# Statistical Machine Learning for Human Behaviour Analysis

# Statistical Machine Learning for Human Behaviour Analysis

Special Issue Editors

**Thomas Moeslund**
**Sergio Escalera**
**Gholamreza Anbarjafari**
**Kamal Nasrollahi**
**Jun Wan**


MDPI • Basel • Beijing • Wuhan • Barcelona • Belgrade • Manchester • Tokyo • Cluj • Tianjin

*Special Issue Editors*

Thomas Moeslund
Visual Analysis of People
Laboratory, Aalborg University
Denmark

Sergio Escalera
Universitat de Barcelona and
Computer Vision Centre
Spain

Gholamreza Anbarjafari
iCV Lab, Institute of Technology,
University of Tartu
Estonia

Kamal Nasrollahi
Visual Analysis of People
Laboratory, Aalborg University,
Research Department of
Milestone Systems A/S
Denmark

Jun Wan
National Laboratory of Pattern
Recognition (NLPR), Institute of
Automation, Chinese Academy
of Sciences
China

*Editorial Office*
MDPI
St. Alban-Anlage 66
4052 Basel, Switzerland

This is a reprint of articles from the Special Issue published online in the open access journal *Entropy* (ISSN 1099-4300) (available at: https://www.mdpi.com/journal/entropy/special_issues/Statistical_Machine_Learning).

For citation purposes, cite each article independently as indicated on the article page online and as indicated below:

LastName, A.A.; LastName, B.B.; LastName, C.C. Article Title. *Journal Name* **Year**, *Article Number*, Page Range.

**ISBN 978-3-03936-228-8 (Pbk)**
**ISBN 978-3-03936-229-5 (PDF)**

# Contents

# About the Special Issue Editors

**Thomas B. Moeslund** received his PhD from Aalborg University in 2003 and is currently Head of the Visual Analysis of People lab at Aalborg University (www.vap.aau.dk). His research covers all aspects of software systems for automatic analysis of people. He has been involved in 14 national and international research projects, both as coordinator, WP leader and researcher. He has published more than 300 peer reviewed journal and conference papers. His awards include the Most Cited Paper in 2009, Best IEEE Paper in 2010, Teacher of the Year in 2010, and the Most Suitable for Commercial Application award in 2012. He serves as Associate Editor and editorial board member for four international journals. He has co-edited two Special Issues and acted as PC member/reviewer for numerous conferences. Professor Moeslund has co-chaired the following eight international conferences/workshops/tutorials: ARTEMIS'12 (ECCV'12), AMDO'12, Looking at People'12 (CVPR12), Looking at People'11 (ICCV'11), Artemis'11 (ICCV'11), Artemis'10 (MM'10), THEMIS'08 (ICCV'09), and THEMIS'08 (BMVC'08).

**Sergio Escalera** obtained his PhD degree on multiclass visual categorization systems for his work at Computer Vision Center, UAB. He obtained the 2008 Best Thesis award for Computer Science at Universitat Autònoma de Barcelona. He is ICREA Academia. He leads the Human Pose Recovery and Behavior Analysis Group at UB, CVC, and the Barcelona Graduate School of Mathematics. He is Full Professor at the Department of Mathematics and Informatics, Universitat de Barcelona. He is also a member of the Computer Vision Center at UAB. He is Series Editor of The Springer Series on Challenges in Machine Learning. He is Vice-President of ChaLearn Challenges in Machine Learning, leading ChaLearn Looking at People events. He is co-creator of Codalab open source platform for organization of challenges. He is also a member of the European Laboratory for Learning and Intelligent Systems ELLIS, the AERFAI Spanish Association on Pattern Recognition, ACIA Catalan Association of Artificial Intelligence, INNS, and Chair of IAPR TC-12: Multimedia and Visual Information Systems. He holds numerous patents and registered models. He has published more than 300 research papers and participated in the organization of scientific events. His research interests include automatic analysis of humans from visual and multimodal data, with special interest in inclusive, transparent, and fair affective computing and characterization of people: personality and psychological profile computing.

**Gholamreza Anbarjafari** (Shahab) is Head of the intelligent computer vision (iCV) lab at the Institute of Technology at the University of Tartu. He was also Deputy Scientific Coordinator of the European Network on Integrating Vision and Language (iV&L Net) ICT COST Action IC1307. He is Associate Editor and Guest Lead Editor of numerous journals, Special Issues, and book projects. He is an IEEE Senior Member and Chair of Signal Processing/Circuits and Systems/Solid-State Circuits Joint Societies Chapter of IEEE Estonian section. He has the recipient of the Estonian Research Council Grant and has been involved in many international industrial projects. He is an expert in computer vision, machine learning, human–robot interaction, graphical models, and artificial intelligence. He has supervised 17 MSc students and 7 PhD students. He has published over 130 scientific works. He has been in the organizing and technical committees of the IEEE Signal Processing and Communications Applications Conference in 2013, 2014, and 2016 and TCP of conferences such as ICOSST, ICGIP, SampTA, and SIU. He has been organizing challenges and workshops in FG17, CVPR17, ICCV17, ECML19, and FG20.

**Kamal Nasrollahi** is Head of Machine Learning at Milestone Systems A/S and Professor of Computer Vision and Machine Learning at Visual Analysis of People (VAP) Laboratory at Aalborg University in Denmark. He has been involved in several national and international research projects. He obtained his MSc and PhD degrees from Amirkabir University of Technology and Aalborg University, in 2007 and 2010, respectively. His main research interest is on facial analysis systems, for which he has published more than 100 peer-reviewed papers on different aspects of such systems in several international conferences and journals. He has won three best conference paper awards.

**Jun Wan** (http://www.cbsr.ia.ac.cn/users/jwan/research.html) received his BS degree from the China University of Geosciences, Beijing, China, in 2008, and PhD degree from the Institute of Information Science, Beijing Jiaotong University, Beijing, China, in 2015. Since January 2015, he has been worked at the National Laboratory of Pattern Recognition (NLPR), Institute of Automation, Chinese Academy of Sciences (CASIA). He received the 2012 ChaLearn One-Shot-Learning Gesture Challenge Award, sponsored by Microsoft, ICPR 2012. He also received the 2013, 2014 Best Paper Awards from the Institute of Information Science, Beijing Jiaotong University. His main research interests include computer vision, machine learning, especially for gesture and action recognition, facial attribution analysis (i.e., age estimation, facial expression, gender and race classification). He has published papers in top journals as the first author or corresponding author, such as JMLR, TPAMI, TIP, TCYB and TOMM. He has served as the reviewer on several top journals and conferences, such as JMLR, TPAMI, TIP, TMM, TSMC, PR, CVPR, ICCV, ECCV, ICRA, ICME, ICPR, FG.

*Editorial*

# Statistical Machine Learning for Human Behaviour Analysis

**Thomas B. Moeslund [1], Sergio Escalera [2,3], Gholamreza Anbarjafari [4,5,6], Kamal Nasrollahi [1,7,*] and Jun Wan [8]**

[1]   Visual Analysis of People Laboratory, Aalborg University, 9000 Aalborg, Denmark; tbm@create.aau.dk
[2]   Computer Vision Centre, Universitat Autònoma de Barcelona, Bellaterra (Cerdanyola),
     08193 Barcelona, Spain; sergio@maia.ub.es
[3]   Department of Mathematics and Informatics, Universitat de Barcelona, 08007 Barcelona, Spain
[4]   iCV Lab, Institute of Technology, University of Tartu, 50411 Tartu, Estonia; shb@ut.ee
[5]   Department of Electrical and Electronic Engineering, Hasan Kalyoncu University, 27900 Gaziantep, Turkey
[6]   PwC Finland, Itämerentori 2, 00100 Helsinki, Finland
[7]   Research Department of Milestone Systems A/S, 2605 Copenhagen, Denmark
[8]   National Laboratory of Pattern Recognition (NLPR), Institute of Automation, Chinese Academy of Sciences,
     Beijing 100190, China; jun.wan@ia.ac.cn
*   Correspondence: kna@milestone.dk

Received: 22 April 2020; Accepted: 6 May 2020; Published: 7 May 2020

**Keywords:** action recognition; emotion recognition; privacy-aware

Human behaviour analysis has introduced several challenges in various fields, such as applied information theory, affective computing, robotics, biometrics and pattern recognition. This Special Issue focused on novel vision-based approaches, mainly related to computer vision and machine learning, for the automatic analysis of human behaviour. We solicited submissions on the following topics: information theory-based pattern classification, biometric recognition, multimodal human analysis, low resolution human activity analysis, face analysis, abnormal behaviour analysis, unsupervised human analysis scenarios, 3D/4D human pose and shape estimation, human analysis in virtual/augmented reality, affective computing, social signal processing, personality computing, activity recognition, human tracking in the wild, and application of information-theoretic concepts for human behaviour analysis. In the end, 15 papers were accepted for this special issue [1–15]. These papers, that are reviewed in this editorial, analyse human behaviour from the aforementioned perspectives, defining in most of the cases the state of the art in their corresponding field.

Most of the included papers are application-based systems, while [15] focuses on the understanding and interpretation of a classification model, which is an important factor for the classifier's credibility. Given a set of categorical data, [15] utilizes multi-objective optimization algorithms, like ENORA and NSGA-II, to produce rule-based classification models that are easy to interpret. Performance of the classifier and its number of rules are optimized during the learning, where the first one is obviously expected to be maximized while the second one is expected to be minimized. Testing on public databases, using 10-fold cross-validation, shows the superiority of the proposed method against classifiers that are generated using other previously published methods like PART, JRip, OneR and ZeroR.

Two published papers ([1,9]) have privacy as their main concern, while they develop their respective systems for biometrics recognition and action recognition. Reference [1] has considered a privacy-aware biometrics system. The idea is that the identity of the users should not be readily revealed from their biometrics, like facial images. Therefore, they have collected a database of foot and hand traits of users while opening a door to grant or deny access, while [9] develops a privacy-aware method for action recognition using recurrent neural networks. The system accumulates reflections of light pulses omitted by a laser, using a single-pixel hybrid photodetector. This includes information about the distance of the objects to the capturing device and their shapes.

Multimodality (RGB-depth) is covered in [14] for sign language recognition; while in [11], multiple domains (spatial and frequency) are used for saliency detection. Reference [14] has applied restricted Boltzmann machine (RBM)s to develop a system for sign language recognition from a given single image, in two modalities of RGB and depth. Two RBMs are designed to process the images coming from the two deployed modalities, while a third RBM fuses the results of the first two RBMs. The inputs to the first two RBMs are hand images that are detected by a convolutional neural network (CNN). The experimental results reported in [14] on two public databases show the state-of-the-art performance of the proposed system. Reference [11] proposes a multi-domain (spatial and frequency)-based system for salient object detection in foggy images. The frequency domain saliency map is extracted using the amplitude spectrum, while the spatial domain saliency map is calculated using the contrast of the local and global super-pixels. These different domain maps are fused using a discrete stationary wavelet transform (DSWT) and are then refined using an encoder-decoder model to pronounce the salient objects. Experimental results on public databases and comparison with state-of-the-art similar methods show the better performance of this system.

Four papers in this special issue have covered action recognition [6,9,12,13]. Reference [12] has proposed a system for toe-off detection using a regular camera. The system extracts the differences between consecutive frames to build silhouettes difference maps, that are then fed into a CNN for feature extraction and classification. Different types of maps are developed and tested in this paper. The experimental results reported in [12] on public databases show state-of-the-art performance. Reference [6] proposes a system for individuals and then crowd condition monitoring and prediction. Individuals participating in this study are grouped into crowds based on their physical locations extracted using GPS on their smartphones. Then, an enhanced context-aware framework using an algorithm for feature selection is used to extract statistical-based time-frequency domain features. Reference [13] focuses on utilizing recurring concepts using adaptive random forests to develop a system that can cope with drastically changing behaviours in dynamic environments, like financial markets. The proposed system is an ensemble-based classifier comprised of trees that are either active or inactive. The inactive ones keep a history of market operators' reactions in previously recorded similar situations, while either an inactive tree or a background tree that has recently been trained replaces the active ones, as a reaction to drift.

In terms of face analysis, in [10] a system is proposed for detecting fuzziness tendencies and utilizing these to design human-machine interfaces. This is motivated by the fact that humans tend to pay more attention to sections of information with fuzziness, which are sections with greater mental entropy. The work of [4] proposes a conditional random field-based system for segmentation of facial images into six facial parts. These are then converted into probability maps, which are used as feature maps for a random decision forest that estimates head-pose, age, and gender.

The method introduced in [3] uses singular value decomposition for removing background of fingerprint images. Then, it finds fingerprints' boundaries and applies an adaptive algorithm based on wavelets extrema and Henry system to detect singular points, which are widely used in applications related to fingerprint, like registration, orientation detection, fingerprint classification, and identification systems.

Three papers have covered emotion recognition, one from body movements [5], and two from speech signals [2,7]. In [2] a committee of classifiers has been applied to a pool of descriptors extracting features from speech signals. Then, it is used as a voting scheme on the classifiers' outputs to get to a conclusion about the emotional status from the used speech signals. The paper in [2] shows that the committee of classifiers outperforms the single individual classifiers in the committee. The system proposed in [7] builds 3D tensors of spectrogram frames that are obtained by extracting 88-dimentional feature vectors from speech signals. These tensors are then used for building a 3D convolutional neural network that is employed for emotion recognition. The system has produced state-of-the-art results on three public databases. The emotional recognition system of [5] does not use facial images or speech signals, but body movements, which are captured by Microsoft Kinect v2 under eight different emotional states. The affective movements are represented by extracting and tracking location and orientation of body joints over time. Experimental results, using different deep learning-based methods, show the state-of-the-art performance of this system.

Finally, two databases have been introduced in this special issue, one for biometric recognition [1] and one for detecting sleeping issues and fatigue [8], the later containing a database of patients suffering from Fibromyalgia, which is a situation resulting in muscle pain and tenderness, accompanied by few other signs including sleep, memory, and mood disorders. It uses similarity functions with configurable convexity or concavity to build a classifier on this collected database in order to predict extreme cases of sleeping issues and fatigue.

**Acknowledgments:** We express our thanks to the authors of the above contributions and to the journal Entropy and MDPI for their support during this work. Kamal Nasrollahi's contribution to this work is partially supported by the EU H2020-funded SafeCare project, grant agreement no. 787002. This work is partially supported by ICREA under the ICREA Academia programme.

**Conflicts of Interest:** The authors declare no conflict of interest.

## References

1. Jahromi, S.M.N.; Buch-Cardona, P.; Avots, E.; Nasrollahi, K.; Escalera, S.; Moeslund, T.B.; Anbarjafari, G. Privacy-Constrained Biometric System for Non-Cooperative Users. *Entropy* **2019**, *21*, 1033. [CrossRef]
2. Kamińska, D. Emotional Speech Recognition Based on the Committee of Classifiers. *Entropy* **2019**, *21*, 920. [CrossRef]
3. Le, N.T.; Le, D.H.; Wang, J.-W.; Wang, C.-C. Entropy-Based Clustering Algorithm for Fingerprint Singular Point Detection. *Entropy* **2019**, *21*, 786. [CrossRef]
4. Khan, K.; Attique, M.; Syed, I.; Sarwar, G.; Irfan, M.A.; Khan, R.U. A Unified Framework for Head Pose, Age and Gender Classification through End-to-End Face Segmentation. *Entropy* **2019**, *21*, 647. [CrossRef]
5. Sapiński, T.; Kamińska, D.; Pelikant, A.; Anbarjafari, G. Emotion Recognition from Skeletal Movements. *Entropy* **2019**, *21*, 646. [CrossRef]
6. Sadiq, F.I.; Selamat, A.; Ibrahim, R.; Krejcar, O. Enhanced Approach Using Reduced SBTFD Features and Modified Individual Behavior Estimation for Crowd Condition Prediction. *Entropy* **2019**, *21*, 487. [CrossRef]
7. Hajarolasvadi, N.; Demirel, H. 3D CNN-Based Speech Emotion Recognition Using K-Means Clustering and Spectrograms. *Entropy* **2019**, *21*, 479. [CrossRef]
8. Sabeti, E.; Gryak, J.; Derksen, H.; Biwer, C.; Ansari, S.; Isenstein, H.; Kratz, A.; Najarian, K. Learning Using Concave and Convex Kernels: Applications in Predicting Quality of Sleep and Level of Fatigue in Fibromyalgia. *Entropy* **2019**, *21*, 442. [CrossRef]
9. Ofodile, I.; Helmi, A.; Clapés, A.; Avots, E.; Peensoo, K.M.; Valdma, S.-M.; Valdmann, A.; Valtna-Lukner, H.; Omelkov, S.; Escalera, S.; et al. Action Recognition Using Single-Pixel Time-of-Flight Detection. *Entropy* **2019**, *21*, 414. [CrossRef]
10. Bao, H.; Fang, W.; Guo, B.; Wang, P. Supervisors' Visual Attention Allocation Modeling Using Hybrid Entropy. *Entropy* **2019**, *21*, 393. [CrossRef]
11. Zhu, X.; Xu, X.; Mu, N. Saliency Detection Based on the Combination of High-Level Knowledge and Low-Level Cues in Foggy Images. *Entropy* **2019**, *21*, 374. [CrossRef]
12. Tang, Y.; Li, Z.; Tian, H.; Ding, J.; Lin, B. Detecting Toe-Off Events Utilizing a Vision-Based Method. *Entropy* **2019**, *21*, 329. [CrossRef]
13. Suárez-Cetrulo, A.L.; Cervantes, A.; Quintana, D. Incremental Market Behavior Classification in Presence of Recurring Concepts. *Entropy* **2019**, *21*, 25. [CrossRef]
14. Rastgoo, R.; Kiani, K.; Escalera, S. Multi-Modal Deep Hand Sign Language Recognition in Still Images Using Restricted Boltzmann Machine. *Entropy* **2018**, *20*, 809. [CrossRef]
15. Jiménez, F.; Martínez, C.; Miralles-Pechuán, L.; Sánchez, G.; Sciavicco, G. Multi-Objective Evolutionary Rule-Based Classification with Categorical Data. *Entropy* **2018**, *20*, 684. [CrossRef]

*Article*

# Privacy-Constrained Biometric System for Non-Cooperative Users

**Mohammad N. S. Jahromi** [1,*], **Pau Buch-Cardona** [2], **Egils Avots** [3], **Kamal Nasrollahi** [1], **Sergio Escalera** [2,4], **Thomas B. Moeslund** [1] **and Gholamreza Anbarjafari** [3,5]

[1]   Visual Analysis of People Laboratory, Aalborg University, 9100 Aalborg, Denmark; kn@create.aau.dk (K.N.); tbm@create.aau.dk (T.B.M.)

[2]   Computer Vision Centre, Universitat Autònoma de Barcelona, 08193 Bellaterra (Cerdanyola), Barcelona, Spain; pbuch@cvc.uab.es (P.B.-C.); sergio@maia.ub.es (S.E.)

[3]   iCV Lab, Institute of Technology, University of Tartu, 50411 Tartu, Estonia; ea@icv.tuit.ut.ee (E.A.); shb@icv.tuit.ut.ee (G.A.)

[4]   Department of Mathematics and Informatics, Universitat de Barcelona, 08007 Barcelona, Spain

[5]   Department of Electrical and Electronic Engineering, Hasan Kalyoncu University, 27900 Gaziantep, Turkey

[*]   Correspondence: mosa@create.aau.dk

Received: 21 September 2019; Accepted: 23 October 2019; Published: 24 October 2019

**Abstract:** With the consolidation of the new data protection regulation paradigm for each individual within the European Union (EU), major biometric technologies are now confronted with many concerns related to user privacy in biometric deployments. When individual biometrics are disclosed, the sensitive information about his/her personal data such as financial or health are at high risk of being misused or compromised. This issue can be escalated considerably over scenarios of non-cooperative users, such as elderly people residing in care homes, with their inability to interact conveniently and securely with the biometric system. The primary goal of this study is to design a novel database to investigate the problem of automatic people recognition under privacy constraints. To do so, the collected data-set contains the subject's hand and foot traits and excludes the face biometrics of individuals in order to protect their privacy. We carried out extensive simulations using different baseline methods, including deep learning. Simulation results show that, with the spatial features extracted from the subject sequence in both individual hand or foot videos, state-of-the-art deep models provide promising recognition performance.

**Keywords:** biometric recognition; multimodal-based human identification; privacy; deep learning

---

## 1. Introduction

Biometric recognition is the science of identification of individuals based on their biological and behavioral traits [1,2]. In the design of a biometrics-based recognition or authentication system, different issues, heavily related to the specific application, must be taken into account. According to the literature, ideally biometrics should be universal, unique, permanent, collectable, and acceptable. In addition, besides the choice of the biometrics to employ, many other issues must be considered in the design stage. The system accuracy, the computational speed, and cost are important design parameters, especially for those systems intended for large populations [3]. Recently, biometric recognition systems have posed new challenges related to personal data protection (e.g., GDPR), which is not often considered by conventional recognition methods [4]. If biometric data are captured or stolen, they may be replicated and misused. In addition, the use of biometrics data may reveal sensitive information about a person's personality and health, which can be stored, processed, and distributed without the user's consent [5]. In fact, GDPR has a distinct category of personal data protection that defines 'biometric data', its privacy, and legal grounds of its processing. According to GDPR, what qualifies as 'Biometric data' is defined as 'personal data resulting from specific technical processing relating to the physical, physiological or behavioural characteristics of a natural person, which allow or confirm the unique identification of that natural person such as facial images' [6]. Furthermore, GDPR attempts to address privacy matters by the preventing of processing any 'sensitive' data revealing information such as health or sexual orientation of individuals. In other words, processing of such sensitive data can be only allowed if it falls under ten exceptions laid down in GDPR [6]. Apart from this privacy concern, in some scenarios, designing and deploying a typical biometric system where any subject has to cooperate and interact with the mechanism may not be practical. In care homes with elderly patients, for example, interaction of the user with typical device-dependent hardware or following specific instruction during biometric scan (e.g., direct contact with a camera, placing a biometric into a specific position, etc.) [7,8]. In other words, the nature of such uncontrolled environments suggest the biometric designer to consider strictly natural and transparent systems that mitigate the user non-cooperativeness behavior, providing an enhanced performance.

This possibility was explored in our earlier work [9] by considering identification of persons when they grab the door handle, which is an unchanged routine, in opening a door and no further user training is required. In our previous work, we designed a bimodal dataset (hand's dorsal, hereafter refer to as hand, and face) by placing two cameras above the door handle and frame, respectively. This was done in order to capture the dorsal hand image of each user while opening the door for multiple times (10 times per user) in a nearly voluntary manner. In addition, face images of users approaching the physical door were collected as a complementary biometric feature. In [9], we concluded that facial images are not always clearly visible due to the nonoperative nature of the environment, but, when visible, it provides complementary features to hand-based identification.

In [9], however, the study disregards the privacy of the users previously mentioned here as all the methods employ the visible face of each subject in the recognition task, which is considered as sensitive information in the new data protection paradigm.

In this paper, we deal with the problem of automatic people recognition under privacy constraints. Due to this constraint, it is crucial to conduct a careful data-collection protocol that excludes any sensitive biometric information that may comprise user's privacy. For instance, to protect the users, acquiring facial or full-body gait information of candidates is not possible. Consequently, we have collected a new data-set containing only the hands and feet of each subject using both RGB and near/infrared cameras. We verified the usefulness of the designed setup for user privacy-constrained classification by performing extensive experiments with both conventional handcrafted methods as well as recent Deep Learning models.

The remainder of this paper is organized as follows: Section 2 discusses related work in the field. In Section 3, the database is presented. In Section 4, the dataset is evaluated with classical and deep learning strategies. Finally, conclusions are drawn in Section 5.

## 2. Related Work

This section reviews the existing methods on hand and the footprint recognition focusing mostly on the use of geometric spatial information. There are a few detailed studies that are reviewing different hand-based biometric recognition systems [10,11]. Visual specifications of hands constitute a paramount criterion for biometric-based identification of persons, owing to the associated respectively low computational requirements and mild memory usages [12]. In addition, they provide superior distinctive representations of persons, which lead to unparalleled recognition success rates. Furthermore, the related procedures can be well adapted into the existing biometric authentication systems, which make them favorable for the foregoing purpose [13–17]. These systems, depending on the type of the features they extract from the hand, can be categorized as follows:

- Group 1: in which the geometric features of the hand are used for the identification. Examples of such features include the length and the width of the hand palm. Conventional methods such as

General Regression Neural Network (GRN) [18], graph theory [18], or later methods like sparse learning [19] are examples of this group.

- Group 2: in which hand vein patterns are used for the identification. These patterns are unique to every individual and are not affected by aging, scars and skin color [20]. Therefore, the vascular patterns of an individual's hand (palm, dorsal or finger) can be used as a feature for biometric recognition systems. An example of this category includes wavelet and Local binary patterns (LBP) based [21,22] or recent deep learning-based methods [23]. Such features have been used in connection with CNNs [24] and extracted using thermal imaging and Hausdorff distance based matching [20,25], and using multi-resolution filtering [26].

- Group 3: in which palm prints are used for identification. Palm prints can be extracted according to texture, appearance, orientations or lines. Besides various conventional techniques, there are dictionary and deep learning methods [27,28] reported in literature. Considering the above categories, the geometry-based hand features are robust to both rotation and translation. However, at the same time, they are not suitable to scale variations. Moreover, in order to achieve high performance for the recognition task, a huge amount of measurements is needed to extract discriminative features of each subject. This will eventually increase the computational complexity. The hand vein features, on the other hand, are robust to varying hand poses and deformation. They may also introduce computational cost if all distances between landmark points are required. Finally, for the palm-print based recognition, some methods that achieve high recognition rates exist, but, in general, acquiring high-resolution palm-print images is challenging due to setup complexities.

*Footprint*

Contrary to many well-established biometric techniques used in the context of automatic human recognition, the human foot features are rarely used as a feature in those solutions. Although the uniqueness property of the human foot is extensively addressed in the forensic studies [29], its commercial solution is considered mostly complicated due to complexity of the data acquisition in the environment [30]. The very early attempt of employing a human foot as means of identification emerged in the forensic study carried out by Kennedy [29] in which he examines the uniqueness of barefoot impression. In [31], the first notion of utilizing the Euclidean distance between a pair of human feet was presented. In [32], the authors propose a static and dynamic footprint-based recognition based on a hidden Markov model. The latter implemented a footprint based biometric system, similar to a hand, which involves exploiting the following foot features:

- Group 1: in which the shape and geometrical information of the human foot are used for identification. Features of this category concentrate on the length, shape and area of the silhouette curve, local foot widths, lengths of toes, eigenfeet features and angles of intertoe valleys [30]. The research works in [33,33–35] are a few examples of this category. In general, a variety of possible features makes shape and geometric-based methods very popular. In addition, these methods are robust to various environmental conditions. The drawback of such a large number of possible features, however, can eventually result in high intrapersonal variability.

- Group 2: in which the texture-based information of the human foot are used for identification. In this group, pressure (soleprint features analogous to palm print based of the hand biometric) and generated heat can be considered as the promising features. Examples in this category can be found in [30,36]. Unlike the shape and geometrical features of feet, acquiring a fine-grained texture of feet requires a high accuracy instrument. For example, skin-texture on palm-print involves extracting rather invisible line patterns as opposed to the similar one in the hands. Similar challenges may exist in recording ridge structure with high resolution. On the other hand, the high-resolution of texture-based features will require higher computational power with respect to shape and geometrical ones.

Minutiae-based ballprint [30] in the foot as well as different distance techniques such as city-block, cosine, and correlation [37] are further examples of the features that are employed in this context. It is also important to mention that gait biometrics [38] are also a potential approach that studies the characteristic of human foot strike.

## 3. Acquisition Setup

In this paper, in order to have a realistic testing environment, an acquisition setup has been designed by employing a standard-size building door with three camera sensors, one mounted above its handle, and two installed at the frame side, respectively.

During data collection, it is important to capture each modal in a clear visible form so that all unique meaningful features can be extracted. In other words, each modal has to be collected by a proper sensor. In this work, for example, each subject approaches a door and grabs its handle to open it. Therefore, each subject's hand should be recorded by a sensor while placed on the door handle. Based on several conducted tests with different available sensors, we choose to employ a near infrared light (NIR) camera (AV MAKO G-223B NIR POE) equipped with a band pass filter to cut off visible light. In this way, for hands, good feature candidates such as veins can be properly extracted. In addition, to guarantee that the hand modals on the door handles are visible in the captured frames, a near infrared light source (SVL BRICK LIGHT S75-850) was also mounted on the door frame. To capture each foot modal, a regular RGB camera (GoPro Hero 3 Black) on the door frame is installed to capture the subject's foot as they approach the door. The third camera in this setup has been used to acquire the face modality of each corresponding subject although it is not used to perform automatic classification. They are collected to conduct alternative studies beyond the scope of this paper and hence excluded. The overall door model together with the installed cameras and the light source are shown in Figure 1.

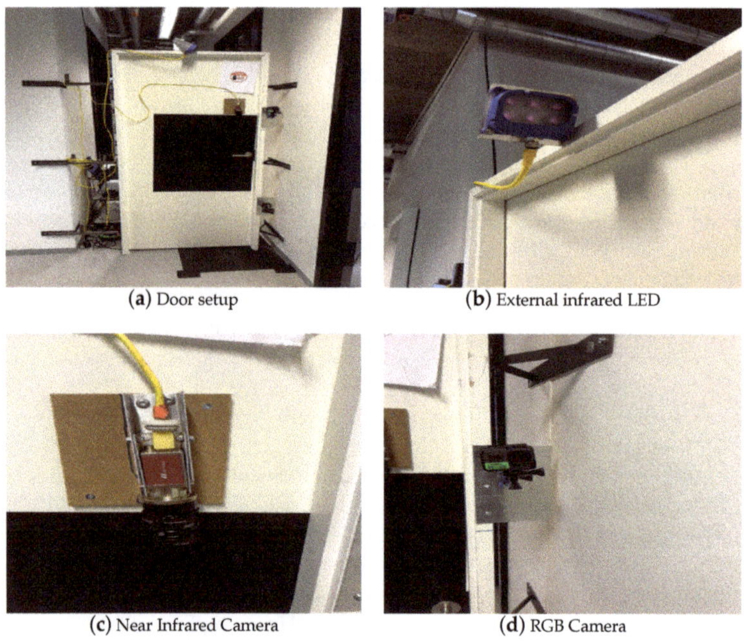

<table>
<tr><td align="center">(a) Door setup</td><td align="center">(b) External infrared LED</td></tr>
<tr><td align="center">(c) Near Infrared Camera</td><td align="center">(d) RGB Camera</td></tr>
</table>

**Figure 1.** Illustration of the main set-up and the sensors.

A total of 77 persons of mixed gender and varying ages from 20 to 55 years participated in the data collection procedure at *Aalborg University*. There exist three paths that each subject can take to approach the door setup. Each person is requested to approach the door from any desired path randomly. These paths can cover both linear and curvature trajectory, making the scenario natural. The participant then walks toward to the door, grabs its handle and then passes through the door. This procedure is repeated two times. During data acquisition, no further instructions were given to the participants. This is done to have the participants grab the door handle as they would naturally perform in any context. As a result, all data are captured in a totally natural scenario where a variety of realistic situations such as occlusion, different pose and partial foot may occur. Furthermore, the lighting condition is not controlled and the data has been collected during different times of day for two months. Figure 2 shows samples acquired by the different cameras.

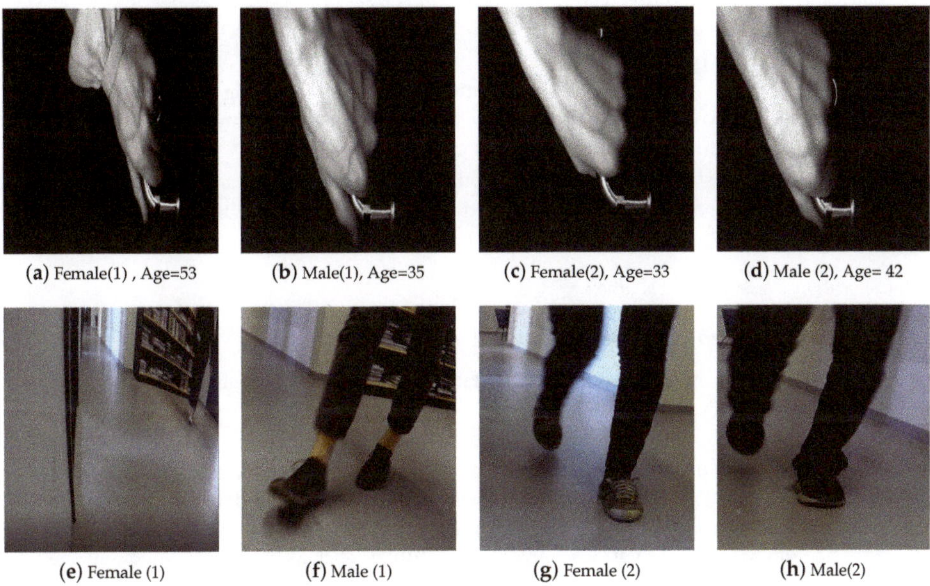

| | | | |
|---|---|---|---|
| (**a**) Female(1) , Age=53 | (**b**) Male(1), Age=35 | (**c**) Female(2), Age=33 | (**d**) Male (2), Age= 42 |
| (**e**) Female (1) | (**f**) Male (1) | (**g**) Female (2) | (**h**) Male(2) |

**Figure 2.** Sample of captured frames of both hand and foot modalities for four subjects.

Each video sequence of the subject's hand/foot is post processed to enhance the quality of captured frames and remove any camera distortion. This is performed by using the well-known chessboard camera calibration tool in the vision library of MATLAB (2019, MathWorks) [39].

Privacy disclaimer: While our proposal moves in the direction of privacy constrained scenarios, we are aware that still some soft biometrics features used in this work could be used in some situations by specific external observers that could be able to identify the user. Without loss of generality, we use privacy-constrained to refer to the scenario where sensitive user information is avoided, making the biometric identification harder in case data are leaked.

## 4. Experimental Results and Discussion

In this section, we first discuss the evaluation protocol of the experiments. Then, we briefly explain the methods used and finally the obtained results and discussions.

### 4.1. Evaluation Protocol

In order to carry out experiments using different methods, we divided the database into mutually disjoint subsets of training, validation, and testing. As there are two cycles of complete action per modality (i.e., each user approaches the door twice), each video sample is divided into two sequences per modality. Next, we use all first sequences from all subjects for both hands and feet to train while utilizing the second sequence of the subjects for validation and testing, respectively. In this manner, we have 77 sample sequences per modality for training, 37 sequences for validation and 40 samples for the test. Each test sample is then associated with a label during simulations.

In this paper, the main focus of all the experiments is around general spatial appearance models. In other words, for all the simulations, the spatial features are extracted through the analysis of each independent frame (uncorrelated frames per same subject). For the evaluated deep learning model, we have further analyzed the contribution of the motion as an input modality. Finally, we have also performed late-fusion on both modalities for all of the experiments.

To summarize, the performed experiments are divided into the following three categories:

- *Independent frame analysis*: The final evaluation is based on frames independently. That is, frames from the same subject are uncorrelated. The maximum output probability from each frame determines its final predicted class.
- *Subject sequence analysis*: The final evaluation is based on grouping frames from the same sequence/subject. We average output probabilities belonging to the same subject, and we finally obtain the maximum output value as its final predicted class.
- *Hand–Feet Late Fusion Analysis*: Averaged output probabilities from hand and feet outputs (aka subject sequence analysis) are averaged together for each subject to finally determine the final predicted class.

*4.2. Conventional Techniques Evaluation*

4.2.1. Local Binary Patterns and Support Vector Machine

Even tough deep neural networks dominate state-of-the-art solutions in image processing, it is still worthwhile to further test conventional methods to create baseline results, in particular in scenarios where a limited amount of annotated data are available. Local binary patterns (LBP) [40–42] are one of the most powerful handcrafted texture descriptors. The core implementation and its variants are extensively used in facial image analysis, including tasks as diverse as face detection, face recognition and facial expression analysis. Benzaoui et al. [43] showed that classification tasks which use LBP for feature extraction can improve various statistical procedures, such as principal component analysis (PCA) and discrete wavelet transform (DWT). For example, by using a combination of DWT, LBP and support vector machine (SVM) [44–46] for classification, it is possible to create a hybrid method for face recognition. Similarly, the same approach can be used for hand and foot classification. The performance of an LBP based feature extractor can be greatly improved, making input data robust against certain image transformations. For example, in the case of face images, this relates to aligned and cropped faces. When considering recordings of human gait, the size and foot orientation is constantly changing, thus adding additional challenges to the description and classification problems. The DWT method is widely used in feature extraction, compression and denoising applications. The process of recognition using DWT is as follows: the wavelet transform of a particular level is applied on the test image and the output is an approximation coefficient matrix, which we consider as a sub-image. Then, we extract rotation invariant LBP feature vectors from the sub-images for SVM training and classification [47]. The regions of interest from the video sequences are extracted using the frame difference method across multiple frames. This approach was robust enough to successfully pre-process all the videos in the database. The system was developed in MATLAB environment, where we used inbuilt functions for single-level 2D wavelet decomposition (dwt2) approximation coefficients matrix, rotation invariant local binary patterns (extractLBPFeatures) with $10 \times 10$ cells [48] and the linear multi-class support vector machine (fitcecoc).

- *Setup*: To acquire the regions of interest (ROI) for the moving object in each frame (hand or foot), in this experiment, we applied simple and fast frame difference. If the difference is greater than 80 pixels for the foot and 30 pixels for the hand videos (these values were found empirically), then the resulted difference frame will be recorded as a binary mask. Figure 3a shows an example of all masks within one sequence, where the color transition from dark gray to white represents transition from the start of the video to the end of the video. For the foot sequences, the bounding box for a particular frame is created by taking that frame and then superimposing the previous 10 frames, and the next 10 frames that contain binary masks (identical to a sliding window). In other words, a binary image was formed by repeating the logical OR operation for 21 consecutive frames after which a bounding box has been found for the detected region as shown in Figure 3b. On the other hand, for the hand sequences, a fixed bounding box was created by using OR operation for all binary frames. After drawing the bounding box, the images were cropped and then resized to $200 \times 200$ pixels.

- *Experiments:* For the same subject, the total number of frames depend on the video length and therefore can not be fixed to a specific amount. The image features were extracted from an approximation coefficients matrix, which is one of the single-level two-dimensional wavelet decomposition method outputs. In particular, we used the Symlet wavelet to find results of the decomposition low-pass filter. Afterwards, the extracted rotation invariant local binary patterns from the output of wavelet decomposition were obtained as feature vectors to train a linear one-versus-one multi-class support vector machine. This process is shown in Figure 4. Finally, the fusion results obtained via majority voting, where a video label was determined by independent foot and hand frames. Results for single frame recognition and fusion can be found in Table 1. Note that, with a limited amount of data, taking into account that a random prediction classifier score in our problem of 77 labels is 1.3% accurate, which can be still considered as a reasonably good performance for the base line method.

(**a**) All frame masks within one sequence obtained by frame difference

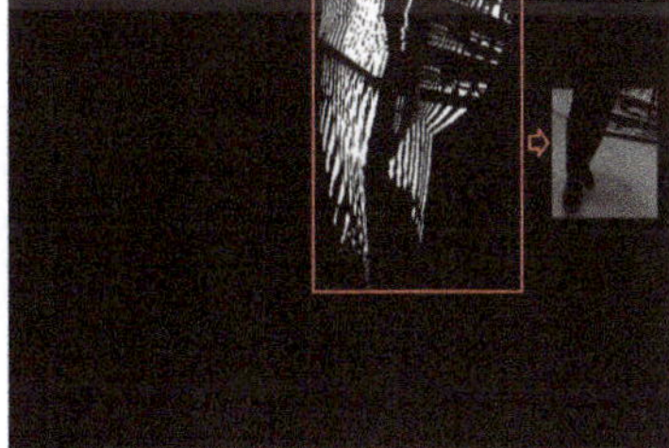

(**b**) Region of interest created by several frame masks

**Figure 3.** Movement detection and bounding box extraction.

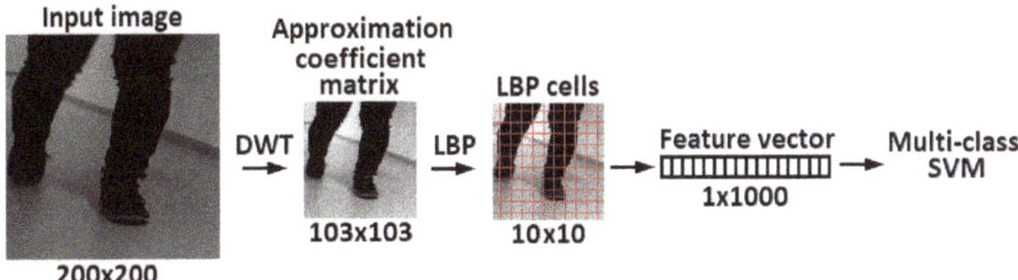


**Figure 4.** Feature extraction flow chart.

**Table 1.** Recognition rate in (%) of the DWT-LBP-SVM approach.

| | Accuracy |
|---|---|
| Modality | Independent frames DWT-LBP-SVM |
| Hand | 37.78% |
| Foot | 34.12% |
| *Hand+Foot (late fusion)* | 57.14% |

### 4.2.2. Dictionary Learning

Sparse based signal processing is a well-established method in the field of Computer Vision. This success is mainly due to the fact that important classes of signals such as audio and images have naturally sparse representations with respect to fixed bases (i.e., Fourier, wavelet), or concatenations of such bases [49]. It has been applied to many Computer Vision tasks such as face recognition [50], image classification [51], denoising [52], etc. In particular, the robust face recognition via sparse representation (SRC) algorithm proposed in [49] uses sparse representation for face recognition. In this method, the basic idea is to form an over-complete dictionary by using the training faces and then classifying a new face by searching the sparsest vector in this dictionary. Hence, this technique is called dictionary learning. Unlike conventional methods such as Eigenface and Fisherface, the dictionary learning can achieve superior results without any explicit feature extraction [53]. This superiority makes the SRC method a convenient method to employ in recognition tasks.

- *Setup:* As a prior step to employ the SRC method for the classifier, we select the frames where both hand and feet are visible. For this, we used the Kalman visual tracker [54], which also defines the ROI within the associated selected images.Figure 5 shows examples of extracted frames to be used in the training stage of Dictionary Learning.
- *Experiments:* For the dictionary learning-based method, we employ the sparse representation classifier (SRC) of [49] for independent frame analysis. We randomly selected 50 extracted frames per subject in the training phase. Therefore, we generate a dictionary of size $100 \times 3850$ with a patch dimension (feature size) of 100 for each subject. This value of feature size has been found experimentally to provide the best performance. Then, at the test stage, using the same feature size, we attempt to recover the sparsest solution ($l_1$-minimization) to linear equation $Ax = b$, where A is the generated dictionary and y is a test image vector, respectively. The obtained results are shown in Table 2. As it can be seen, from both Table 1 and Table 2, conventional spatial appearance models provide poor classification results in all evaluated scenarios. This suggests that a more effective feature extractor is needed in this context.

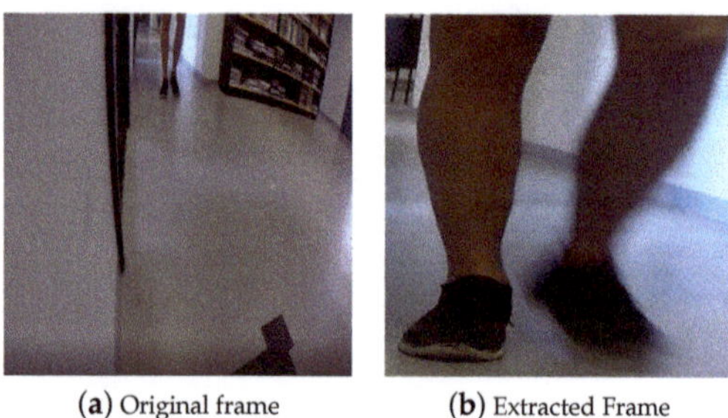

(**a**) Original frame      (**b**) Extracted Frame

**Figure 5.** Sample of the extracted frame using a Kalman tracker.

**Table 2.** Average recognition rate in (%) of the network for sparse representation classifier.

| Accuracy (Dictionary Learning) | |
| --- | --- |
| Modality | Independent frames SRC |
| Hand | 49.1% |
| Foot | 41.3% |
| *Hand+Foot (late fusion)* | 54.1% |

## 4.3. Deep Learning

Deep Neural Networks, and especially Convolutional Neural Networks (CNN), have gained a lot of attention due to their state-of-the-art classification performance in many Computer Vision tasks since the breakthrough of AlexNet architecture [55] in the 2012 ILSVRC (ImageNet Large-Scale Visual Recognition Challenge).

In the context of this work, we opted to process the video frames as individual RGB images from both hand and feet datasets given the limited amount of data. However, for completeness, we have also considered motion maps as network inputs since some motion features may be unique to each subject regardless of their clothing. To that end, we have extracted the Optical Flow (OF) values (u,v) of each pair of consecutive video frames for each subject. The resulting OF values can then be used to generate a heat map that may potentially describe the motion features. Figure 6 shows a sample heat map generated by the Optical Flow vector of consecutive video frames of both hand and foot modality per subject. The rest of the simulations are arranged as follows:

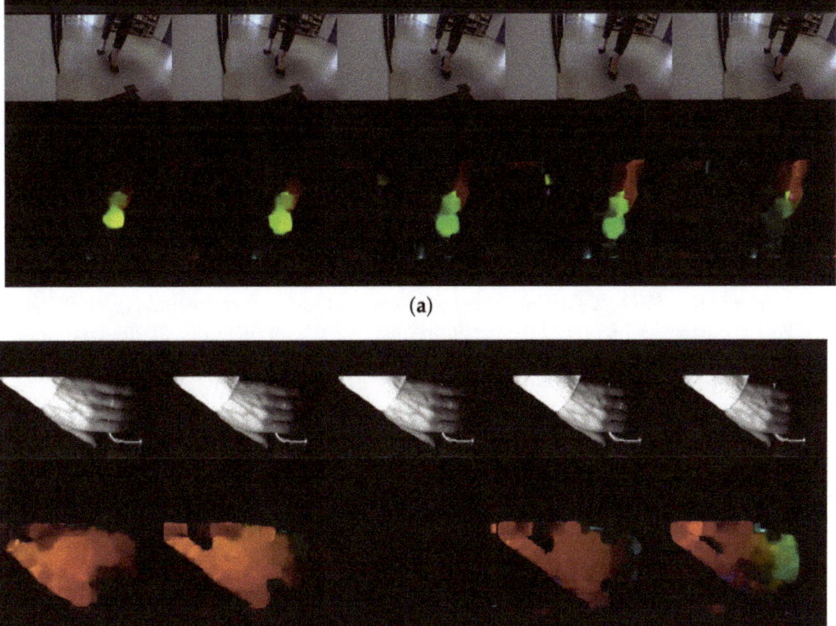

(a)

(b)

**Figure 6.** The heat map generated from an optical flow vector of consecutive video frames per each subject's modality (**a**) and the corresponding heat map (**b**).

- *Setup:* Since the dataset under study can be clearly linked to a classification problem, we have found it convenient to conduct our experiments on a standard ResNet-50 neural network architecture as shown in Figure 7. ResNet-50 has been proved to have a faster performance and lower computational cost compared to those of standard classification architectures such as VGG-16 due to its skip connection configuration [56]. For this purpose, we have constrained the input data (frames) to a [224 × 224] image size, batch size of 32 and output classes to 77 (number of eligible subjects) during the training phase. We left the number of input channels as a degree of freedom that will be set according to the different experiments we conducted.

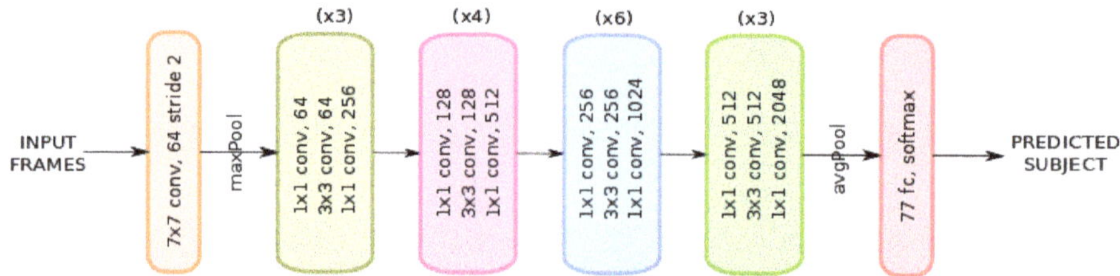


**Figure 7.** ResNet-50 neural network architecture [56].

- *Experiments:* In order to utilize the network, we have arranged the input to the network in the following four fashions:

1. *Appearance*—In this setting, the extracted frames, as before, are fed to the network for both the hand and foot datasets. Hence, input channel dimension is set to 3 due to the RGB nature of the frames. The recognition accuracy rate of this network is reported in Table 3.

**Table 3.** Average recognition rate in (%) of the ResNet network for the appearance model.

| | | Accuracy (Deep Learning) | |
|---|---|---|---|
| No. of Samples | Modality | Independent Frames (appearance) | Subject Sequence (appearance) |
| 2889 | Hand | 70.0% | 80.5% |
| 10195 | Foot | 58.6% | 70.1% |
| | *Hand+Foot (late fusion)* | — | 84.4% |

2. *Optical Flow (OF)*—OF values (u,v) are extracted for each consecutive pair of frames for both the hand and foot in the dataset. In this case, we set the input channel parameter to 2 due to the OF dimensionality. Table 4 summarizes the results of this setting.

**Table 4.** Average recognition rate in (%) of the ResNet network for the OF model.

| | | Accuracy (Deep Learning) | |
|---|---|---|---|
| No. of Samples | Modality | Independent Frames (Optical Flow) | Subject Sequence (Optical Flow) |
| 2812 | Hand | 32.6% | 62.3% |
| 10118 | Foot | 35.1% | 48.1% |
| | *Hand+Foot (late fusion)* | — | 59.7% |

3. *Appearance + OF*—We apply an early fusion to the extracted frames and OF calculated values for both hand and feet datasets. That is, a 5-channel input parameter is set in order to match the RGB(3) + OF(2) new dimension. The results of this simulation are tabulated in Table 5.

**Table 5.** Average recognition rate in (%) of the ResNet network for both the appearance and OF model.

| | | Accuracy (Deep Learning) | |
|---|---|---|---|
| No.of Sample | Modality | Independent frames (appearance+ OF) | Subject Sequence (appearance+ OF) |
| 2812 | Hand | 34.7% | 62.3% |
| 10118 | Foot | 54.8% | 71.4% |
| | *Hand+Foot (late fusion)* | — | 71.4% |

4.  *Appearance + Optical Flow (late fusion)*—We finally bring appearance analysis and OF analysis, computed separately, together. From each 'branch' output, we apply the same principle from the late fusion modality and study its performance. The result of this mode category can be seen in Table 6.

**Table 6.** Average recognition rate in (%) of the ResNet network for the late-fused of both the appearance and OF model.

| | Accuracy (Deep Learning) |
|---|---|
| Modality | Subject Sequence Appearance+OF/ Late Fusion |
| *hand+foot* | 83.1% |

As it can be seen from the results in all of the tables, one can observe that the conventional techniques for both modalities and the late fusion can not effectively utilize the spatial information of the modalities and hence they may not be good candidate methods to be used in the context of a real solution for several reasons: on one hand, handcraft features can not properly model fine-grain information present in the data. Those small details are indeed the ones that may identify properly the subject in this challenging scenario. On the other hand, even for handcraft methods, the limited amount of data per subject in this dataset reduces the generalization capability of handcraft strategies. Still, please note that a random prediction guesser in this scenario will achieve 1.3% accuracy. Thus, an accuracy over 50% and a better result of the combined hand–foot model shows that handcraft methods, up to some degree, are able to learn some discriminative features and their complementary nature.

In Deep Learning, however, we find the best setup classification results (84.4% accuracy) when analyzing the appearance per subject sequence modality—that is, when we use the whole sequence of frames per subject to determine the resulting class. This makes sense because some uncertain frame predictions do not normally contribute too much to the subject's final estimation. We could imagine these misclassification outcomes as noisy samples, which are mostly cancelled out when averaging multiple data. Only the hand model achieves better performance than the only foot one. It was somehow expected because of the more controlled recording of hands and the freedom of the subject in terms of walking, i.e., different walking paths, different point of view, and different scales because of the distance to the camera. Interestingly, the late fusion combination increases around four points the results of the hand, suggesting that complementary and discriminative features are captured by the deep approaches. Some visual misclassified examples are shown in Figure 8. It can be seen that the frame on the left was misclassified as belonging to the same subject on the right. Some explainability can be found by just visual inspection (resemblance between subjects appearance). On the other hand, we find the worst performance when analyzing OF per independent frame analysis modality (32.6% and 35.1% accuracy). We believe motion can produce complementary features to the appearance ones and benefit from its appearance invariant descriptor. However, in order to obtain an increase in performance because of the use of motion, additional data and further strategies to mitigate the overfitting effect (e.g., data augmentation) should be considered.

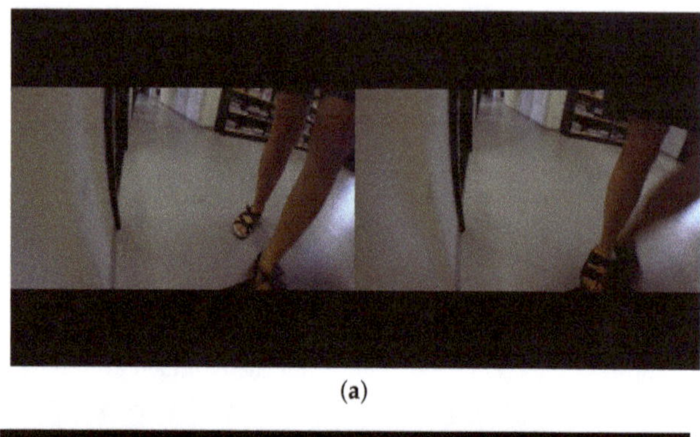

(**a**)

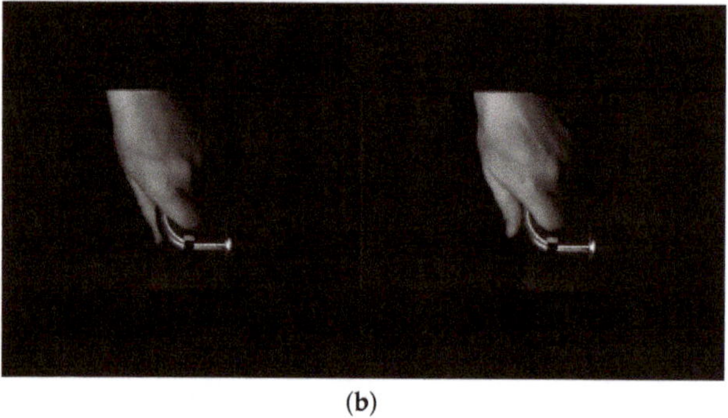

(**b**)

**Figure 8.** Appearance of misclassified examples. Left frames are being misclassified as belonging to the subject on the right. Misclassified hand modality of a subject (**a**). Misclassified foot modality of a subject (**b**).

## 5. Conclusions

In this paper, we presented a dataset containing hand and foot sequences for 77 subjects with the goal of performing automatic people recognition under privacy constraints. The dataset was collected using both RGB and near/infrared camera. We carried out extensive simulations using: (1) handcraft conventional techniques such as LBP, DTW, SRC, and SVM, and (2) deep learning. The results show that poor recognition performance is achieved when applying handcraft techniques, independently of the usage of hand or foot modality. On the other hand, the ResNet-50 deep model evaluated achieves a recognition rate of over 70% for feet and 80% for hands, further improved when fused, showing their complementary nature, and obtaining a final score of 84.4%. Interestingly, the inclusion of optical flow maps to enrich the appearance network channel did not show any improvement. This could have happened because of the limited amount of training data available per participant in the data set. All in all, spatial appearance deep learning showed a high generation performance to recognize users by the combination of hand and foot data.

**Author Contributions:** Formal analysis—K.N., S.E., T.B.M., and G.A.; Methodology—M.N.S.J., P.B.-C., and E.A.; Writing—original draft, M.N.S.J.; Writing—review and editing, P.B.-C., E.A., K.N., S.E., T.B.M., and G.A.

**Funding:** This work has been partially supported by the Spanish project TIN2016-74946-P (MINECO/FEDER, UE), CERCA Programme/Generalitat de Catalunya, and the Estonian Centre of Excellence in IT (EXCITE).

**Acknowledgments:** We gratefully acknowledge the support of NVIDIA with the donation of the GPU used for this research. This work is partially supported by ICREA under the ICREA Academia program.

**Conflicts of Interest:** The authors declare no conflict of interest.

## References


1. Crisan, S. A Novel Perspective on Hand Vein Patterns for Biometric Recognition: Problems, Challenges, and Implementations. In *Biometric Security and Privacy*; Springer: Cham, Switzerland, 2017; pp. 21–49.
2. Litvin, A.; Nasrollahi, K.; Escalera, S.; Ozcinar, C.; Moeslund, T.B.; Anbarjafari, G. A novel deep network architecture for reconstructing RGB facial images from thermal for face recognition. *Multimed. Tools Appl.* **2019**, *78*, 1–13. [CrossRef]
3. Leyvand, T.; Li, J.; Meekhof, C.; Keosababian, T.; Stachniak, S.; Gunn, R.; Stuart, A.; Glaser, R.; Mays, E.; Huynh, T.; et al. Biometric recognition. U.S. Patent 9,539,500, 1 October 2017.
4. Rath, A.; Spasic, B.; Boucart, N.; Thiran, P. Security Pattern for Cloud SaaS: From System and Data Security to Privacy Case Study in AWS and Azure. *Computers* **2019**, *8*, 34. [CrossRef]
5. Campisi, P. *Security and Privacy in Biometrics*; Springer: Cham, Switzerland, 2013.
6. Regulation Protection. Regulation (EU) 2016/679 of the European Parliament and of the Council, April 2016. Available online: http://www.gkdm.co.il/wp-content/uploads/2018/02/GDPR-Israel.pdf (accessed on 23 October 2019).
7. Ofodile, I.; Helmi, A.; Clapés, A.; Avots, E.; Peensoo, K.M.; Valdma, S.M.; Valdmann, A.; Valtna-Lukner, H.; Omelkov, S.; Escalera, S.; et al. Action Recognition Using Single-Pixel Time-of-Flight Detection. *Entropy* **2019**, *21*, 414. [CrossRef]
8. Sapiński, T.; Kamińska, D.; Pelikant, A.; Anbarjafari, G. Emotion recognition from skeletal movements. *Entropy* **2019**, *21*, 646. [CrossRef]
9. Sabet Jahromi, M.N.; Bonderup, M.B.; Asadi, M.; Avots, E.; Nasrollahi, K.; Escalera, S.; Kasaei, S.; Moeslund, T.; Anbarjafari, G. Automatic Access Control Based on Face and Hand Biometrics in A Non-Cooperative Context. In Proceedings of the IEEE Winter Conference on Applications of Computer Vision - Cross Domain Biometric Recognition Workshop, Hawaii, HI, USA, 7 January 2018, pp. 1–9.
10. Ishihara, T.; Kitani, K.M.; Ma, W.C.; Takagi, H.; Asakawa, C. Recognizing hand-object interactions in wearable camera videos. In Proceedings of the IEEE International Conference on Image Processing (ICIP), Quebec City, QC, Canada, 27–30 September 2015; pp. 1349–1353.
11. Cheng, H.; Yang, L.; Liu, Z. Survey on 3D hand gesture recognition. *IEEE Trans. Circuits Syst. Video Technol.* **2016**, *26*, 1659–1673. [CrossRef]
12. Duta, N. A survey of biometric technology based on hand shape. *Pattern Recognit.* **2001**, *42*, 2797–2806. [CrossRef]
13. Bobick, A.F.; Davis, J.W. The recognition of human movement using temporal templates. *IEEE Trans. Pattern Anal. Mach. Intell.* **2001**, *23*, 257–267. [CrossRef]
14. Kumar, A.; Wong, D.C.; Shen, H.C.; Jain, A.K. Personal authentication using hand images. *Pattern Recognit. Lett.* **2006**, *27*, 1478–1486. [CrossRef]
15. Amayeh, G.; Bebis, G.; Erol, A.; Nicolescu, M. Peg-free hand shape verification using high order Zernike moments. In Proceedings of the Conference on Computer Vision and Pattern Recognition Workshop, CVPRW'06, New York, NY, USA, 17–22 June 2006; p. 40.
16. Pavešić, N.; Ribarić, S.; Ribarić, D. Personal authentication using hand-geometry and palmprint features–The state of the art. *Hand* **2004**, *11*, 12.
17. Zheng, J.; Feng, Z.; Xu, C.; Hu, J.; Ge, W. Fusing shape and spatio-temporal features for depth-based dynamic hand gesture recognition. *Multimed. Tools Appl.* **2017**, *76*, 20525–20544. [CrossRef]
18. Polat, Ö.; Yıldırım, T. Hand geometry identification without feature extraction by general regression neural network. *Expert Syst. Appl.* **2008**, *34*, 845–849. [CrossRef]
19. Goswami, G.; Mittal, P.; Majumdar, A.; Vatsa, M.; Singh, R. Group sparse representation based classification for multi-feature multimodal biometrics. *Inf. Fusion* **2016**, *32*, 3–12. [CrossRef]
20. Kumar, A.; Prathyusha, K.V. Personal authentication using hand vein triangulation and knuckle shape. *IEEE Trans. Image Process.* **2009**, *18*, 2127–2136. [CrossRef] [PubMed]
21. Malutan, R.; Emerich, S.; Crisan, S.; Pop, O.; Lefkovits, L. Dorsal hand vein recognition based on Riesz Wavelet Transform and Local Line Binary Pattern. In Proceedings of the 3rd International Conference on Frontiers of Signal Processing (ICFSP), Paris, France, 6–8 September 2017; pp. 146–150.
22. Anbarjafari, G.; Izadpanahi, S.; Demirel, H. Video resolution enhancement by using discrete and stationary wavelet transforms with illumination compensation. *Signal Image Video Process.* **2015**, *9*, 87–92. [CrossRef]

23. Li, X.; Huang, D.; Wang, Y. Comparative study of deep learning methods on dorsal hand vein recognition. In Proceedings of the Chinese Conference on Biometric Recognition, Chengdu, China, 14–16 October 2016, pp. 296–306.

24. Qin, H.; El-Yacoubi, M.A. Deep representation-based feature extraction and recovering for finger-vein verification. *IEEE Trans. Inf. Forensics Secur.* **2017**, *12*, 1816–1829. [CrossRef]

25. Wang, L.; Leedham, G. A thermal hand vein pattern verification system. In Proceedings of the International Conference on Pattern Recognition and Image Analysis, Bath, UK, 22–25 August 2005; pp. 58–65.

26. Wang, L.; Leedham, G.; Cho, D.S.Y. Minutiae feature analysis for infrared hand vein pattern biometrics. *Pattern Recognit.* **2008**, *41*, 920–929. [CrossRef]

27. Xu, Y.; Fan, Z.; Qiu, M.; Zhang, D.; Yang, J.Y. A sparse representation method of bimodal biometrics and palmprint recognition experiments. *Neurocomputing* **2013**, *103*, 164–171. [CrossRef]

28. Wan, H.; Chen, L.; Song, H.; Yang, J. Dorsal hand vein recognition based on convolutional neural networks. In Proceedings of the IEEE International Conference on Bioinformatics and Biomedicine (BIBM), Kansas City, MO, USA, 13–16 November 2017; pp. 1215–1221.

29. Kennedy, R.B. Uniqueness of bare feet and its use as a possible means of identification. *Forensic Sci. Int.* **1996**, *82*, 81–87. [CrossRef]

30. Uhl, A. Footprint-based biometric verification. *J. Electron. Imaging* **2008**, *17*, 011016. [CrossRef]

31. Nakajima, K.; Mizukami, Y.; Tanaka, K.; Tamura, T. Footprint-based personal recognition. *IEEE Trans. Biomed. Eng.* **2000**, *47*, 1534–1537. [CrossRef]

32. Jung, J.W.; Bien, Z.; Lee, S.W.; Sato, T. Dynamic-footprint based person identification using mat-type pressure sensor. In Proceedings of the 25th Annual International Conference of the IEEE Engineering in Medicine and Biology Society (IEEE Cat. No. 03CH37439), Cancun, Mexico, 17–21 September 2003; pp. 2937–2940.

33. Kumar, V.A.; Ramakrishan, M. Employment of footprint recognition system. *Indian J. Comput. Sci. Eng. (IJCSE)* **2013**, *3*, 774–778.

34. Barker, S.; Scheuer, J. Predictive value of human footprints in a forensic context. *Med. Sci. Law* **1998**, *38*, 341–346. [CrossRef] [PubMed]

35. Kumar, V.A.; Ramakrishnan, M. Manifold feature extraction for foot print image. *Indian J. Bioinform. Biotechnol.* **2012**, *1*, 28–31.

36. Kushwaha, R.; Nain, N.; Singal, G. Detailed analysis of footprint geometry for person identification. In Proceedings of the 13th International Conference on Signal-Image Technology & Internet-Based Systems (SITIS), Jaipur, India, 4–7 December 2017; pp. 229–236.

37. Rohit Khokher, R.C.S. Footprint-based personal recognition using scanning technique. *Indian J. Sci. Technol.* **2016**, *9*. [CrossRef]

38. Boyd, J.E.; Little, J.J. Biometric gait recognition. *Advanced Studies in Biometrics*; Springer: Cham, Switzerland 2005; pp. 19–42.

39. MathWorks. Single Camera Calibrator App, 2018. Available online: https://www.mathworks.com/help/vision/ug/single-camera-calibrator-app.html (accessed on 23 October 2019).

40. Ahonen, T.; Hadid, A.; Pietikäinen, M. Face recognition with local binary patterns. In Proceedings of the European Conference on Computer Vision, Prague, Czech Republic, 11–14 May 2004; pp. 469–481.

41. Ahonen, T.; Hadid, A.; Pietikainen, M. Face description with local binary patterns: Application to face recognition. *IEEE Trans. Pattern Anal. Mach. Intell.* **2006**, *28*, 2037–2041.

42. Demirel, H.; Anbarjafari, G. Data fusion boosted face recognition based on probability distribution functions in different colour channels. *EURASIP J. Adv. Signal Process.* **2009**, *2009*, 25. [CrossRef]

43. Benzaoui, A.; Boukrouche, A.; Doghmane, H.; Bourouba, H. Face recognition using 1DLBP, DWT and SVM. In Proceedings of the 3rd International Conference on Control, Engineering & Information Technology (CEIT), Tlemcen, Algeria, 25–27 May 2015; pp. 1–6.

44. Drucker, H.; Burges, C.J.; Kaufman, L.; Smola, A.J.; Vapnik, V. Support vector regression machines. In Proceedings of the 9th International Conference on Neural Information Processing Systems, Denver, CO, USA, 3–5 December 1997; pp. 155–161.

45. Schölkopf, B.; Burges, C.; Vapnik, V. Incorporating invariances in support vector learning machines. In Proceedings of the International Conference on Artificial Neural Networks, Bochum, Germany, 16–19 July 1996; pp. 47–52.

46. Elshatoury, H.; Avots, E.; Anbarjafari, G.; Initiative, A.D.N. Volumetric Histogram-Based Alzheimer's Disease Detection Using Support Vector Machine. Available online: https://content.iospress.com/articles/journal-of-alzheimers-disease/jad190704 (accessed on 23 October 2019).
47. Cherifi, D.; Cherfaoui, F.; Yacini, S.N.; Nait-Ali, A. Fusion of face recognition methods at score level. In Proceedings of the International Conference on Bio-engineering for Smart Technologies (BioSMART), Dubai, UAE, 4–7 December 2016; pp. 1–5.
48. Ojala, T.; Pietikäinen, M.; Mäenpää, T. Multiresolution gray-scale and rotation invariant texture classification with local binary patterns. *IEEE Trans. Pattern Anal. Mach. Intell.* **2002**, *24*, 971–987.
49. Wright, J.; Ma, Y.; Mairal, J.; Sapiro, G.; Huang, T.S.; Yan, S. Sparse representation for computer vision and pattern recognition. *Proc. IEEE* **2010**, *98*, 1031–1044. [CrossRef]
50. Li, X.; Jia, T.; Zhang, H. Expression-insensitive 3D face recognition using sparse representation. In Proceedings of the IEEE Conference on Computer Vision and Pattern Recognition, Miami, FL, USA, 20–25 June 2009; pp. 2575–2582.
51. Mairal, J.; Bach, F.; Ponce, J.; Sapiro, G.; Zisserman, A. *Discriminative Learned Dictionaries for Local Image Analysis*; Minnesota Univ. Minneapolis Inst. for Mathematics and Its Applications: Minneapolis, MN, USA; 2008.
52. Cai, J.F.; Ji, H.; Liu, C.; Shen, Z. Blind motion deblurring from a single image using sparse approximation. In Proceedings of the Conference on Computer Vision and Pattern Recognition, Miami, FL, USA, 20–25 June 2009; pp. 104–111.
53. Zhang, Q.; Li, B. Discriminative K-SVD for dictionary learning in face recognition. In Proceedings of the Computer Society Conference on Computer Vision and Pattern Recognition, San Francisco, CA, USA, 13–18 June 2010; pp. 2691–2698.
54. Azarbayejani, A.; Pentland, A.P. Recursive estimation of motion, structure, and focal length. *IEEE Trans. Pattern Anal. Mach. Intell.* **1995**, *6*, 562–575.
55. Krizhevsky, A.; Sutskever, I.; Hinton, G.E. Imagenet classification with deep convolutional neural networks. In Proceedings of the Neural Information Processing Systems Conference, Lake Tahoe, NV, USA, 3–8 December 2012; pp. 1097–1105.
56. He, K.; Zhang, X.; Ren, S.; Sun, J. Deep residual learning for image recognition. In Proceedings of the Conference on Computer Vision and Pattern Recognition, Las Vegas, NV, USA, 26 June–1 July 2016; pp. 770–778.

*Article*

# Emotional Speech Recognition Based on the Committee of Classifiers

**Dorota Kamińska**

Lodz University of Technology, Institute of Mechatronics and Information Systems, Stefanowskiego 18/22 Str. 90-924 Lodz, Poland; dorota.kaminska@p.lodz.pl

Received: 12 August 2019; Accepted: 19 September 2019; Published: 21 September 2019

**Abstract:** This article presents the novel method for emotion recognition from speech based on committee of classifiers. Different classification methods were juxtaposed in order to compare several alternative approaches for final voting. The research is conducted on three different types of Polish emotional speech: acted out with the same content, acted out with different content, and spontaneous. A pool of descriptors, commonly utilized for emotional speech recognition, expanded with sets of various perceptual coefficients, is used as input features. This research shows that presented approach improve the performance with respect to a single classifier.

**Keywords:** emotion recognition; speech; committee of classifiers

---

## 1. Introduction

During a conversation people are constantly sending and receiving different nonverbal clues, communicated through speech signal (paralanguage), body movements, facial expressions, and physiological changes. The discrepancy between the words spoken and the interpretation of their actual content relies on nonverbal communication. Emotions are a medium of information regarding feelings of an individual and one's expected feedback. The ability to recognize the attitude and thoughts from one's behaviour was the original system of communication prior to spoken language. Understanding the emotional state enhances interaction. Although computers are now a part of human life, the relation between human and machine is far from being natural [1]. Proper identification of emotional state can significantly improve quality of human-computer interfaces. It can be applied for monitoring of psycho-physiological states of individuals e.g., to assess the level of stress or fatigue, forensic data analysis [2], advertisement [3], social robotic [4], video conferencing [5], violence detection [6], animation or synthesis of life-like agents xue2018voice, and many others. Automatic emotion recognition methods utilize various input types i.e., facial expressions [7–9], speech [10–12], gesture and body language [13,14], physical signals such as electrocardiogram (ECG), electromyography (EMG), electrodermal activity, skin temperature, galvanic resistance, blood volume pulse (BVP), and respiration [15]. Facial expressions have been studied most extensively and about 95% of literature dedicated to this topic focuses on faces as a source, at the expense of other modalities [16]. Speech is one of the most accessible form the above mentioned signals, thus recently it is increasingly significant research direction in emotion recognition. Despite an enormous amount of research, the issue is still far from its satisfactory solution. Analysis of emotional content embedded in speech is an issue that presents multiple difficulties. The main problem is gathering and compiling a database of viable and relevant experimental material. Most available corpora comprise speech samples uttered by professional actors, which are not guaranteed to reflect the real environment with its background noise or overlapping voices. Additionally, individual features of the speaker such as gender, age, origin and social influence can greatly affect universal consistency in emotional speech. The first most important work published before 20th century studying emotions was *The Expression of the Emotions in Man and*

*Animals* by Charles Darwin [17]. Darwin made the first description of the paralanguage conveying emotional states of the speaker. Based on the study of people and different species of animals, he came to the conclusion that there is a direct connection between the modulation of speech signal and the internal state of the individual. He also observed that acoustic signals could trigger emotional reactions of the listener. The theoretical and practical approach suggests that specific paralinguistic cues such as loudness, rate, pitch, pitch contour and formant frequencies contribute to the emotive quality of an utterance. Emotions may cause changes in the way of breathing, phonation or articulation, which are reflected in the speech. For example, states like anger or fear are characterized by fast pace, high values of pitch, wide range of intonation, sudden acceleration of heart rate, increased blood pressure and, in some cases, dry mouth and muscle tremor. The opposite phenomena occur in case of sadness and boredom. Speech becomes slow and monotonous, pitch is reduced without any major changes in intonation. This is caused partially due to activation of the parasympathetic system, relief of cardiac rhythm, blood pressure drop and increased secretion of saliva. Consequently, paralinguistic cues relating to emotion have a huge effect on ultimate meaning of the message [18]. This paper refers to my previous research [19], where the novel method for emotional speech recognition based on committee of classifiers was presented. This method is based on a set of classifiers (nodes) whose individual predictions are combined to make the final decision. Current paper is an extension of the previous approach. I investigated three different type of Polish corpora: acted out, in which the actors repeat the same sentence while expressing different emotional states [20]; acted out, in which the actors repeat several different sentences while expressing different emotional states [2]; spontaneous speech samples collected from live shows and programs such as reality shows [21]. I combined different classification methods as nodes (k-NN, MLP, SL, SMO, Bagging, RC, j48, LMT, NBTree, RF) and juxtaposed several alternative approaches to final voting. This research shows that some of presented approaches improve the performance with respect to a single classifier. A pool of descriptors, commonly utilized for emotional speech recognition, expanded with sets of various perceptual coefficients, is used as input feature vectors. The following list summarises the contributions of this work:

1. This research is carried out on three different types of Polish corpora, which allows the analysis of impact of various type of database on the final result. The classifiers were tested by using mixed sets (corpora-dependent and corpora-independent tests) to verify if acted out database can be used as a training set for application operating in real environment.
2. In comparison to similar research where each classifier is trained with the exact same data, in this paper the whole feature set is divided into subsets before classification process. Despite their similarity (e.g., MFCC and BFCC) different models provide varied results on specific feature subsets, affecting the final assessment. Thus, the most effective model may be selected appropriately for a specific subset (in similar research voting is performed on different classifiers working on the same features). This approach significantly increases accuracy of results, in comparison to related works. Presented algorithm was verified using different voting methods.
3. It presents a thorough analysis of extensive set of features on the recognition of several emotional classes-groups of features are examined separately as well as a whole collection.

The structure of the paper is as following. Next section presents a brief review of works related to speech emotion recognition (SER). Section 3 describes proposed research methodology: relevant corporas of emotional voice, speech signal descriptors and outline of adopted strategy for emotion recognition. Section 4 presents obtained results followed by their discussion. Finally, Section 5 gives the conclusion and future directions of this research.

## 2. Related Works

Since emotion recognition from speech signal is a pattern recognition problem, standard approach consisting of three processes: feature extraction, feature selection, and classification is used to solve the task. The main research issue is selection of an optimal feature set that efficiently characterizes the emotional content of the utterance. The number of acoustic parameters proven to contain emotional information is still increasing. Generally, the most commonly used features can be divided into three groups: prosodic features (e.g., fundamental frequency, energy, speed of speech) [22], quality characteristics (e.g., formants, brightness) [23] and spectrum characteristics (e.g., mel-frequency cepstral coefficients) [24,25]. The final features vector is based on their statistics such as mean, maximum, minimum, change rate, kurtosis, skewness, zero-crossing rate, variance etc., [26,27]. However, a vector of too many features may give rise to high dimension and redundancy, making the learning process complicated and increasing the likelihood of overfitting [28]. Therefore prior to classification, methods of balancing a numerous features vector, feature selection or extraction are studied to speed up the learning process and minimize the curse of dimensionality problem [29,30]. Emotion classification is generally performed using standard techniques such as SVM [31–33], various types of artificial neural networks (NN) [34–37], different types of the k-NN classifier [19,38] or using Hidden Markov Model (HMM) and its variations [39]. However, it is a complex task with many unresolved issues. Therefore, hybrids and multilevel classifiers [40,41] or ensemble models [42] have been widely used to enhance the performance of single classifiers. Classifying committees (Ensemble, Committee, Multiple Classifier Systems) are based on the principle of *divide and conquer*: they consist of a set of classifiers (nodes) whose individual predictions are combined. A necessary condition for this approach is that member classifiers should have a substantial level of disagreement, i.e., mistakes made by nodes should be independent, regardless of the others. The most commonly used and most intuitive technique consists of several models $C$ (Figure 1a) working separately on the same or similar feature set, with their results merged on decision $D$ level

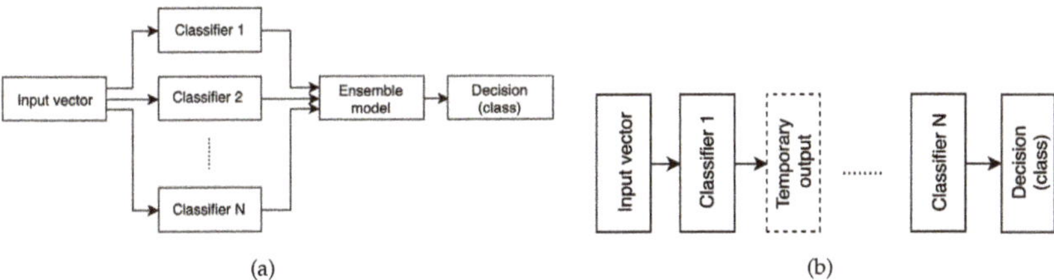



**Figure 1.** (a) Combining the results via simple voting, weighted or highest confidence voting, or other methods. (b) Multilevel classification.

This kind of approach was used in [43], where the authors present a multiple classifier system for 5 emotional states (anger, happiness, sadness, boredom and neutral) and task is performed on Mandarin speech. They investigated several classifiers such as k-NN, weighted k-NN, Weighted Average Patterns of Categorical k-NN, Weighted Discrete k-NN and SVM. To combine results, majority voting, minimum misclassification and maximum accuracy methods were compared. The experimental results have shown that classifier combination schemes perform better than the single classifiers with the improvement ranging from 0.9–6.5%. The improvement of the automatic perception of vocal emotion using ensemble methods over traditional classification is shown in [44]. The authors compared two emotional speech data sources: natural, spontaneous emotional speech and acted or portrayed emotional speech to demonstrate the advantages and disadvantages of both. Basing on prosodic features (namely: fundamental frequency, energy, rhythm, and formant frequencies) two ensemble methods (stacked generalisation and unweighted vote) were applied. These techniques shown a modest improvement in prediction accuracy. In [45], the authors analysed the effectiveness of employing five ensemble models such as Bagging, Adaboost, Logitboost, Random Subspace and Random Committee, estimating emotional Arabic speech. The system recognizes happy, angry, and surprise emotion from natural speech samples. The highest improvement in accuracy in relation to the classical approach (19.09%) was obtained by the Boosting technique having the Naïve Bayes

Multinomial as the node. Multilevel approach (see Figure 1b) is predicated on splitting the classification process into several consecutive stages. For example in [46] the authors propose a hierarchical classification, which achieves greater accuracy of SER than corresponding classical methods. In the first stage of this algorithm, features vector is used to separate anger and neutral (group 1) from happiness and sadness (group 2). Finally, group 1 is classified into anger and neutrality, and group 2 into happiness and sadness. Similar approach is presented in [47]. First, the emotional states are categorized according to the dimensional model into positive or negative valence and high or low arousal using Gaussian Mixture Model and Support Vector Machines. Final decisions are made inside subsets with fewer categories using spectral representation. Studies were performed using the Berlin Emotional database [48] and the Surrey Audio-Visual Expressed Emotion corpus. In [49], the authors studied the effect of age and gender of the speaker on the effectiveness of emotion recognition system. They proposed a hierarchical classification model to investigate the importance of identifying those features before identifying the emotional label. They compared the performance of four different models and presented the relationship between the age gender and the emotion recognition accuracy. The results proved that using a separate emotion model for each gender and age category gives a higher accuracy compared with using one classifier for all the data. Similarly, in [50], gender is identified on the first level. Next, the dimensional reduction using PCA, LDA and mixed algorithm is performed according to particular gender-set. In [51], the authors underline a fuzzy nature of particular emotional states (e.g., sadness and boredom) and suggest that global classifier cannot obtain effective results. Thus, they proposed a hierarchical approach, which divides the set of utterances into *active* and *passive* on the first level, in order to classify them into emotional categories on the second one. The experiments were conducted on two different corpora: Berlin and DES [52] database. Obtained results outperform those obtained via single classifier.

## 3. Methods

### 3.1. Database

As mentioned in Section 1, for the purpose of this project three different types of Polish datasets were investigated. They will be briefly described below and summarised in Table 1.

**Table 1.** Main characteristics of databases investigated in this research.

| Database | No. of Samples/Per Emotion | Female/Male | Type | No. of Emotions |
| --- | --- | --- | --- | --- |
| MERIP | 560/unsp | 8/8 | acted | 7: Ne, Sa, Su, Fe, Di, An, Ha |
| PESD | 240/40 | 4/4 | acted | 6: Ha, Bo, Fe, An, Sa, Ne |
| PSSD | 748/80 | nd/nd | natural | 8: Ha, Sa, An, Fe, Di, Su, An, Ne |

#### 3.1.1. MERIP Database

MERIP emotional speech database is a subset of the Multimodal Emotion Recognition in Polish project [20]. The database consists of 560 samples recorded in the rehearsal room of *Teatr Nowy im. Kazimierza Dejmka w Łodzi*. Samples were collected from separate utterances of 16 professional actors/actresses (8 male and 8 female) aged from 25 to 64. The subjects were asked to utter a sentence *Każdy z nas odczuwa emocje na swój sposób* (English translation: *Each of us perceives emotions in a different manner*) while expressing different emotional states in the following order: neutral, sadness, surprise, fear, disgust, anger, and happiness (this set of discrete emotions was based on examination conducted by Ekman in [53]). All emotions were acted out 5 times, without any guidelines or prompts from the researchers. This allowed to gather 80 samples per each emotional state. Audio files were captured using dictaphone Roland R-26 in the form of wav audio files 44.1 kHz, 16 bit, stereo). The samples were evaluated by 12 subjects (6 male and 6 female) who were allowed to listen each sample only once and determine the emotional state. The average emotion recognition rate was 90% (ranging from 84% to 96% for different emotional state).

### 3.1.2. Polish Emotional Speech Database

The Polish Polish Emotional Speech Database (PESD) [2] was prepared and shared by the Medical Electronics Division, Lodz University of Technology. The database consists of 240 samples recorded in the aula of the Polish National Film Television and Theater School in Lodz. Samples were collected from separate utterances of 8 professional actors/actresses (4 male and 4 female). Each speaker was asked to utter five different sentences (*They have bought a new car today, His girlfriend is coming here by plane, Johnny was today at the hairdresser's, This lamp is on the desk today* and *I stop to shave from today on*) with six types of emotional load: joy, boredom, fear, anger, sadness, and neutral (no emotion). Audio data was collected in the form of wav audio files (44.1 kHz, 16 bit). The samples were evaluated by 50 subjects through a procedure of classification of 60 randomly generated samples (10 samples per particular emotion). Listeners were asked to classify each utterance into emotional categories. The average emotion recognition rate was 72% (ranging from 60 to 84% for different subjects).

### 3.1.3. Polish Spontaneous Speech Database

The spontaneous Polish Speech Database (PSSD) [21] consists of 748 samples containing emotional carrier of seven basic states, from the Plutchik's wheel of emotions [54]: joy, sadness, anger, fear, disgust, surprise, anticipation and neutral. Speech samples were collected from discussions in TV programs, live shows or reality shows and the proportion of speakers' gender and age was maintained. Each utterance was unique and varied from one-word articulations such as *Yes* or *No*, single words, phrases to short sentences. Occasionally additional sounds such as screaming, squealing, laughing or crying are featured in the corpora. The data was collected in the form of wav audio files of varied quality. The samples were evaluated by 15 male and female volunteers aged from 21 to 58. All listeners were presented random samples that consisted of at least half of each prequalified basic emotions recordings. The evaluators listened to audio samples one by one, each assessment was recorded in the database. Every sample could have been played any number of times before the final decision, but after the classification, it was not possible to return to the recording. Average emotion recognition was 82.66% (ranging from 63% to 93% for different subjects).

To juxtapose these three different databases for the purpose of this project, an equal number of emotional sets was selected, which means that utterances expressing surprise and anticipation were omitted. Additionally, in case of PSSD, the number of samples for emotions has been unified to 80.

### 3.2. Extracted Features

Representation of speech signal in time or frequency domain is too complex to analyze, thus usually high-level statistical features (HLS) are sought to determine its properties. In most cases a large number of HLS features are extracted at the utterance level, which is followed by dimension reduction techniques to obtain a robust representation of the problem. Feature extraction comprises of two different stages. First, a number of low level (LL) features are extracted from short frames. Next, HLS features such as mean, max, min, variance, std, are applied to each of the LLs over the whole utterance, and the results are concatenated into a final feature vector. The role of the HLS is to describe temporal variations and contours of the different LLs during particular speech chunk [55]. Most commonly used LLs, for the purpose of emotional speech recognition, can be divided into two groups: prosodies and spectrum characteristics, both of them described below.

### 3.2.1. Prosodies

Speech prosodic features are associated with larger units such as syllables, words, phrases, and sentences, thus are considered as supra-segmental information. They represent the perceptual properties of speech, which are commonly used by humans to carry various information [56]. As it has been repeatedly emphasised in the literature, prosodic features such as energy, duration, intonation

(*F0* contour) and their derivatives are commonly used as important information sources for describing emotional states.

*F0*, which is the frequency of vocal folds, is inextricably linked with the scale of the human voice, accent and intonation, all of which have a considerable impact on the nature of speech. *F0* does change during utterances and rate of those changes is dependent on the speaker's intended intonation [22]. For the purpose of this research *F0* was extracted using autocorrelation technique. The analysis window was set to 20 ms with 50% overlap.

Another feature that provides information useful in distinguishing emotions is signal energy, which describes the volume or intensity of speech. For example, some emotional states, like joy or anger, have increased energy levels in comparison to other emotional states.

### 3.2.2. Spectrum Characteristics

Nowadays, perceptual features are a standard in voice recognition. They are also used in emotional speech analysis. Perceptual approach is based on frequency conversion, corresponding to subjective reception of the human auditory system. For this purpose, the perceptual scales such as Mel or Bark are used. In this paper Mel Frequency Cepstral Coefficients *MFCC* [57], Human Factor Cepstral Coefficients *HFCC* [58], Bark Frequency Cepstral Coefficients *BFCC* [59], Perceptual Linear Prediction *PLP* [60] and Revised Perceptual Linear Prediction *RPLP* [59] coefficients are employed. Additionally, Linear Prediction Coefficients (LPC) [61] were taken into consideration, as they are the most frequently used features for speech recognition. Initially, for all particular perceptual features sets, the number of coefficients has been specified to 12. For all above mentioned LLs sets. HLS such as maximum, minimum, range, mean and standard deviation were determined for all LLs.

Another important feature type, describing properties of vocal tract, are formant frequencies, at which local maxima of the speech signal spectrum envelope occur. They can be utilized to determine the speaker's identity and the form and content of their utterance [62]. Usually 3 to 5 formants are applied in practice, thus this paper estimates 3 of them and on their basis HLS such as mean, median, standard deviation, maximum and minimum are determined, giving a total of 15 features.

### 3.2.3. Features Selection

Initially, the number of extracted HLS features amounted to 407. Correlation-based Feature Selection (CFS) algorithm [63] has been applied on the whole set of features as well as on all subsets separately in order to remove redundancy and select descriptors most relevant for analysis.

This procedure resulted in a significant reduction of the feature vector dimension, after CFS the final vectors length was: 93 in case of MERIP, 88 for PESD and 91 for PSSD. Distribution of features before and after the selection process applied on a particular subset is presented in Figure 2. Selected features are presented in Appendix A, Tables A1 and A2.

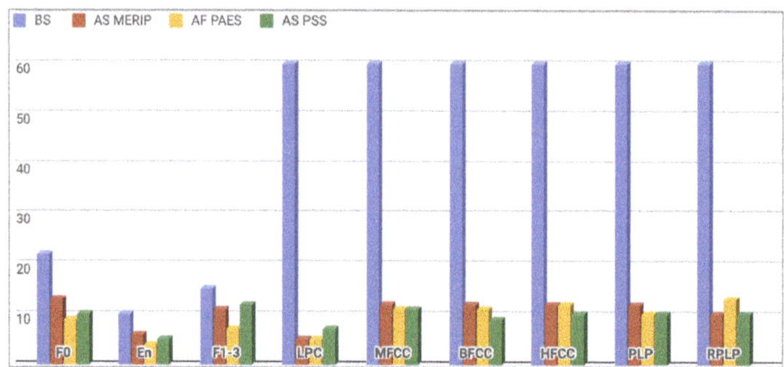


**Figure 2.** Distribution of features count for particular sets before and after selection process for each database. BS—before selection, AS—after selection.

### 3.3. Classification Model

Proposed algorithm, presented in see Figure 3, starts with division of the HFL feature vector, describing speech samples, into separate sub-vectors of particular group of features (i.e., sub-vector with MFCC coefficients). Each sub-vector is subjected to the selection process, followed by classification using different models $M$ (e.g., M1: k-NN, M2: MLP etc.). Subsequently, among the models operating on particular sub-vector, one model with the lowest error rate is selected for further analysis. The error rate is calculated according to Equation (1). Final voting is done among the highest scoring models for particular sub-vectors.

$$err = 1 - accuracy = 1 - \frac{(\#classified\_correct)}{(\#classified\_total)} = \frac{(\#classified\_incorrect)}{(\#classified\_total)} \tag{1}$$

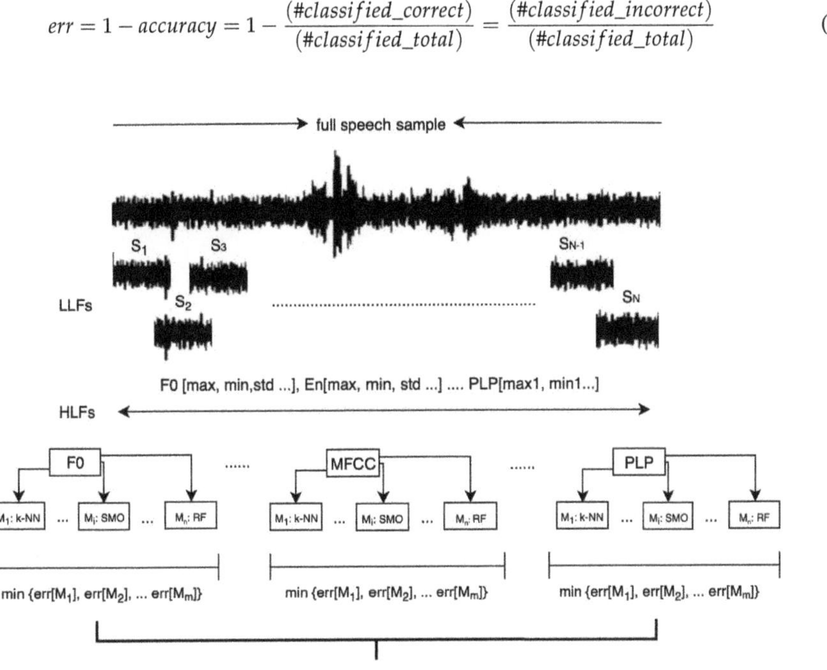


**Figure 3.** Proposed algorithm for emotion recognition using committee of classifiers.

In the basic algorithm, the final decision is made using equal voting. This method does not require additional calculations, only votes of individual models, rendering this process simple and effective. A decision is made collectively, using the following equations:

$$r_i = \sum_{j=1}^{m} d_{ji} \tag{2}$$

$$Z = arg \max_{i=1}^{l} [r_i] \tag{3}$$

where: $m$—number of classifiers (models), $l$—number of different classes, $d_{ji}$—decision of $j$ classifier for $i$ class.

Unequal impact of particular descriptors on the recognition provides the basis for replacing equal with the weighted voting. For each model different weights $w_1, w_2, \ldots w_m$ are determined, which allows to prioritize more precise models. In this case the Equation (2) is replaced by the following:

$$r_i = \sum_k^{j=1} d_{ji} w_j \qquad (4)$$

This approach requires the assessment (or at least comparison) of all models. In this study weights were selected experimentally, based on the error rate of individual classifiers. Appropriate weight $w_i$ for individual model was calculated based on the error rate $err_i$ according to the following equations.

$$w_i = 1 - err_i \qquad (5)$$

$$w_i = \frac{1}{err_i} \qquad (6)$$

$$w_i = \left(\frac{1}{err_i}\right)^2 \qquad (7)$$

## 4. Results and Discussion

### 4.1. Efficiency of Features Subsets

The verification of efficiency of feature subsets is carried out using several types of classifiers such as k-NN, Multilayer Perceptron (MLP), Simple Logistic (SL), SMO, Bagging, Random Cometee (RC), j48, LMT, NBTree and Random Forest (RF) using Weka [64], with 10-fold cross-validation. This approach allows to evaluate the efficacy of particular features set and determine the most efficient ones. Tables 2–4 present the efficiency of above mentioned feature subsets obtained for three independent speech corpora. In the course of research, the parameters for each classifier were identified and selected to achieve the highest recognition results.

**Table 2.** Average recognition results [%] of features subsets for MERIP database.

|  | k-NN | MLP | SL | SMO | Bagging | RC | j48 | LMT | NBTree | RF |
|---|---|---|---|---|---|---|---|---|---|---|
| F0 | 34,02 | 32,39 | 34,75 | 34,75 | 34,75 | 34,75 | 30,26 | 33,33 | 26,47 | **37,35** |
| En | 30,91 | 29,03 | 26,93 | 25,99 | 33,49 | 28,57 | 26,22 | **31,62** | 29,03 | 28,1 |
| F1-F3 | 36,06 | 35,36 | 39,11 | 36,06 | 34,43 | 33,02 | 29,03 | **40,04** | 33,26 | 33,3 |
| LPC | 42,39 | 45,2 | 44,73 | 41,22 | 39,11 | 42,39 | 32,55 | 44,26 | 34,43 | **45,9** |
| MFCC | **59,33** | 57,37 | 51,52 | 53,63 | 50,58 | 54,1 | 44,73 | 51,05 | 44,96 | 51,3 |
| BFCC | 57,21 | **58,39** | 52,69 | 54,56 | 51,05 | 53,63 | 45,19 | 57,61 | 46,37 | 56,9 |
| HFCC | 39,11 | 33,02 | 37 | 37 | 35,12 | 39,81 | 33,72 | 37,23 | 37 | **42,86** |
| PLP | 56,44 | 49,18 | 43,09 | 40,75 | 47,3 | 54,1 | 44,26 | 53,16 | 41,69 | **54,8** |
| RPLP | 55,5 | 43,09 | 39,11 | 43,32 | 50,11 | 52,46 | 46,37 | 47,54 | 40,28 | **56,9** |

**Table 3.** Average recognition results [%] of features subsets for PESD database.

|       | k-NN  | MLP   | SL    | SMO   | Bagging | RC    | j48   | LMT   | NBTree | RF    |
|-------|-------|-------|-------|-------|---------|-------|-------|-------|--------|-------|
| F0    | 33,76 | 34,18 | 38,82 | 34,18 | 30,8    | 32,91 | 30,8  | 37,55 | 32,49  | **40,08** |
| En    | 35,02 | 34,17 | 34,17 | 29,95 | **37,97** | 32,06 | 29,95 | 37,55 | 35,86  | 33,75 |
| F1-F3 | 38,82 | 36,71 | 34,18 | 36,29 | 38,4    | 37,55 | 37,97 | 34,17 | 32,91  | **39,24** |
| LPC   | 38,82 | 32,06 | 38,37 | 37,55 | 32,49   | 33,75 | 32,07 | **38,96** | 31,22 | 31,22 |
| MFCC  | 38,82 | **58,22** | 51,89 | 57,38 | 45,99 | 49,36 | 41,77 | 51,89 | 39,24  | 54,43 |
| BFCC  | 54    | **61,6** | 49,37 | 59,49 | 42,19  | 49,36 | 41,35 | 48,1  | 43,46  | 55,69 |
| HFCC  | 55,27 | 51,05 | 56,19 | **56,96** | 47,26 | 50,21 | 42,62 | 55,27 | 41,77  | 52,74 |
| PLP   | **57,74** | 53,58 | 56,96 | 49,36 | 47,68 | 54,85 | 44,72 | 55,69 | 43,04  | 56,11 |
| RPLP  | 54,43 | **60,34** | 54,43 | 53,59 | 48,95 | 47,26 | 42,19 | 53,16 | 41,77  | 56,96 |

**Table 4.** Average recognition results [%] of features subsets for PSSD database.

| Features: | k-NN  | MLP   | SL    | SMO   | Bagging | RC    | j48   | LMT   | NBTree | RF    |
|-----------|-------|-------|-------|-------|---------|-------|-------|-------|--------|-------|
| F0        | 48,86 | 48,57 | 50,28 | 48,01 | **51,42** | 49,14 | 41,76 | 51,13 | 42,04  | 53,69 |
| En        | 55,39 | 52,84 | 45,45 | 42,61 | **57,95** | 50    | 50,85 | 55,11 | 51,7   | 55,96 |
| F1-F3     | 53,97 | 52,56 | 57,95 | **59,94** | 54,26 | 53,41 | 50,28 | 57,67 | 45,45  | 58,52 |
| LPC       | 65,9  | **70,73** | 67,89 | 66,19 | 61,93 | 66,19 | 63,92 | 67,89 | 57,38  | 68,18 |
| MFCC      | 76,7  | 76,98 | 71,85 | 76,13 | 64,2    | 73,57 | 63,35 | 71,87 | 64,2   | **80,68** |
| BFCC      | 76,32 | 72,29 | 77,55 | 77,55 | 69,66   | 71,36 | 62,85 | 77,7  | 63,46  | **79,87** |
| HFCC      | 72,29 | 69,97 | 72,91 | 71,82 | 74,67   | 71,39 | 61,76 | 72,91 | 60,22  | **74,61** |
| PLP       | **76,42** | 74,14 | 71,59 | 71,02 | 67,89 | 74,43 | 65,34 | 71,59 | 67,05  | 74,43 |
| RPLP      | 72,29 | 69,45 | 68,11 | 69,67 | 66,09   | 71,21 | 55,72 | 68,11 | 54,48  | **74,45** |

It is clearly visible that the best results are achieved for the subsets containing perceptual coefficients (MERIP: 59.33% using MFCC, PESD: 61.5% using BFCC, PSSD: 80.68% using MFCC). In each case, these results are obtained using a different classification algorithms: k-NN, MLP, RF, for MERIP, PESD and PSSD respectively. The lowest results are collected in case of F0, formants and energy and this is noticeable for all datasets.

Analyzing results retrieved from different models, in most cases, a significant recognition rate improvement when using the RF classifier can be observed. it is very evident especially for MERIP and PSSD corpora, where the best results were gathered using RF for 6 out of 10 models in case of MERIP and 5 out of 10 in case of PSSD. When it comes to PESD, MLP gives the best recognition results for 4 out of 10 models. Other classifiers (k-NN, SMO, Bagging or LMT) give best results in individual cases, but without any repeatable pattern. SL, RC, NBTree and j48 algorithms did not take the lead in any model and thus will be omitted in further analysis.

There is a discrepancy between different types of databases (acted out: MERIP and PESD, and spontaneous PSSD) as well as between the same type of databases (MERIP and PESD). Thus, it can be assumed that recognition is affected not only by the type of database, but also by its size and by the type of samples such as uttered sentences and individual features of the speaker. Such varied results and the lack of repeatability indicates the necessity of conducting efficiency tests and selection of appropriate methods every time the corpora is modified.

*4.2. Efficiency of Proposed Algorithm*

Based on the results presented in the previous section, classifiers providing highest results on specific feature sets are selected to be part of the proposed algorithm. Thus, for example, in case of MERIP, the final algorithm consists of: RF for F0, LPC, HFCC, PLP, and RPLP LMT for energy and formants, k-NN for MFCC, MLP for BFCC. Next, the error rate of each model is taken into account to calculate the weights for weighted voting (see Figure 4). To assess the proposed method, the results are compared with those obtained using classical approach: using common classifiers on the whole feature set (see Table A1).

According to Tables 5 and 6 an improvement of the overall accuracy using proposed algorithm can be observed in comparison to commonly used classifiers for all datasets. The lowest increase of results is observed for MERIP: MLP gives 66.9% and the third method of weighted voting 69.38%. It is important to note that equal voting among best models gives lower recognition than MLP. Significantly improved recognition quality can be observed in case of PESD and PSSD, where the proposed method boost the overall accuracy from 66.83% (k-NN) and 83.52% to 76.25% and 86.14% for weighted voting respectively. In case of PSSD dataset equal voting gives the same results as MLP. The average accuracy on MERIP, PESD and PSSD databases is illustrated as a confusion matrix in Figure 5.

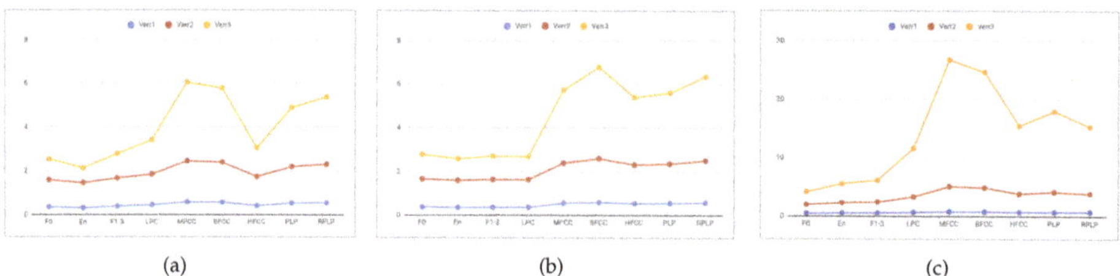

(a)        (b)        (c)

**Figure 4.** The discrete distribution of the error rate obtained by selected models for each features sets for (**a**) MERIP (**b**) PESD (**c**) PSSD.

**Table 5.** Average recognition results [%] of all features subsets for MERIP, PESD and PSSD database using commonly known classifiers.

|       | k-NN  | MLP   | SL    | SMO   | Bagging | RC    | j48   | LMT   | NBTree | RF    |
|-------|-------|-------|-------|-------|---------|-------|-------|-------|--------|-------|
| MERIP | 60,99 | 66,9  | 60,75 | 61,22 | 48,22   | 56,26 | 49,4  | 60,28 | 43,97  | 59,1  |
| PESD  | 66,83 | 62,82 | 59,4  | 57,27 | 59,15   | 57    | 41,88 | 60,25 | 47,86  | 65,55 |
| PSSD  | 78,97 | 83,52 | 80,96 | 83,23 | 74,43   | 77,84 | 65,9  | 80,68 | 69,03  | 81,53 |

**Table 6.** Average recognition results [%] of all features subsets for MERIP, PESD and PSSD database using proposed algorithm with equal voting (EV) juxtaposed with three different approaches for weighted voting.

|       | EV    | $w_i = 1 - err_i$ | $w_i = \frac{1}{err_i}$ | $w_i = \left(\frac{1}{err_i}\right)^2$ |
|-------|-------|-------------------|-------------------------|----------------------------------------|
| MERIP | 60,99 | 66.43             | 67.61                   | **69,38**                              |
| PESD  | 64,58 | **76,25**         | 72,39                   | 74,58                                  |
| PSSD  | 83,52 | 84,94             | 85,22                   | **86,14**                              |

|        | An | Di | Fe | Ha | Ne | Sa |
|--------|----|----|----|----|----|----|
| **An** | 56 | 6  | 5  | 10 | 2  | 1  |
| **Di** | 6  | 53 | 9  | 3  | 6  | 3  |
| **Fe** | 0  | 10 | 43 | 8  | 9  | 10 |
| **Ha** | 11 | 4  | 5  | 59 | 0  | 1  |
| **Ne** | 0  | 0  | 1  | 0  | 70 | 9  |
| **Sa** | 0  | 3  | 5  | 0  | 20 | 52 |

(a)

|        | An | Bo | Fe | Ha | Ne | Sa |
|--------|----|----|----|----|----|----|
| **An** | 35 | 0  | 0  | 4  | 0  | 1  |
| **B**  | 0  | 24 | 4  | 0  | 7  | 5  |
| **Fe** | 2  | 2  | 29 | 2  | 2  | 3  |
| **Ha** | 3  | 0  | 1  | 36 | 0  | 0  |
| **Ne** | 2  | 8  | 1  | 0  | 27 | 2  |
| **Sa** | 1  | 1  | 1  | 0  | 5  | 32 |

(b)

|        | An | Di | Fe | Ha | Ne | Sa |
|--------|----|----|----|----|----|----|
| **An** | 68 | 0  | 0  | 0  | 0  | 7  |
| **Di** | 0  | 50 | 12 | 0  | 13 | 0  |
| **Fe** | 0  | 6  | 72 | 0  | 4  | 0  |
| **Ha** | 0  | 0  | 0  | 75 | 2  | 3  |
| **Ne** | 0  | 10 | 2  | 0  | 63 | 0  |
| **Sa** | 5  | 0  | 0  | 0  | 0  | 78 |

(c)

**Figure 5.** Confusion matrices presenting the best results obtained for (**a**) MERIP (**b**) PESD (**c**) PSSD. Emotional states: An—anger, Di—disgust, Fe—fear, Ha—happiness, Ne—neutral, Sa—sadness.

The analysis of confusion matrix illustrates that the mistakes are different for each database. For example in case of MERIP anger and happiness are most confused and the same issue occurs for PESD. However, in case of PSSD misrecognition of anger and sadness is more clearly visible. Additionally, it can be observed that confusion between boredom-sadness-neutral is a common mistake for all datasets.

Table 7 presents the accuracy achieved in state-of-the-art research on PESD and PSSD datasets, which has been improved using the algorithm proposed in this paper. It is impossible to compare results for MERIP, since the database has been released recently and, up to now, there has been no research carried out on it.

**Table 7.** Comparison with similar works.

| Database | Reference | Method | Accuracy |
|----------|-----------|--------|----------|
| PESD | [65] | SVM | 75,42 |
| PESD | [66] | Binary Tree/SVM | 56,25 |
| PESD | [19] | EC (k-NN) | 70,9 |
| PSSD | [19] | EC (k-NN) | 84,7 |
| PSSD | [67] | k-NN/SVM | 83,95 |

In order to verify if acted out database can be used as a training set for application operating in real environment, selected classifiers were tested using mixed sets. In the first experiment, the training set consists of one of acted out databases (MERIP or PESD). In the second experiment both sets are connected, creating a larger training set. PSSD is a testing set in both cases. Obtained results are presented in Table 8.

**Table 8.** The average emotion recognition rates for mixed database. Columns named $EV$, $Verr_1$, $Verr_2$, $Verr_3$ represent the voting methods proposed in this paper.

| Training | Testing | k-NN | MLP | SL | Bagging | RC | RF | EV | $Verr_1$ | $Verr_2$ | $Verr_3$ |
|----------|---------|------|-----|----|---------|----|----|----|----------|----------|----------|
| MERIP | PSSD | 31,94 | 29,94 | 32,84 | 30,67 | 29,03 | 30,12 | 31,65 | 32,91 | 33,02 | 33,26 |
| PESD | PSSD | 29,47 | 28,2 | 32,73 | 32 | 38,52 | 34,72 | 35,02 | 38,4 | 38,82 | 39,24 |
| MERIP + PESD | PSSD | 37,97 | 42,89 | 32,07 | 32,07 | 34,6 | 45,99 | 44,3 | 47,68 | 48,57 | 47,72 |

As expected, the effectiveness of classifiers whose testing and training sets comprised different datasets is much lower in comparison to those operating on one particular database. When the acted out database is the training one, the average emotion recognition rate barely exceeds 30%. Increasing the number of samples in the training set by combining both acted out datasets, increased the quality of the classification. However, even in this case, the results do not exceed 50%. It should be noted that, as in previous cases, the proposed algorithm gives better results.

## 5. Conclusions

In this paper, performance of a committee of classifiers working on small subsets of features was studied and competitive performance in speech-based emotion recognition was shown. The proposed algorithm was tested on three different types of databases and in every case it achieved performance equal or better than current state-of-the-art methods. Although obtained results look promising when working within one particular database. When it comes to mixed database classification, the results are much lower and require further study. The research indicates that using the acted out database as a training set of a model that is supposed to operate in real conditions is not the perfect approach. To achieve higher results, it is recommended either to use a training set with bigger number of samples than a test set or train the model using spontaneous speech samples. This is crucial to create a system operating in real-world environment. Future works may include adding a gender recognition module right before emotional states classification, since a huge impact of gender on SER is noticed in many papers. It is also worth to explore and examine robust features, which have an impact on differentiation between emotional states with similar resonance such as anger and happiness, as well as neutral, sad and boredom. Additionally, replacing classic algorithm models with deep learning e.g., CNN or LSTM-RNN can be considered on the grounds that the use of neural networks provides good results in SER. At the same time, it must be emphasized that deep learning requires a large number of training samples whereas widely used and accessible databases still have their limitations.

**Funding:** This research received no external funding.

**Conflicts of Interest:** The author declares no conflict of interest.

## Abbreviations

The following abbreviations are used in this manuscript:

| | |
|---|---|
| BFCC | Bark Frequency Cepstral Coefficients |
| CNN | Convolutional Neural Network |
| EC | Ensemble Committee |
| EV | Equal Voting |
| HFCC | Human Factor Cepstral Coefficients |
| HMM | Hidden Markov Models |
| LSTM | Long Short-Term Memory |
| k-NN | k Nearest Neighbours |
| LDA | Linear Discriminant Analysis |
| LMT | Logistic Model Trees |
| MERIP | Multimodal Emotion Recognition in Polish |
| MFCC | Mel Frequency Cepstral Coefficients |
| MLP | Multilayer Perceptron |
| NBTree | Naive Bayes Tree |
| PCA | Principal Component Analysis |
| PESD | Polish Emotional Speech Database, |
| PLP | Perceptual Linear Prediction |
| PSSD | Polish Spontaneous Speech, |
| RC | Random Cometee |
| RF | Random Forest |
| RNN | Recurrent Neural Network |
| RPLP | Revised Perceptual Linear Prediction |
| SER | Speech Emotion Recognition |
| SMO | Sequential Minimal Optimization |
| SL | Simple Logistic |
| SVM | Support Vector Machines |

## Appendix A

This section provides the details about selected features for each set.

**Table A1.** Feature selected applied on the whole sets for MERIP, PESD and PSSD corpora.

| Database | Features |
| --- | --- |
| *MERIP* | F0: mean; energy: median, std; F1, F2: mean, median, F2, F3: max, min, F3:std, LPC mean: 2–11; MFCC mean: 1,6, 10, 11,12; MFCC std: 12, MFCC median: 6,11; MFCC min: 1; MFCC max: 7,11; BFCC mean: 1,6,10; BFCC median: 1,6; BFCC min: 1,6; BFCC max: 11; PLP mean: 5–8; PLP median: 6,10,11; PLP std: 3–10, RPLP mean: 1,3,5,9; RPLP median: 3; RPLP std: 5,9–10; |
| *PESD* | F0: min, upper quartile, variation rate, rising-ranege max; energy: min; F1-F2: mean; F3: median; F3: min; LPC mean: 2,3,7,8,11; LPC median: 2,7,8; LPC max: 7; MFCC mean: 3,4,6,7,12; MFCC std: 1,4,5,9; MFCC median: 4,6,8–11; MFCC min: 1,3–5; MFCC max: 4; BFCC mean: 1,4; BFCC median: 5–8; BFCC min: 4,6; BFCC max: 5–7; PLP max: 1; PLP min: 1–3,4,10; PLP std: 4–7,10; RPLP mean: 1,4,7,9; RPLP median: 4–7; RPLP std: 1,4; |
| *PSSD* | F0: mean, kurtosis; energy: median, std; F1:mean, median; F3: min, std; LPC mean: 5–7,10,11; MFCC mean: 1,3,9–11; MFCC std: 11,12, MFCC median: 1,7; MFCC min: 7; MFCC max: 1; BFCC mean: 1–4,10; BFCC median: 2,6,9; BFCC min: 1,5–8; BFCC max: 1–3; PLP mean: 5,6,10,12; PLP std: 3,7,10; PLP min: 1; PLP max: 8,9; RPLP mean: 1,3,5,9; RPLP median: 3; RPLP std: 5,9–10; |

**Table A2.** Feature sets after subsets selection obtained for MERIP, PESD and PSSD corpora.

| | *MERIP* | *PESD* | *PSSD* |
| --- | --- | --- | --- |
| *F0* | mean *F0*, median *F0*, std *F0*, max *F0*, range *F0*, upper quartile *F0*, lower quartile *F0*, interquartile range *F0*, skewness *F0*, kurtosis *F0*, *F0*, rising-slope max *F0*, falling-slope max *F0*; | mean *F0*, median *F0*, max *F0*, upper quartile *F0*, lower quartile *F0*, interquartile range *F0*, kurtosis *F0*, rising-range min *F0*, falling-range min *F0*; | mean *F0*, median *F0*, max *F0*, min *F0*, rane *F0*, upper quartile *F0*, lower quartile *F0*, interquartile range *F0*, skewness *F0*, falling-range max *F0*; |
| *F1-F3* | mean: *F1*, *F3*; max: *F1*, *F3*; median: *F1*, *F2*, *F3*, min: *F1*, *F3*; standard deviation *F1*, *F3* | mean: *F1*, *F3*; median: *F2*, *F3*, min: *F1*; standard deviation *F1*, *F3*; | mean: *F1*, *F2*, *F3*; max: *F1*, *F3*; median: *F1*, *F2*, min: *F1*, *F3*; standard deviation *F1*, *F2*, *F3* |
| *Energy* | max, min, median, std, range, mean | max, min, median, std | max, min, median, std, range |
| *LPC* | *LPC* mean: 2, 4–6,12 | *LPC* mean: 5,7–11 | *LPC* mean: 2–10 |
| *MFCC* | *MFCC*: 1,3,6 -mean *MFCC*: 2,4,6 -median *MFCC*: 5,10 -std *MFCC*: 3,10 -max *MFCC*: 2,12 -min | *MFCC*: 1,3,9 -mean *MFCC*: 1 -median *MFCC*: 3,5,12,14 -std *MFCC*: 1,12 -max *MFCC*: 14 -min | *MFCC*: 2,6,7 -mean *MFCC*: 2 -median *MFCC*: 1,3,7 -std *MFCC*: 1,4 -max *MFCC*: 1,4 -min |
| *BFCC* | *BFCC*: 2,7,8 -mean *BFCC*: 7 - std *BFCC*: 1,2,3,11 -median *BFCC*: 1,4 -max *BFCC*: 2,5 -min | *BFCC*: 3,9 -mean *BFCC*: 4,10 -std *BFCC*: 2,3,5,7, -median *BFCC*: 1 -max *BFCC*: 2,3 -min | *BFCC*: 8 -mean *BFCC*: 1,2 -std *BFCC*: 2,6,12 -median *BFCC*: 1,2 -max *BFCC*: 4 -min |
| *HFCC* | *HFCC*: 1,2 -mean *HFCC*: 1,2,4,6 -std *HFCC*: 2,4 median *HFCC*: 1,7,10 -max *HFCC*: 5,6 -min | *HFCC*: 1,2,4 -mean *HFCC*: 1,3,4 -std *HFCC*: 1,4 median *HFCC*: 2,3 -max *HFCC*: 1,2 -min | *HFCC*: 3,5 -mean *HFCC*: 2,5,7 -std *HFCC*: 2,4 -max *HFCC*: 1,4,7 -min |

**Table A2.** *Cont.*

|        | *MERIP* | *PESD* | *PSSD* |
| --- | --- | --- | --- |
| *PLP* | *PLP*: 3,5 -mean *PLP*: 2,3,6 -median *PLP*: 7,10 -std *PLP*: 1,4,7 -max *PLP*: 1,4 -min | *PLP*: 1,5,9 -mean *PLP*: 4,6 -median *PLP*: 3,7 -std *PLP*: 1,2 -max *PLP*: 6 -min | *PLP*: 1 -mean *PLP*: 8,10 -median *PLP*: 1,4,6 -std *PLP*: 1,5 -max *PLP*: 1,9 -min |
| *RPLP* | *RPLP*: 1,2,3 -mean *RPLP*: 5,6,8 -median *RPLP*: 2 -std *RPLP*: 9–11 -max | *RPLP*: 1,12 -mean *RPLP*: 3–5 -median *RPLP*: 2,7 -std *RPLP*: 1–3,6, -max *RPLP*: 2,4 -min | *RPLP*: 2,3 -mean *RPLP*: 1 -median *RPLP*: 2,3 -std *RPLP*: 1,3,8 -max *RPLP*: 1,4 -min |

## References


1. Noroozi, F.; Kaminska, D.; Corneanu, C.; Sapinski, T.; Escalera, S.; Anbarjafari, G. Survey on emotional body gesture recognition. *IEEE Trans. Affect. Comput.* **2018**. [CrossRef]
2. Ślot, K.; Cichosz, J.; Bronakowski, L. Emotion recognition with poincare mapping of voiced-speech segments of utterances. In Proceedings of the International Conference on Artificial Intelligence and Soft Computing, Zakopane, Poland, 16–20 June 2019; pp. 886–895.
3. McDuff, D.; Kaliouby, R.; Senechal, T.; Amr, M.; Cohn, J.; Picard, R. Affectiva-mit facial expression dataset (am-fed): Naturalistic and spontaneous facial expressions collected. In Proceedings of the IEEE Conference on Computer Vision and Pattern Recognition Workshops, Portland, OR, USA, 23–28 June 2013; pp. 881–888.
4. Ofodile, I.; Helmi, A.; Clapés, A.; Avots, E.; Peensoo, K.M.; Valdma, S.M.; Valdmann, A.; Valtna-Lukner, H.; Omelkov, S.; Escalera, S.; et al. Action Recognition Using Single-Pixel Time-of-Flight Detection. *Entropy* **2019**, *21*, 414. [CrossRef]
5. Shaburov, V.; Monastyrshyn, Y. Emotion Recognition in Video Conferencing. U.S. Patent 9,576,190, 2018.
6. Datta, A.; Shah, M.; Lobo, N.D.V. Person-on-person violence detection in video data. In *Object Recognition Supported by User Interaction for Service Robots*; IEEE: Quebec, QC, Canada, 2002; Volume 1, pp. 433–438.
7. Kaliouby, R.; Robinson, P. Mind Reading Machines Automated Inference of Cognitive Mental States from Video. In Proceedings of the IEEE International Conference on Systems, Man and Cybernetics The Hague, The Netherlands, 10–13 October 2004; pp. 682–688.
8. Ofodile, I.; Kulkarni, K.; Corneanu, C.A.; Escalera, S.; Baro, X.; Hyniewska, S.; Allik, J.; Anbarjafari, G. Automatic Recognition of Deceptive Facial Expressions of Emotion. *arXiv* **2017**, arxiv:1707.04061.
9. Ekman, P.; Wallace, F. *Facial Action Coding System: A Technique for the Measurement of Facial Movement*; Consulting Psychologist Press: Washington, DC, USA, 1978.
10. Silva, P.; Madurapperuma, A.; Marasinghe, A.; Osano, M. A Multi-Agent Based Interactive System Towards Childs Emotion Performances Quantified Through Affective Body Gestures. In Proceedings of the 18th International Conference on Pattern Recognition (ICPR'06), Hong Kong, China, 20–24 August 2006; pp. 1236–1239.
11. Noroozi, F.; Kamińska, D.; Sapiński, T.; Anbarjafari, G. Supervised Vocal-Based Emotion Recognition Using Multiclass Support Vector Machine, Random Forests, and Adaboost. *J. Audio Eng. Soc.* **2017**, *65*, 562–572. [CrossRef]
12. Noroozi, F.; Kamińska, D.; Sapiński, T.; Anbarjafari, G. Vocal-based emotion recognition using random forests and decision tree. *Int. J. Speech Technol.* **2017**, *9*, 239–246. [CrossRef]
13. Kleinsmith, A.; Bianchi-Berthouze, N. Affective Body Expression Perception and Recognition: A Survey. *IEEE Trans. Affect. Comput.* **2013**, *4*, 15–33. [CrossRef]
14. Karg, M.; Samadani, A.A.; Gorbet, R.; Kuhnlenz, K.; Hoey, J.; Kulic, D. Body Movements for Affective Expression: A Survey of Automatic Recognition and Generation. *IEEE Trans. Affect. Comput.* **2013**, *4*, 341–359. [CrossRef]
15. Garay, N.; Cearreta, I.; López, J.; Fajardo, I. Assistive Technology and Affective Mediation. *Interdiscip. J. Humans Ict Environ.* **2006**, *2*, 55–83. [CrossRef]
16. Gelder, B.D. Why Bodies? Twelve Reasons for Including Bodily Expressions in Affective Neuroscience. *Hilosophical Trans. R. Soc. Biol. Sci.* **2009**, *364*, 3475–3484. doi:doi:10.1098/rstb.2009.0190. [CrossRef]
17. Darwin, C. *The Expression of the Emotions in Man and Animals*; John Murray: London, UK, 1872.

18. Izdebski, K. *Emotion in the Human Voice, Volume I Fundations*; Plural Publishing: San Diego, CA, USA, 2008.

19. Kamińska, D.; Sapiński, T. Polish emotional speech recognition based on the committee of classifiers. *Przegląd Elektrotechniczny* **2017**, *93*, 101–106. [CrossRef]

20. Sapiński, T.; Kamińska, D.; Pelikant, A.; Ozcinar, C.; Avots, E.; Anbarjafari, G. Multimodal Database of Emotional Speech, Video and Gestures. In Proceedings of the International Conference on Pattern Recognition, Beijing, China, 20–24 August 2018; pp. 153–163.

21. Kaminska, D.; Sapinski, T.; Pelikant, A. Polish Emotional Natural Speech Database. In Proceedings of the Conference: Signal Processing Symposium 2015, Debe, Poland, 10–12 June 2015.

22. Liu, Z.T.; Wu, M.; Cao, W.H.; Mao, J.W.; Xu, J.P.; Tan, G.Z. Speech emotion recognition based on feature selection and extreme learning machine decision tree. *Neurocomputing* **2018**, *273*, 271–280. [CrossRef]

23. Mannepalli, K.; Sastry, P.N.; Suman, M. Analysis of Emotion Recognition System for Telugu Using Prosodic and Formant Features. In *Speech and Language Processing for Human-Machine Communications*; Springer: Berlin/Heidelberg, Germany, 2018; pp. 137–144.

24. Nancy, A.M.; Kumar, G.S.; Doshi, P.; Shaw, S. Audio Based Emotion Recognition Using Mel Frequency Cepstral Coefficient and Support Vector Machine. *J. Comput. Theor. Nanosci.* **2018**, *15*, 2255–2258. [CrossRef]

25. Zamil, A.A.A.; Hasan, S.; Baki, S.M.J.; Adam, J.M.; Zaman, I. Emotion Detection from Speech Signals using Voting Mechanism on Classified Frames. In Proceedings of the 2019 International Conference on Robotics, Electrical and Signal Processing Techniques (ICREST), Dhaka, Bangladesh, 10–12 January 2019; pp. 281–285.

26. Anagnostopoulos, C.N.; Iliou, T.; Giannoukos, I. Features and classifiers for emotion recognition from speech: A survey from 2000 to 2011. *Artif. Intell. Rev.* **2015**, *43*, 155–177. [CrossRef]

27. El Ayadi, M.; Kamel, M.S.; Karray, F. Survey on speech emotion recognition: Features, classification schemes, and databases. *Pattern Recognit.* **2011**, *44*, 572–587. [CrossRef]

28. Fewzee, P.; Karray, F. Dimensionality Reduction for Emotional Speech Recognition. In Proceedings of the 2012 ASE/IEEE International Confer-ence on Social Computing and 2012 ASE/IEEE International Conference on Privacy, Security, Risk and Trust, Amsterdam, The Netherlands, 3–5 September 2012; pp. 532–537.

29. Arruti, A.; Cearreta, I.; Álvarez, A.; Lazkano, E.; Sierra, B. Feature Selection for Speech Emotion Recognition in Spanish and Basque: On the Use of Machine Learning to Improve Human-Computer Interaction. *PLoS ONE* **2014**, *9*, e108975. [CrossRef] [PubMed]

30. Han, W.; Zhang, Z.; Deng, J.; Wöllmer, M.; Weninger, F.; Schuller, B. Towards Distributed Recognition of Emotion From Speech. In Proceedings of the 5th International Symposium on Communications, Control and Signal Processing, Rome, Italy, 2–4 May 2012.

31. Ke, X.; Zhu, Y.; Wen, L.; Zhang, W. Speech Emotion Recognition Based on SVM and ANN. *Int. J. Mach. Learn. Comput.* **2018**, *8*, 198–202. [CrossRef]

32. Avots, E.; Sapiński, T.; Bachmann, M.; Kamińska, D. Audiovisual emotion recognition in wild. *Mach. Vis. Appl.* **2018**, *30*, 975–985. [CrossRef]

33. Sun, L.; Fu, S.; Wang, F. Decision tree SVM model with Fisher feature selection for speech emotion recognition. *Eurasip J. Audio Speech Music. Process.* **2019**, *2019*, 2. [CrossRef]

34. Zhao, J.; Mao, X.; Chen, L. Speech emotion recognition using deep 1D & 2D CNN LSTM networks. *Biomed. Signal Process. Control* **2019**, *47*, 312–323.

35. Zhao, J.; Mao, X.; Chen, L. Learning deep features to recognise speech emotion using merged deep CNN. *IET Signal Process.* **2018**, *12*, 713–721. [CrossRef]

36. Han, K.; Yu, D.; Tashev, I. Speech emotion recognition using deep neural network and extreme learning machine. In Proceedings of the Fifteenth Annual Conference of the International Speech Communication Association, Singapore, 14–18 September 2014.

37. Hajarolasvadi, N.; Demirel, H. 3D CNN-Based Speech Emotion Recognition Using K-Means Clustering and Spectrograms. *Entropy* **2019**, *21*, 479. [CrossRef]

38. Swain, M.; Routray, A.; Kabisatpathy, P. Databases, features and classifiers for speech emotion recognition: A review. *Int. J. Speech Technol.* **2018**, *21*, 93–120. [CrossRef]

39. Swain, M.; Sahoo, S.; Routray, A.; Kabisatpathy, P.; Kundu, J.N. Study of feature combination using HMM and SVM for multilingual Odiya speech emotion recognition. *Int. J. Speech Technol.* **2015**, *18*, 387–393. [CrossRef]

40. Rathor, S.; Jadon, R. Acoustic domain classification and recognition through ensemble based multilevel classification. *J. Ambient. Intell. Humaniz. Comput.* **2019**, *10*, 3617–3627. [CrossRef]

41. Wu, C.H.; Liang, W.B.; Cheng, K.C.; Lin, J.C. Hierarchical modeling of temporal course in emotional expression for speech emotion recognition. In Proceedings of the 2015 International Conference on Affective Computing and Intelligent Interaction (ACII), Xi'an, China, 21–24 September 2015; pp. 810–814.

42. Shih, P.Y.; Chen, C.P.; Wu, C.H. Speech emotion recognition with ensemble learning methods. In Proceedings of the 2017 IEEE International Conference on Acoustics, Speech and Signal Processing (ICASSP), New Orleans, LA, USA, 5–9 March 2017; pp. 2756–2760.

43. Pao, T.L.; Chien, C.S.; Chen, Y.T.; Yeh, J.H.; Cheng, Y.M.; Liao, W.Y. Combination of multiple classifiers for improving emotion recognition in Mandarin speech. In Proceedings of the Third International Conference on Intelligent Information Hiding and Multimedia Signal Processing (IIH-MSP 2007), Kaohsiung, Taiwan, 26–28 November 2007; Volume 1, pp. 35–38.

44. Morrison, D.; Wang, R.; De Silva, L.C. Ensemble methods for spoken emotion recognition in call-centres. *Speech Commun.* **2007**, *49*, 98–112. [CrossRef]

45. Zantout, R.; Klaylat, S.; Hamandi, L.; Osman, Z. Ensemble Models for Enhancement of an Arabic Speech Emotion Recognition System. In Proceedings of the Future of Information and Communication Conference, San Francisco, CA, USA, 14–15 March 2019, pp. 174–187.

46. Sultana, S.; Shahnaz, C. A non-hierarchical approach of speech emotion recognition based on enhanced wavelet coefficients and K-means clustering. In Proceedings of the 2014 International Conference on Informatics, Electronics & Vision (ICIEV), Dhaka, Bangladesh, 23–24 May 2014; pp. 1–5.

47. Trabelsi, I.; Ayed, D.B.; Ellouze, N. Evaluation of influence of arousal-valence primitives on speech emotion recognition. *Int. Arab J. Inf. Technol.* **2018**, *15*, 756–762.

48. Xiao, Z.; Dellandrea, E.; Dou, W.; Chen, L. Automatic hierarchical classification of emotional speech. In Proceedings of the Ninth IEEE International Symposium on Multimedia Workshops (ISMW 2007), Beijing, China, 10–12 December 2007; pp. 291–296.

49. Shaqra, F.A.; Duwairi, R.; Al-Ayyoub, M. Recognizing Emotion from Speech Based on Age and Gender Using Hierarchical Models. *Procedia Comput. Sci.* **2019**, *151*, 37–44. [CrossRef]

50. Xiao, Z.; Dellandréa, E.; Chen, L.; Dou, W. Recognition of emotions in speech by a hierarchical approach. In Proceedings of the 2009 3rd International Conference on Affective Computing and Intelligent Interaction and Workshops, Amsterdam, The Netherlands, 10–12 September 2009; pp. 1–8.

51. You, M.; Chen, C.; Bu, J.; Liu, J.; Tao, J. A hierarchical framework for speech emotion recognition. In Proceedings of the 2006 IEEE International Symposium on Industrial Electronics, Montreal, QC, Canada, 9–13 July 2006; Volume 1, pp. 515–519.

52. Engberg, I.S.; Hansen, A.V. *Documentation of the Danish Emotional Speech Database (DES)*; Internal AAU Report; Center for Person Kommunikation: Aalborg, Denmark, 1996; p. 22.

53. Ekman, P.; Friesen, W.V. Constants across cultures in the face and emotion. *J. Personal. Soc. Psychol.* **1971**, *17*, 124. [CrossRef]

54. Plutchik, R. The nature of emotions: Human emotions have deep evolutionary roots, a fact that may explain their complexity and provide tools for clinical practice. *Am. Sci.* **2001**, *89*, 344–350. [CrossRef]

55. Mirsamadi, S.; Barsoum, E.; Zhang, C. Automatic speech emotion recognition using recurrent neural networks with local attention. In Proceedings of the 2017 IEEE International Conference on Acoustics, Speech and Signal Processing (ICASSP), New Orleans, LA, USA, 5–9 March 2017; pp. 2227–2231.

56. Rao, K.S.; Koolagudi, S.G.; Vempada, R.R. Emotion recognition from speech using global and local prosodic features. *Int. J. Speech Technol.* **2013**, *16*, 143–160. [CrossRef]

57. Zieliński, T. *Cyfrowe Przetwarzanie Sygnałów*; Wydawnictwa Komunikacji i Łączności: Warsaw, Poland, 2013.

58. Skowronski, M.; Harris, J. Increased mfcc filter bandwidth for noise-robust phoneme recognition. In Proceedings of the IEEE International Conference on Acoustics, Speech, and Signal Processing, Orlando, FL, USA, 13–17 May 2002; pp. 801–804.

59. Kumar, P.; Biswas, A.; Mishra, A.; Chandra, M. Spoken Language Identification Using Hybrid Feature Extraction Methods. *J. Telecommun.* **2010**, *1*, 11–15.

60. Hermansky, H. Perceptual Linear Predictive (PLP) Analysis of Speech. *J. Acoust. Soc. Am.* **1989**, *87*, 1738–1752. [CrossRef] [PubMed]

61. O'Shaughnessy, D. Linear predictive coding. *IEEE Potentials* **1988**, *7*, 29–32. [CrossRef]

62. Mermelstein, P. Determination of the vocal-tract shape from measured formant frequencies. *J. Acoust. Soc. Am.* **1967**, *41*, 1283–1294. [CrossRef] [PubMed]

63. Hall, M.A. Correlation-Based Feature Selection for Machine Learning. Available online:https://www.cs.waikato.ac.nz/~mhall/thesis.pdf (accessed on 20 September 1999).

64. Hall, M.; Frank, E.; Holmes, G.; Pfahringer, B.; Reutemann, P.; Witten, I.H. The WEKA data mining software: An update. *ACM SIGKDD Explor. Newsl.* **2009**, *11*, 10–18. [CrossRef]

65. Hook, J.; Noroozi, F.; Toygar, O.; Anbarjafari, G. Automatic speech based emotion recognition using paralinguistics features. *Bull. Pol. Acad. Sci. Tech. Sci.* **2019**. *67*, 479–488.

66. Yüncü, E.; Hacihabiboglu, H.; Bozsahin, C. Automatic speech emotion recognition using auditory models with binary decision tree and svm. In Proceedings of the 2014 22nd International Conference on Pattern Recognition, Stockholm, Sweden, 24–28 August 2014; pp. 773–778.

67. Kamińska, D.; Sapiński, T.; Anbarjafari, G. Efficiency of chosen speech descriptors in relation to emotion recognition. *EURASIP J. Audio Speech Music. Process.* **2017**, *2017*, 3. [CrossRef]

*Article*

# Entropy-Based Clustering Algorithm for Fingerprint Singular Point Detection

Ngoc Tuyen Le [1], Duc Huy Le [2], Jing-Wein Wang [1,*] and Chih-Chiang Wang [3]

1   Institute of Photonic Engineering, National Kaohsiung University of Science and Technology, Kaohsiung 80778, Taiwan
2   Department of Electronic Engineering, National Kaohsiung University of Science and Technology, Kaohsiung 80778, Taiwan
3   Department of Computer Science and Information Engineering, National Kaohsiung University of Science and Technology, Kaohsiung 80778, Taiwan
*   Correspondence: jwwang@nkust.edu.tw; Tel.: +886-930943143

Received: 19 July 2019; Accepted: 9 August 2019; Published: 12 August 2019

**Abstract:** Fingerprints have long been used in automated fingerprint identification or verification systems. Singular points (SPs), namely the core and delta point, are the basic features widely used for fingerprint registration, orientation field estimation, and fingerprint classification. In this study, we propose an adaptive method to detect SPs in a fingerprint image. The algorithm consists of three stages. First, an innovative enhancement method based on singular value decomposition is applied to remove the background of the fingerprint image. Second, a blurring detection and boundary segmentation algorithm based on the innovative image enhancement is proposed to detect the region of impression. Finally, an adaptive method based on wavelet extrema and the Henry system for core point detection is proposed. Experiments conducted using the FVC2002 DB1 and DB2 databases prove that our method can detect SPs reliably.

**Keywords:** singular point detection; boundary segmentation; blurring detection; fingerprint image enhancement; fingerprint quality

---

## 1. Introduction

Fingerprint biometrics is increasingly being used in the commercial, civilian, physiological, and financial domains based on two important characteristics of fingerprints: (1) fingerprints do not change with time and (2) every individual's fingerprints are unique [1–5]. Owing to these characteristics, fingerprints have long been used in automated fingerprint identification or verification systems. These systems rely on accurate recognition of fingerprint features. At the global level, fingerprints have ridge flows assembled in a specific formation, resulting in different ridge topology patterns such as core and delta (singular points (SPs)), as shown in Figure 1a. These SPs are the basic features required for fingerprint classification and indexing. Local fingerprint features are carried by local ridge details such as ridge endings and bifurcations (minutiae), as shown in Figure 1b. Fingerprint minutiae are often used to conduct matching tasks because they are generally stable and highly distinctive [6].

Most previous SP extraction algorithms were performed directly over fingerprint orientation images. The most popular method is based on the Poincaré index [7], which typically computes the accumulated rotation of the vector field along a closed curve surrounding a local point. Wang et al. [8] proposed a fingerprint orientation model based on 2D Fourier expansions to extract SPs independently. Nilsson and Bigun [9] as well as Liu [10] used the symmetry properties of SPs to extract them by first applying a complex filter to the orientation field in multiple resolution scales by detecting the parabolic and triangular symmetry associated with core and delta points. Zhou et al. [11] proposed a feature of differences of the orientation values along a circle (DORIC) in addition to the Poincaré index to effectively remove spurious detections, take the topological relations of SPs as a global constraint for fingerprints, and use the global orientation field for SP detection. Chen et al. [12] obtained candidate SPs by the multiscale analysis of orientation entropy and then applied some post-processing steps to filter the spurious core and delta points.

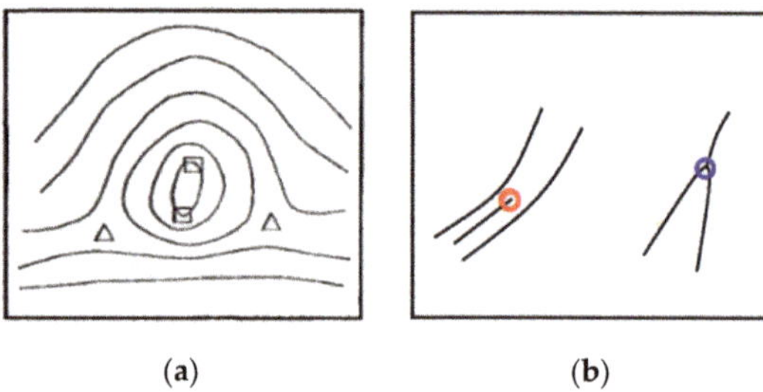

(a)        (b)

**Figure 1.** The global and local features in the fingerprint. (**a**) Singular points (SPs) (square: core; triangle: delta) and (**b**)minutiae (red circle: ridges ending; blue circle: bifurcation).

However, SP detection is sensitive to noise, and extracting SPs reliably is a very challenging problem. When input fingerprint images have poor quality, the performance of these methods degrades rapidly. Noise in fingerprint images makes SP extraction unreliable and may result in a missed or wrong detection. Therefore, fingerprint image enhancement is a key step before extracting SPs.

Fingerprint image enhancement remains an active area of research. Researchers have attempted to reduce noise and improve the contrast between ridges and valleys in fingerprint images. Most fingerprint image enhancement algorithms are based on the estimation of an orientation field [13–15]. Some methods use variations of Gabor filters to enhance fingerprint images [16,17]. These methods are based on the estimation of a single orientation and a single frequency; they can remove undesired noise and preserve and improve the clarity of ridge and valley structures in images. However, they are not suitable for enhancing ridges in regions with high curvature. Wang and Wang [18] first detected the SP area and then improved it by applying a bandpass filter in the Fourier domain. However, detecting the SP region when the fingerprint image has extremely poor quality is highly difficult. Yang et al. [19] first enhanced fingerprint images in the spatial domain with a spatial ridge-compensation filter by learning from the images and then used a frequency bandpass filter that is separable in the radial- and angular-frequency domains. Yun and Cho [20] analyzed fingerprint images, divided them into oily, neutral, and dry according to their properties, and then applied a specific enhancement strategy for each type. To enhance fingerprint images, Fronthaler et al. [21] used a Laplacian-like image pyramid to decompose the original fingerprint into subbands corresponding to different spatial scales and then performed contextual smoothing on these pyramid levels, where the corresponding filtering directions stem from the frequency-adapted structure tensor. Bennet and Perumal [22] transformed fingerprint images into the wavelet domain and then used singular value decomposition (SVD) to decompose the low subband coefficient matrix. Fingerprint images were enhanced by multiplying the singular value matrix of the low-low(LL) subband with the ratio of the largest singular value of the generated normalized matrix with mean of 0 and variance of 1 and the largest singular value of the LL subband. However, the resulting images were sometimes uneven. This is because SVD was applied only to the low subband and a generated normalized matrix was used. To overcome this problem, Wang et al. [23] introduced a novel lighting compensation scheme involving the use of adaptive SVD on wavelet coefficients. First, they decomposed the input fingerprint image into four subbands by 2D discrete wavelet transform (DWT). Subsequently, they compensated fingerprint images by adaptively obtaining the compensation coefficients for each subband based on the referred Gaussian template.

The aforementioned methods for enhancing fingerprint images can reduce noise and improve the contrast between ridges and valleys in the images. However, they are not really effective with fingerprint images having very poor quality, particularly blurring. To overcome this problem, we need to segment the fingerprint foreground with the interleaved ridge and valley structure from the complex background with non-fingerprint patterns for more accurate and efficient feature extraction and identification. Many studies have investigated segmentation on rolled and plain fingerprint images. Mehtreet al. [24] partitioned a fingerprint image into blocks and then performed block classification based on gradient and variance information to segment fingerprint images into blocks. This method was further extended to a composite method [25] that takes advantage of both the directional and the variance approaches. Zhang et al. [26] proposed an adaptive total variation decomposition model by incorporating the orientation field and local orientation coherence for latent fingerprint segmentation. Based on a ridge quality measure that was defined as the structural similarity between the fingerprint patch and its dictionary-based reconstruction, Cao et al. [27] proposed a learning-based method for latent fingerprint image segmentation.

This study proposes an efficient approach by combining the novel adaptive image enhancement, compact boundary segmentation, and a novel clustering algorithm by integrating wavelet frame entropy with region growing to evaluate the fingerprint image quality so as to validate the SPs. Experiments were conducted on FVC2002 DB1 and FVC2002 DB2 databases [28]. The experimental results indicate the excellent performance of the proposed method.

The rest of this paper is organized as follows. Section 2 introduces the proposed image enhancement, precise boundary segmentation, and blurring detection based on wavelet entropy clustering algorithm. Section 3 describes the proposed algorithm for SP detection. Section 4 presents experimental results to verify the proposed approach. Finally, Section 5 presents the conclusions of this study.

## 2. Blurring Detection for Fingerprint Impression

### 2.1. Fingerprint Background Removal

SVD has been widely used in digital image processing [29–31]. Without loss of generality, we suppose that $f$ is a fingerprint image with a resolution of $M \times N$ ($M \geq N$). The SVD of a fingerprint image $f$ can be written as follows:

$$f = U\Sigma V^T, \tag{1}$$

where $U = [u_1, u_2, \ldots, u_N]$ and $V = [v_1, v_2, \ldots, v_N]$ are orthogonal matrices containing singular vectors and $\Sigma = [D, O]$ contains the sorted singular values on its main diagonal. $D = \mathrm{diag}\,(\lambda_1, \lambda_2, \ldots, \lambda_k)$ with singular values $\lambda_i$, $i = 1, 2, \ldots, k$ in a non-increasing order, O is a $M \times (M - k)$ zero matrix, and $k$ is the rank of $f$. We also can expand the fingerprint image as follows:

$$f = \lambda_1 u_1 v_1^T + \lambda_2 u_2 v_2^T + \ldots + \lambda_k u_k v_k^T, \;\; \lambda_1 \geq \lambda_2 \geq \ldots \geq \lambda_k. \tag{2}$$

The terms $\lambda_i u_i v_i^T$ containing the vector outer-product in Equation (2) are the principal images. The Frobenius norm of the fingerprint image is preserved in SVD transformation:

$$\|f\|_F^2 = \sum_{i=1}^{k} \lambda_i^2. \tag{3}$$

Equation (3) shows how the signal energy of f can be partitioned by the singular values in the sense of the Frobenius norm. It is common to discard the small singular values in SVD to obtain matrix approximations whose rank equals the number of remaining singular values. Good matrix approximations can always be obtained with a small fraction of the singular values. The highly concentrated property of SVD helps remove background noise from the foreground ridges.

We performed some experiments to observe the effects of singular values on a fingerprint image. Figure 2a shows a fingerprint image in the FVC 2002 DB2 database. First, all singular values of the fingerprint image, as shown in Figure 2a, were set to 1 and the fingerprint image was then reconstructed. Figure 2b shows the reconstructed fingerprint image without the effect of singular values, implying that the singular vectors represent the background information of the given fingerprint image. Next, all singular values of the fingerprint image shown in Figure 2a were multiplied by 2 and the fingerprint image was then reconstructed. As shown in Figure 2c, the fingerprint image looks clearer and the background of the fingerprint image was removed. It suggests that the singular values represent the foreground ridges of the given fingerprint image. Thus, SVD can be used for enhancing the ridge structure and removing noise from the background of the fingerprint image. In addition, if the fingerprint image is a low-contrast image, this problem can be corrected by replacing $\Sigma$ with an equalized singular matrix obtained from a normalized image, which is considered to be that with a probability density function involving a Gaussian distribution with a mean and variance calculated using the available dataset. This normalized image is called a Gaussian template image.

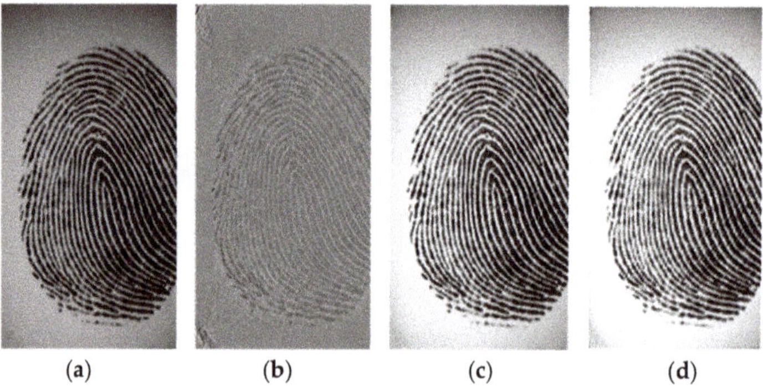

(a)      (b)      (c)      (d)

**Figure 2.** Effects of singular values on a fingerprint image. (a) Fingerprint image in FVC 2002 DB2 database; (b) reconstructed fingerprint image when all singular values of Figure 2a are set to 1; (c) reconstructed fingerprint image when all singular values of Figure 2a are multiplied by 2; (d) equalized fingerprint images of Figure 2a.

Based on observations of the effects of SVD on a fingerprint image, and to effectively remove the background, we examined the singular values of the fingerprint image, which contains most of the foreground information. We automatically adjusted the illumination of an image to obtain an equalized image that has a normal distribution. If the fingerprint image had low contrast, the singular values were multiplied with a scalar larger than 1. A normalized intensity image with no illumination problem can be considered an image that has a Gaussian distribution and that can easily be obtained by generating random pixel values with Gaussian distribution. Moreover, the first singular value contributes 99.72% of energy to the original image and the first two singular values contribute 99.88% of the total energy [31]. The larger singular value represents the energy of the fingerprint pattern and the smaller one, the energy of the background and noise. To effectively remove the background, we set a compensation weight, $\alpha$, that enhanced the image contrast. It is easy to remove the ridge of images when the compensation weight is larger than 1, and the image contrast is reduced when the compensation weight is smaller than 1. Therefore, we compared the maximum singular value of the Gaussian template with the maximum singular value of the original fingerprint image to compute the compensation weight as follows:

$$\alpha = \begin{cases} max\left(\frac{\max(\Sigma_G)}{\max(\Sigma)}, \frac{\max(\Sigma)}{\max(\Sigma_G)}\right) & , \ \max(\Sigma) < \eta \\ 1 & , \ max(\Sigma) \geq \eta \end{cases}, \tag{4}$$

where the threshold value $\eta$ is experimentally set as 90,000, and $\Sigma_G$ is the singular value matrix of the Gaussian template image with mean and variance calculated from the adopted database as shown in Table 1. The equalized image, $f_{eq}$, having the same size as the original fingerprint image can be generated by the following:

$$f_{eq} = U(\alpha\Sigma)V^T. \tag{5}$$

This task that actually equalizes the fingerprint image can eliminate the undesired background noise. As shown in Figure 2d, the background of the fingerprint image has been removed, thereby providing an image with nearly normal distribution. It also improves the clarity and continuity of ridge structures in the fingerprint image.

**Table 1.** Mean and standard deviation of Gaussian distribution function in each database.

| Database | Mean | Standard Deviation |
|---|---|---|
| FVC2002 DB1 | 0.84 | 0.24 |
| FVC2002 DB2 | 0.50 | 0.18 |

## 2.2. Impression Region Detection and Boundary Segmentation

The fingerprint texture should be distinguished from the background by a suitable binary threshold obtained from the energy analysis as a very useful and distinctive preprocessing for boundary segmentation. An analysis of the energy distribution of fingerprint images from the public fingerprint image database indicates a prominent distinction between the fingerprint object and the undesired background owing to the construction of ridges and valleys. In this section, we propose an impression region detection approach based on the energy difference between the impression contour and the background scene. The most obvious feature of the fingerprint ridge is the texture; it exhibits variances in the energy roughness of the impression region. Roughness corresponds to the perception that our sense of touch can feel with an object, and it can be characterized in two-dimensional scans by depth (energy strength) and width (separation between ridges). Before ridge object extraction, a smoothing filter is used to smooth the image and enhance the desired local ridge. The local standard average $\mu$ and energy $\varepsilon$ of the $7 \times 7$ pixels defined by the mask are given by the following expressions:

$$\mu(x, y) = \frac{1}{N} \sum_{i=-3}^{3} \sum_{j=-3}^{3} f_{eq}(x + i, y + j), \tag{6}$$

$$\varepsilon(x, y) = \frac{1}{N} \sum_{i=-3}^{3} \sum_{j=-3}^{3} \left(f_{eq}(x + i, y + j) - \mu(x, y)\right)^2, \tag{7}$$

where $f_{eq}(x, y)$ is the equalized image, as discussed in Section 2.1, and $N = 49$ is a normalizing constant. For transforming the grayscale intensity image in Figure 3a into a logical map, a binarized image of the equalized image, $f_b(x, y)$, is obtained by extracting the interesting object from the background as follows:

$$f_b(x,y) = \begin{cases} 255, & if \ \varepsilon(x, y) \geq 255 \\ 0, & if \ \varepsilon(x, y) < 255 \end{cases}, \tag{8}$$

where $f_b(x, y)$ is a binarized image; pixel values labeled 255 are objects of interest, whereas pixel values labeled 0 are undesired ones.

Figure 3b shows the binarized image obtained by applying Equation (8) to the equalized image. Based on the binary images, as shown in Figure 3b, we can detect the region of impression (ROI), which is very useful as a distinctive preprocessing of boundary segmentation. Figure 3b shows that the proposed algorithm can perform very well for discriminating the blur region. Pixel $(x, y)$ with energy $\varepsilon(x,y) \geq 255$ is an object of the ROI; therefore, we can detect the ROI, $f_{ROI}(x, y)$, as follows:

$$f_{ROI} = \{(x,y)|\varepsilon(x,y) \geq 255\}. \tag{9}$$

To define the fingerprint contour, we determine the boundary location of the fingerprint. Most human fingerprint contours have elliptical shapes. Thus, the left, right, and horizontal projections for an elliptical fingerprint contour are divided to search for landmarks by commencing from two sides in every 15 pixels from top to down. Based on the located landmarks, the contour of the fingerprint is acquired in a polygon. As illustrated in Figure 3c, the blue, green, and red lines present the contours received by using left, right, and horizontal projections, respectively. This method is advantageous because it is simple and is less influenced by finger pressure.

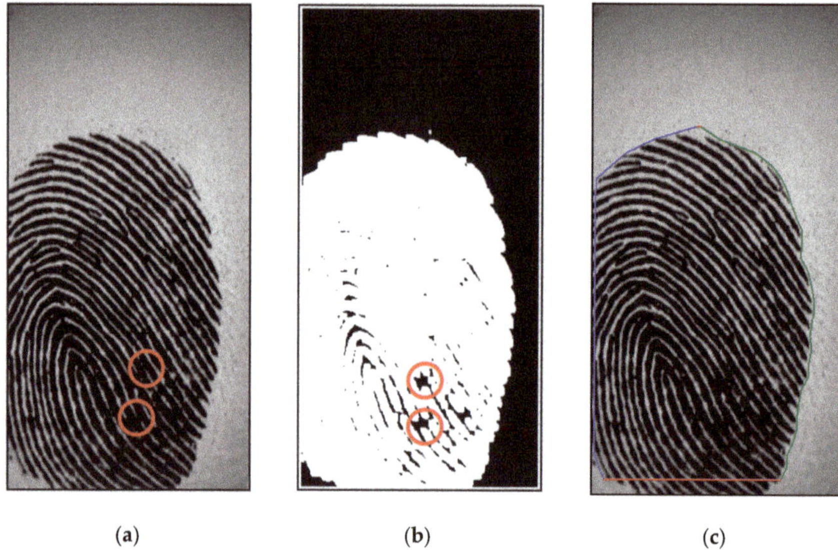

(a)          (b)          (c)

**Figure 3.** (**a**) Original fingerprint image in FVC 2002 DB2 database; (**b**) binary image by using energy transformation and blur detection obtained with2D non-separable wavelet entropy filtering for Figure 3a; (**c**) segmented image of Figure 3a.

### 2.3. Blurring Detection

Our proposed method improves the fingerprint image quality, as discussed in Section 2.1, and the ROI is defined, as discussed in Section 2.2. However, the fingerprint image still contains a blur region within the ROI, leading to the false detection of SPs. In this section, we propose a method for detecting the blur region in a fingerprint image and then ignoring it during detection to reduce the time and improve the accuracy of SP detection.

To locate the blur region, we perform region segmentation by finding a meaningful boundary based on a point aggregation procedure. Choosing the center pixel of the region is a natural starting point. Grouping points to form the region of interest, while focusing on 4-connectivity, would yield a clustering result when there are no more pixels for inclusion in the region. After region growing, the region is measured to determine the size of the blur region. Entropy filtering for blur detection of pixels in the $11 \times 11$ ($N = 11$) neighborhood defined by the mask is given by the following:

$$e_{NSDWF} = \frac{-1}{N^2} \sum_{x,\,y=0}^{N-1} \left| d^{HH}(x,\,y) \right| \log \left| d^{HH}(x,\,y) \right|, \tag{10}$$

where $d^{HH}$ is the coefficient of a non-subsampled version of the 2D non-separable discrete wavelet transform (NSDWT) [32,33] in the high-frequency subband decomposed at the first level ($d_{j+1}^{HH}$), $j = 0$, as shown in Figure 4. Figure 3b shows that the proposed algorithm can perform very well for discriminating the blur region.

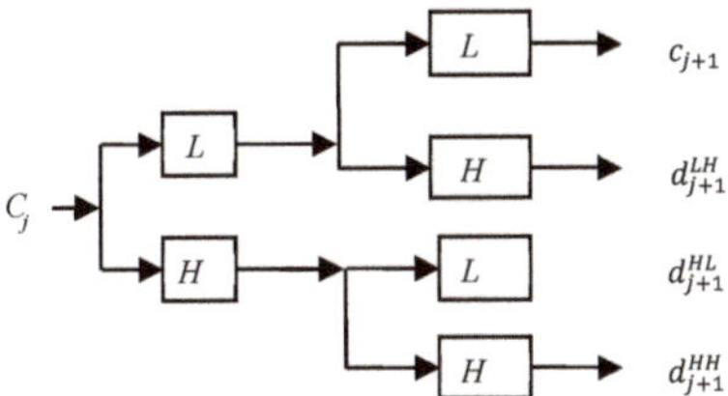


**Figure 4.** Filter bank implementation of 2D non-separable discrete wavelet transform (NSDWT), $j$: level.

## 3. SP Detection

In general, SPs of a fingerprint are detected by a Poincaré index-based algorithm. However, the Poincaré index method usually results in considerable spurious detection, particularly for low-quality fingerprint images. This is because the conventional Poincaré index along the boundary of a given region equals the sum of the Poincaré indices of the core points within this region, and it contains no information about the characteristics and cannot describe the core point completely. To overcome the shortcoming of the Poincaré index method, we propose an adaptive method based on wavelet extrema for core point detection. Wavelet extrema contain information on both the transform modulus maxima and minima in the image, considered to be among the most meaningful features for signal characterization.

First, we align the ROI based on the Poincaré's core points and the local orientation field. The Poincaré index at pixel $(x,y)$, which is enclosed by 12 direction fields taken in a counterclockwise direction, is calculated as follows:

$$Poincare(x,y) = \frac{1}{2\pi} \sum_{k=0}^{M-1} \Delta(k), \tag{11}$$

where

$$\Delta(k) = \begin{cases} \delta(k), & if \left| \delta(k) \right| < \pi/2 \\ \pi + \delta(k), & if \left| \delta(k) \right| < -\pi/2 \\ \pi - \delta(k), & otherwise \end{cases} \tag{12}$$

and

$$\delta(k) = \theta(x(k'), y(k')) - \theta(x(k), y(k)); \; k' = (k+1) \bmod M; \; M = 12, \tag{13}$$

where $(x(k'), y(k'))$ and $(x(k), y(k))$ are the paired neighboring coordinates of the direction fields. A core point has a Poincaré index of +1/2. By contrast, a delta point has a Poincaré index of -1/2. The core pointsdetected in this step are called rough core points.

Next, we align the fingerprint image under the right-angle coordinate system based on the number and location of preliminary core points. Because fingerprints may have different numbers of cores, the first step in alignment is to adopt the preliminary Poincaré indexed positions as a reference. If the number of preliminary cores is 2, the image is rotated along the orientation calculated from the midpoint between the two cores. If the number of cores is equal to 1, the image is rotated along the direction calculated from the neighboring orientation of the core. If the number of cores is zero, the fingerprint is kept intact. The rotation angle is calculated as follows:

$$\varnothing_{j<y_c} = \frac{1}{2} tan^{-1} \frac{\sum_{i\in\zeta} sin2O_{i,j}}{\sum_{i\in\zeta} cos2O_{i,j}},$$ (14)

where $O_{i,j}$ is the local orientation around a pixel and $\zeta$ is the core subregion of interest (COI) centered at the Poincaré index core point $(x_c, y_c)$ with size of $60 \times 60$ pixels, which was determined to avoid possible variability near the boundary while one is fingerprinted by the reader. Fingerprint alignment is performed to make the pattern rotation-invariant and to reduce the false rejection rate. The rotations are given by the following Equation:

$$\begin{cases} y' = xsin\phi + ycos\phi \\ x' = xcos\phi - ysin\phi \end{cases},$$ (15)

and point $(x, y)$ with orientation angle $\phi$ is mapped to point $(x', y')$. Figure 5 shows some fingerprint alignment by our method with different numbers of cores.

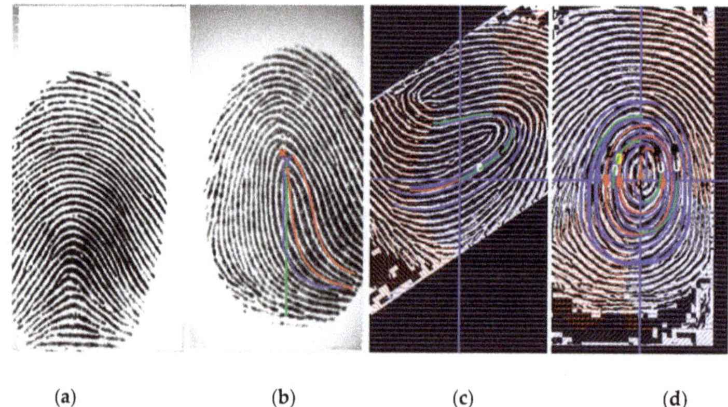

(a)      (b)      (c)      (d)

**Figure 5.** Fingerprint alignment: (**a**) number of cores = 0; (**b**) number of cores = 1; (**c,d**) number of cores = 2.

After alignment, the COI subregion with size of $60 \times 60$ pixels centered at the Poincaré's detected point is further segmented from the aligned image. The COI then goes through a skeletonization process to peel off as many ridge pixels as possible without affecting the general shape of the ridge, as shown in Figure 6a, and is then transformed to a skeletonized ridge image, as shown in Figure 6b. The skeletonized ridge image is used to compute the wavelet extrema, as shown in Figure 6c.

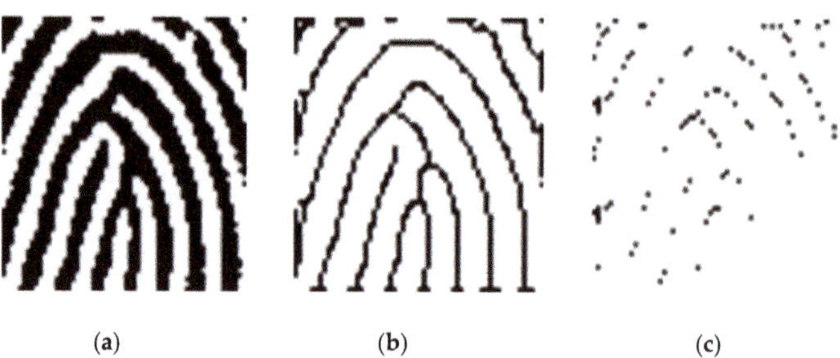

(a)      (b)      (c)

**Figure 6.** (**a**) COI subregion; (**b**)skeletonized ridges; (**c**) 2D wavelet extrema.

Wavelet modulus maxima representations for two-dimensional signals were proposed by Mallat [33] as a tool for extracting information on singularities, which were considered to be among the most meaningful features for signal characterization. Most wavelet transform local extrema are actually modulus maxima (there are examples of signals for which the wavelet extrema and modulus representations are the same). The set of indices and the local maximum, denoted as $M(f)$, and local minimum, denoted as $m(f)$, of skeletonized ridge image $f$ are defined as follows:

$$M(f) = \{(z, f(z)) : f(z-1) \leq f(z) \text{ and } f(z+1) \leq f(z)\}, \tag{16}$$

$$m(f) = \{(z, f(z)) : f(z-1) \geq f(z) \text{ and } f(z+1) \geq f(z)\}, \tag{17}$$

Where $z \in Z$. Similarly, the indices and values of wavelet transform extrema for an image $f$ is defined as follows:

$$E(f) = \{\{M(w_j(f))\} \cup \{m(w_j(f))\}\}; j = 1, 2, \ldots, J\}, \tag{18}$$

where $w_j(f)$ is the 2D non-separable wavelet transform of image $f$ at scale $j$, $j = 1, 2, \ldots, J$. The SP of a fingerprint image can be found by extracting curvature primitives and discovering the location of these primitives in the subregion, as shown in Figure 6c.

We find the exact location of the core point defined by the Henry system and trace the skeletonized ridge curves with 8-adjacency to explore wavelet extrema in 1-pixel increments by starting at 10 pixels apart from two sides. The highest extrema in the ridge curve correspond to core point candidates. We devise two 8-adjacency grids to locate the wavelet extrema (Figure 7a,b). Beginning from two opposite ends and moving toward the center of the subregion, the black-colored pixel of each grid is designated as the central point to trace. Based on this central point, the moving guideline is as follows: if the gray-level of the adjacent pixel is 0, then move toward that pixel, where the number shown in the grid indicates the moving sequence. This method enables one to follow the real track of the ridge curve. Whenever a singularity is detected, its location is noted. Figure 7c shows that it is common to find multiple core point candidates with small vertical displacements, and the area below the lowest ridge curve is circumscribed for locating the core point. In the Henry system, exact core point location can be performed as follows: (a) locate the topmost extrema in the innermost ridge curve if there is no rod; (b) otherwise, locate the top of the rods. The following equation summarizes this process:

$$s = \begin{cases} \omega_{e,0}, & i = 0 \\ \omega_{e,(i/2)+(i \bmod 2)}, & i \geq 1 \end{cases}, \tag{19}$$

Where $s$ is the determined core point, $i$ is the number of rods below the innermost ridge curve, $\omega_{e,0}$ is the topmost extrema in the innermost ridge curve, and $\omega_{e,(i/2)+(i \bmod 2)}$ is the located rod extrema below the innermost ridge curve. Figure 7d presents an example marked with the blue cross.

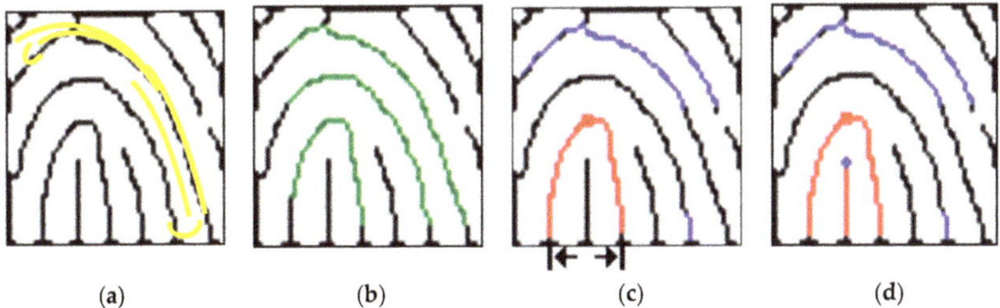

(a)   (b)   (c)   (d)

**Figure 7.** Core point detection based on wavelet extrema and Henry system. (**a**) Two 8-adjacency grids moving toward each other along the ridge curve indicated in yellow; (**b**) traced path of the ridge curve (green line: from left to right); (**c**) SP located at the lowest ridge curve (red square) and the area beneath (blue line: searching extrema from right to left); (**d**) SP detection in accordance with the Henry system (blue cross).

## 4. Experimental Results and Discussion

In this section, to illustrate the effectiveness of our proposed method, we present some of the performed experiments using both FVC2002 DB1 and DB2 fingerprint databases. FVC2002 includes four databases, namely, DB1, DB2, DB3, and DB4, collected using different sensors or technologies that are widely used in practice. Each database is 110 fingers wide (w) and 8 impressions per finger deep (d) (880 fingerprints in all). Fingerprints from 101 to 110 (set B) have been made available to the participants to allow for parameter tuning before the submission of the algorithms. The benchmark is then constituted by fingers numbered from 1 to 100 (set A). Volunteers were randomly partitioned into three groups (30 persons each); each group was associated with a database and therefore to a different fingerprint scanner. Each volunteer was invited to present themselves at the collection place in three distinct sessions, with at least two weeks between each session. The forefinger and middle finger of both hands (in total, four fingers) of each volunteer were acquired by interleaving the acquisition of the different fingers to maximize differences in finger placement. No efforts were made to control image quality and the sensor platens were not systematically cleaned. In each session, four impressions were acquired of each of the four fingers of each volunteer. During the second session, individuals were requested to exaggerate the displacement (impressions 1 and 2) and rotation (impressions 3 and 4) of the finger without exceeding 35°. During the third session, fingers were alternatively dried (impressions 1 and 2) and moistened (impressions 3 and 4). The SPs of all fingerprints in the testing database were manually labeled beforehand to obtain the ground truth. For a ground-truth $SP(x_0, y_0)$, if a detected $SP(x, y)$ satisfies $\sqrt{(x - x_0)^2 - (y - y_0)^2} < 10$, it is said to be truly detected; otherwise, it is called a miss.

The singular point detection rate (SDR) is defined as the ratio of truly detected SPs to all ground-truth SPs:

$$\text{SDR} = \frac{Num(truly\ detected\ SPs)}{Num(groundtruth\ SPs)} \times 100\%. \tag{20}$$

The singular point miss rate (SMR) is defined as the ratio of the number of missed SPs to the number of all ground-truth SPs. The sum of the detection rate and miss rate is 100%:

$$\text{SMR} = \frac{Num(missed\ SPs)}{Num(groundtruth\ SPs)} \times 100\% = 100 - \text{SDR}. \tag{21}$$

The singular point false alarm rate (SFR) is defined as the ratio of the number of falsely detected SPs to the total number of ground-truth $SPs$:

$$\text{SFR} = \frac{Num(falsely\ detected\ SPs)}{Num(groundtruth\ SPs)} \times 100\%. \tag{22}$$

The singular point correctly detected rate (SCR) is defined as the ratio of all truly detected SPs to all detected SPs in a fingerprint of all fingerprint images:

$$\text{SCR} = \frac{Num(truely\ detected\ SPs)}{Num(detected\ SPs)} \times 100\%. \tag{23}$$

First, the compensation weight coefficients are calculated by using Equation (4) and the equalized image, $f_{eq}$, having the same size as the original fingerprint image can be generated by Equation (5). Figures 8 and 9 show example image results of the proposed method for FVC2002 DB1 and DB2, respectively. As shown in Figures 8c and 9b, the background of the fingerprint image has been removed, thereby providing an image with nearly normal distribution. It also improves the clarity and continuity of ridge structures in the fingerprint image.

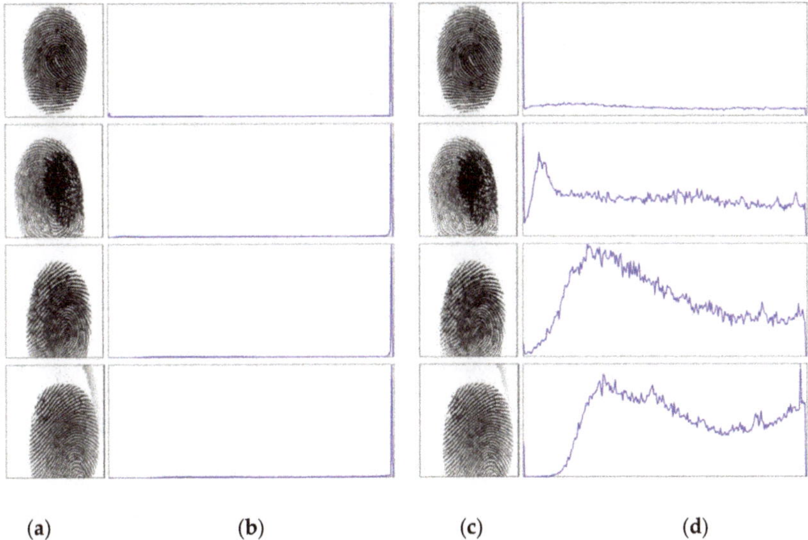

(a)   (b)   (c)   (d)

**Figure 8.** Results of our proposed method for the FVC2002 DB1 database. (**a**) Original fingerprint images; (**b**) histogram of Figure 8a; (**c**) equalized fingerprint images of Figure 8a; (**d**) histogram of Figure 8c.

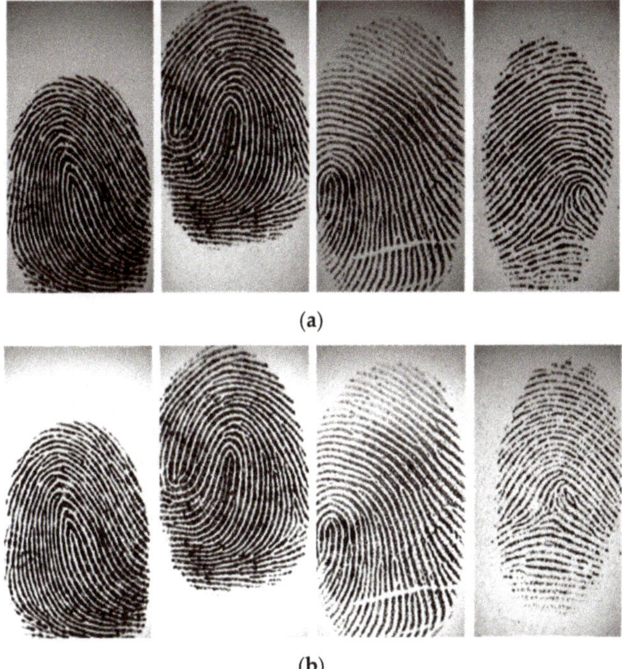

(a)

(b)

**Figure 9.** Results of our proposed method for the FVC2002 DB2 database. (**a**) Original fingerprint images and (**b**) equalized fingerprint images of Figure 9a.

Then, we show the effectiveness by comparing the amount of information in our method and in the original fingerprint images by using the entropy of an image. The entropy of information $H$ was introduced by Shannon [34] in 1948, and it can be calculated by the following equation:

$$H = -\sum_{v=0}^{255} p_i \log_2 p_i, \tag{24}$$

where $p_i$ denotes the probability mass function of gray level $i$, and it is calculated as follows:

$$p_i = \frac{Number\ of\ occurrences\ of\ intensity\ levels}{Number\ of\ intensity\ levels}. \tag{25}$$

In digital image processing, entropy is a measure of an image's information content, which is interpreted as the average uncertainty of the information source. The entropy of an image can be used for measuring image visual aspects [35] or for gathering information to be used as parameters in some systems [36]. Entropy is widely used for measuring the amount of information within an image. Higher entropy implies that an image contains more information.

Entropy is measured to quantify the information produced from the enhanced image. For good enhancement, the entropy of the enhanced image should be close to that of the original image. This small difference between entropies of the original and the enhanced images indicates that the image details are preserved. It also shows that the histogram shape is maintained; thus, the saturation case can be avoided. Table 2 shows the entropy of equalized images compared with original images for each image shown in Figures 8 and 9. The result shows that the equalized fingerprint images have smaller entropy while they are still close to the entropy of the original image. It means that our method can remove noise from the original image while retaining the structure of the fingerprint image.

**Table 2.** Entropy of equalized images compared with original images for each database.

| | The Entropy of Image | | | | | | | |
| --- | --- | --- | --- | --- | --- | --- | --- | --- |
| | FVC2002 DB1 | | | | FVC2002 DB2 | | | |
| Original image | 5.1222 | 5.5262 | 5.4171 | 5.3983 | 7.2446 | 7.1543 | 7.7012 | 7.5129 |
| Equalized image | 4.8028 | 5.3939 | 5.0496 | 4.9844 | 6.8401 | 7.0603 | 6.3322 | 6.2088 |

Next, the equalized fingerprint image was used to determine the contour and detected the blur region of the fingerprint, as discussed in Sections 2.2 and 2.3. Figure 10 shows the binarized image obtained by applying Equation (8) to the equalized image. Based on the binary images, as shown in Figure 10b, we can detect the region of impression (ROI), and the contour of the fingerprint is acquired in a polygon, as shown in Figure 10c. Figure 11b presents the blur detection result obtained by 2D non-separable wavelet entropy filtering for low-quality images, as discussed in Section 2.3. In what follows, an ROI with a 30% blur region is considered to have bad quality, and its SP detection is not good enough.

Our experiments were tested on the FVC2002 DB1_A and FVC2002 DB2_A databases. We compared the results of our proposed SP detection with results obtained using other methods, including a rule-based algorithm [5], Zhou's algorithm [11], Tico's algorithm [37], Ramo's algorithm [38], and Chikkerur and Ratha's algorithm [39]. In these methods, the singular points were measured on Euclidean distance. While no standard terms exist to define a correct detection, we devoted our attention in this research to a method for detecting a singular point precisely and followed the convention for adopting the 10-pixel deviation on the distance between the expected and the detected singular points to validate the performance of the proposed method. In addition, the singular point detection based on the Poincaré index method is sensitive for low-quality fingerprints. In this paper, we show that by combining a novel adaptive image enhancement, compact boundary segmentation, and NSDWT for localization, the detection of singular points is more robust. Moreover, a novel clustering algorithm by integrating wavelet frame entropy with region growing is introduced to evaluate the fingerprint image quality to validate the detected singular points. Tables 3 and 4 show the correctly detected rate, detection rate, miss rate, and false alarm rate. The results in the tables indicate that our method not only has a higher correctly detected rate than other methods but also has a low false alarm rate. Figure 12 presents the results of truly detected SPs on the FVC2002 database; the core points and the delta points are closer as ground truth SPs. Figure 13 presents some comparison results of SP detection for the FVC2002 database using our proposed method and the Poincaré index method. In this figure, blue and green crosses indicate the core and delta points, respectively, detected by our proposed method, and the red cross indicates the core point detected by the Poincaré index method. The results show that the location of the SPs detected using our method is more accurate than those of the SPs detected using the Poincaré index method.

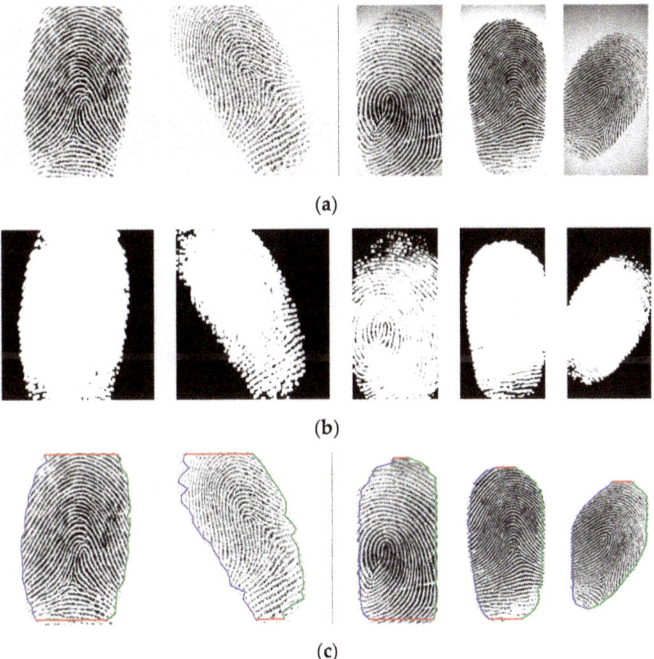

(a)

(b)

(c)

**Figure 10.** Binary images by using energy transformation for the FVC 2002 DB1 and DB2 databases. (a) Equalized images of five fingerprint images in the FVC 2002 database; (b) binary images of Figure 10a; (c) segmented images of Figure 10a.

**Table 3.** Comparison results of various detection algorithms for the FVC2002 DB1-A fingerprint database.

| Algorithm | SCR | SDR | | SMR | | SFR | |
|---|---|---|---|---|---|---|---|
| | | Core | Delta | Core | Delta | Core | Delta |
| Tico's [37] | 58.50 | 90.27 | 55.49 | 9.83 | 44.51 | 10.78 | 80.20 |
| Ramo's [38] | 53.54 | 92.19 | 68.42 | 7.81 | 31.58 | 8.47 | 46.15 |
| Zhou's [11] | 88.88 | 95.78 | 96.98 | 4.22 | 3.02 | 2.27 | 9.97 |
| Chikkerur's [39] | 85.13 | 95.89 | 92.75 | 4.11 | 7.25 | 6.93 | 8.16 |
| Rule-based [5] | 50.00 | 86.26 | 55.24 | 13.74 | 44.76 | 15.92 | 81.04 |
| Proposed | 90.72 | 92.43 | 97.25 | 7.57 | 2.75 | 1.41 | 3.07 |

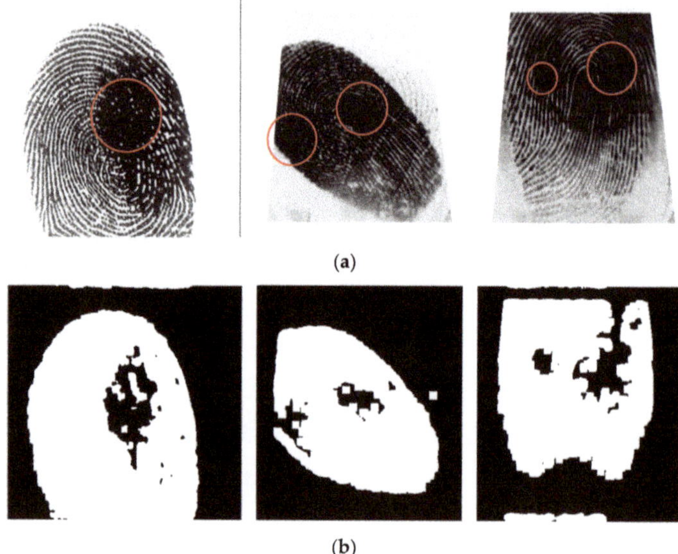

**Figure 11.** Blur detection result obtained by 2D non-separable wavelet entropy filtering for low-quality images: (**a**) original images and (**b**) blur detection results.

**Table 4.** Comparison results of different detection algorithms for the FVC2002 DB2-A fingerprint database.

| Algorithm | SCR | SDR | | SMR | | SFR | |
|---|---|---|---|---|---|---|---|
| | | Core | Delta | Core | Delta | Core | Delta |
| Tico's [37] | 32.32 | 65.38 | 34.75 | 34.62 | 65.25 | 52.94 | 187.80 |
| Ramo's [38] | 49.49 | 80.72 | 37.50 | 19.28 | 62.50 | 23.88 | 166.67 |
| Zhou's [11] | 81.25 | 95.95 | 90.88 | 4.05 | 9.12 | 8.45 | 12.54 |
| Chikkerur's [39] | 73.25 | 93.23 | 94.20 | 6.77 | 5.80 | 13.87 | 28.62 |
| Rule-based [5] | 56.57 | 73.86 | 37.61 | 26.14 | 62.39 | 35.40 | 165.85 |
| Proposed | 89.92 | 95.54 | 95.21 | 4.46 | 4.79 | 1.51 | 2.76 |

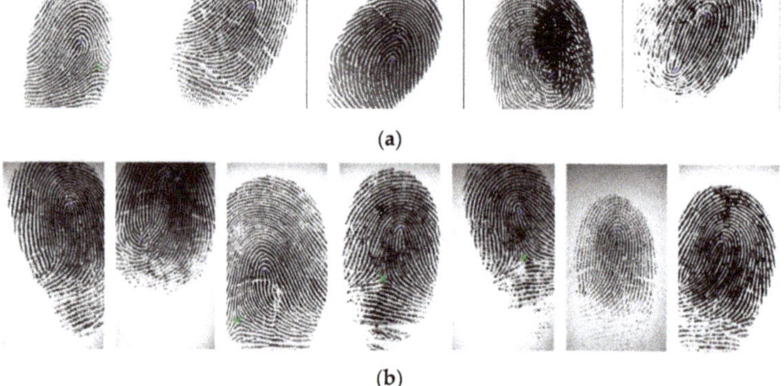

**Figure 12.** Truly detected SPs for the FVC2002 database (blue: core point; green: delta point) by our proposed method: (**a**) FVC2002 DB1 and (**b**) FVC2002 DB2 databases.

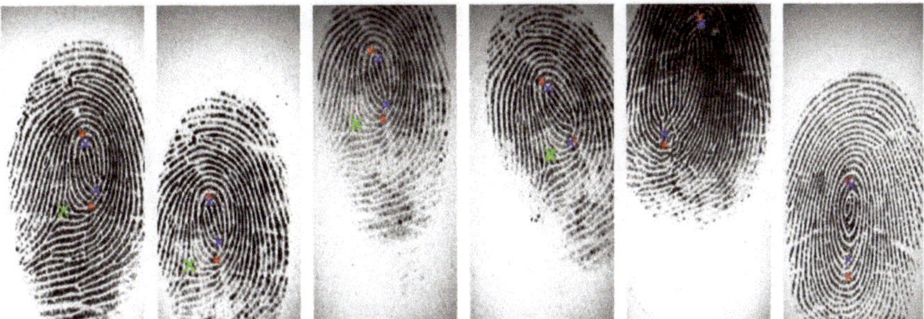

**Figure 13.** Some comparison results of SP detection for the FVC2002 database. The blue and green crosses indicate the core and delta points, respectively, detected by our proposed method, and the red cross indicates the core point detected by the Poincaré index method.

## 5. Conclusions

Because the conventional Poincaré index along the boundary of a given region equals the sum of the Poincaré indices of the core points within this region, it contains no information about the characteristics and cannot describe the core point completely. To solve this problem, we proposed an adaptive method to detect SPs in a fingerprint image. First, a novel fingerprint enhancement algorithm was proposed to considerably eliminate the background, thereby improving the clarity and continuity of ridge structures. Second, we demonstrated that the proposed algorithm could effectively detect low-quality regions with a high correct rate. Third, based on the threshold value, the proposed algorithm inspected and made a True/False decision about whether a detected SP was accepted. Experimental results demonstrate that the proposed algorithm effectively detects SPs and the results are better than those obtained by rule-based [5], Zhou [11], Tico [37], Ramo [38], and Chikkerur [39].

**Author Contributions:** N.T.L. developed the fingerprint hardware and software coding, and wrote the original draft. J.-W.W. guided the research direction and edited the paper. D.H.L. designed the experiments. C.-C.W. contributed to editing the paper. All authors discussed the results and contributed to the final manuscript.

**Funding:** This research was funded in part by MOST 107-2218-E-992-310 and 108-2221-E-992-076 from the Ministry of Science and Technology, Taiwan.

**Acknowledgments:** The authors appreciate the support from National Kaohsiung University of Science and Technology in Taiwan.



## References

1. Henry, E.R. *Classification and Uses of Finger Prints*; George Rutledge & Sons: London, UK, 1900.
2. Jain, A.; Lin, H.; Bolle, R. On-line fingerprint verification. *IEEE Trans. Pattern Anal. Mach. Intell.* **1997**, *19*, 302–314. [CrossRef]
3. Ratha, N.; Bolle, R. *Automatic Fingerprint Recognition Systems*; Springer: New York, NY, USA, 2004.
4. Wang, C.-N.; Wang, J.-W.; Lin, M.-H.; Chang, Y.-L.; Kuo, C.-M. Optical methods in fingerprint imaging for medical and personality applications. *Sensors* **2017**, *17*, 2418. [CrossRef] [PubMed]
5. Maltoni, D.; Maio, D.; Jain, A.K.; Prabhakar, S. *Handbook of Fingerprint Recognition*; Springer Science & Business Media: New York, NY, USA, 2009.
6. Pankanti, S.; Prabhakar, S.; Jain, A.K. On the individuality of fingerprints. *IEEE Trans. Pattern Anal. Mach. Intell.* **2002**, *24*, 1010–1025. [CrossRef]
7. Kawagoe, M.; Tojo, A. Fingerprint pattern classification. *Pattern Recognit.* **1984**, *17*, 295–303. [CrossRef]
8. Wang, Y.; Hu, J.; Phillips, D. A Fingerprint Orientation Model Based on 2D Fourier Expansion (FOMFE) and Its Application to Singular-Point Detection and Fingerprint Indexing. *IEEE Trans. Pattern Anal. Mach. Intell.* **2007**, *29*, 573–585. [CrossRef] [PubMed]

9. Nilsson, K.; Bigun, J. Localization of corresponding points in fingerprints by complex filtering. *Pattern Recognit. Lett.* **2003**, *24*, 2135–2144. [CrossRef]

10. Liu, M. Fingerprint classification based on Adaboost learning from singularity features. *Pattern Recognit.* **2010**, *43*, 1062–1070. [CrossRef]

11. Zhou, J.; Chen, F.; Gu, J. A novel algorithm for detecting singular points from fingerprint images. *IEEE Trans. Pattern Anal. Mach. Intell.* **2009**, *31*, 1239–1250. [CrossRef]

12. Chen, H.; Liang, J.; Liu, E.; Tian, J. Fingerprint singular point detection based on multiple-scale orientation entropy. *IEEE Signal Process. Lett.* **2011**, *18*, 679–682. [CrossRef]

13. Hung, D.D. Enhancement and feature purification of fingerprint images. *Pattern Recognit.* **1993**, *26*, 1661–1671. [CrossRef]

14. He, Y.; Tian, J.; Lou, X.; Zhang, T. Image enhancement and minutiae matching in fingerprint verification. *IEEE Signal Process. Lett.* **2003**, *24*, 1349–1360. [CrossRef]

15. Jirachaweng, S.; Hou, Z.; Yau, W.Y.; Areekul, V. Residual orientation modeling for fingerprint enhancement and singular point detection. *Pattern Recognit.* **2011**, *44*, 431–442. [CrossRef]

16. Wang, W.; Li, J.; Huang, F.; Feng, H. Design and implementation of Log-Gabor filter in fingerprint image enhancement. *Pattern Recognit. Lett.* **2008**, *29*, 301–308. [CrossRef]

17. Gottschlich, C. Curved-region-based ridge frequency estimation and curved Gabor filters for fingerprint image enhancement. *IEEE Trans. Image Process.* **2011**, *21*, 2220–2228. [CrossRef]

18. Wang, S.; Wang, Y. Fingerprint enhancement in the singular point area. *IEEE Signal Process. Lett.* **2004**, *11*, 16–19. [CrossRef]

19. Yang, J.; Xiong, N.; Vasilakos, A.V. Two-stage enhancement scheme for low-quality fingerprint images by learning from the images. *IEEE Trans. Human-Mach. Syst.* **2013**, *43*, 235–259. [CrossRef]

20. Yun, E.K.; Cho, S.B. Adaptive fingerprint image enhancement with fingerprint image quality analysis. *Image Vis. Comput.* **2006**, *24*, 101–110. [CrossRef]

21. Fronthaler, H.; Kollreider, K.; Bigun, J. Local features for enhancement and minutiae extraction in fingerprint. *IEEE Trans. Image Process.* **2008**, *17*, 354–363. [CrossRef]

22. Bennet, D.; Perumal, D.S.A. Fingerprint: DWT, SVD based enhancement and significant contrast for ridges and valleys using fuzzy measures. *J. Comput. Sci. Eng.* **2011**, *6*, 28–32.

23. Wang, J.; Le, N.T.; Wang, C.C.; Lee, J.S. Enhanced ridge structure for improving fingerprint image quality based on a wavelet domain. *IEEE Signal Process. Lett.* **2015**, *22*, 390–395. [CrossRef]

24. Mehtre, B.M.; Murthy, N.N.; Kapoor, S.; Chatterjee, B. Segmentation of fingerprint images using the directional image. *Pattern Recognit.* **1987**, *20*, 429–435. [CrossRef]

25. Mehtre, B.M.; Chatterjee, B. Segmentation of fingerprint images – a composition method. *Pattern Recognit.* **1989**, *22*, 381–385. [CrossRef]

26. Zhang, J.; Lai, R.; Kou, C.C.J. Adaptive directional total – variation model for latent fingerprint segmentation. *IEEE Trans. Inf. Forensics Secur.* **2013**, *8*, 1261–1273. [CrossRef]

27. Cao, K.; Liu, E.; Jain, A.K. Segmentation and enhancement of latent fingerprint: A coarse to fine ridge structure dictionary. *IEEE Trans. Pattern Anal. Mach. Intell.* **2014**, *36*, 1847–1859. [CrossRef]

28. Maio, D.; Maltoni, D.; Cappelli, R.; Wayman, J.L.; Jain, A.K. FVC2002: Second fingerprint verification competition. Proceeding of the Object Recognition Supported by User Interaction for Service Robots, Quebec City, QC, Canada, 11–15 August 2002; pp. 811–814.

29. Andrews, H.; Patterson, C. Singular value decompositions and digital image processing. *IEEE Trans. Acoust. Speech Signal Process.* **1976**, *24*, 26–53. [CrossRef]

30. Wang, J.W.; Le, N.T.; Wang, C.C. Color face image enhancement using adaptive singular value decomposition in Fourier domain for face recognition. *Pattern Recognit.* **2016**, *57*, 31–49. [CrossRef]

31. Wang, J.W.; Le, N.T.; Lee, J.S.; Wang, C.C. Illumination compensation for face recognition using adaptive singular value decomposition in the wavelet domain. *Inf. Sci.* **2018**, *435*, 69–93. [CrossRef]

32. Unser, M. Texture classification and segmentation using wavelet frames. *IEEE Trans. Image Process.* **1995**, *4*, 1546–1560. [CrossRef]

33. Mallat, S. *A Wavelet Tour of Signal Processing*, 3rd ed.; The Sparse Way; Academic Press, Inc.: Cambridge, MA, USA, 2008.

34. Shannon, C.E. A mathematical theory of communication. *Bell Syst. Tech. J.* **1948**, *27*, 379–423/623–656. [CrossRef]

35. Tsai, D.Y.; Lee, Y.; Matsuyama, E. Information entropy measure for evaluation of image quality. *J. Digital Imaging* **2008**, *21*, 338–347. [CrossRef]
36. Min, B.S.; Lim, D.K.; Kim, S.J.; Lee, J.H. A novel method of determining parameters of CLAHE based on image entropy. *Int. J. Softw. Eng. Its Appl.* **2013**, *7*, 113–120. [CrossRef]
37. Tico, M. and Kuosmanen, P. A multiresolution method for singular points detection in fingerprint images. *IEEE Int. Symp. Circuit Syst.* **1999**, *4*, 183–186.
38. Ramo, P.; Tico, M.; Onnia, V.; Saarinen, J. Optimized singular point detection algorithm for fingerprint images. *IEEE Trans. Image Process.* **2001**, *3*, 242–245.
39. Chikkerur, S.; Ratha, N.K. Impact of singular point detection on fingerprint matching performance. In Proceedings of the Fourth IEEE Workshop on Automatic Identification Advanced Technologies, Buffalo, NY, USA, 17–18 October 2005; pp. 207–212.

Article

# A Unified Framework for Head Pose, Age and Gender Classification through End-to-End Face Segmentation

Khalil Khan [1]*, Muhammad Attique [2,*], Ikram Syed [3], Ghulam Sarwar [3], Muhammad Abeer Irfan [4] and Rehan Ullah Khan [5]

1   Department of Electrical Engineering, University of Azad Jammu and Kashmir, Muzafarabbad 13100, Pakistan
2   Department of Software Engineering, Sejong University, Seoul 05006, Korea
3   Department of Software Engineering, University of Azad Jammu and Kashmir, Muzafarabbad 13100, Pakistan
4   Dipartimento di Elettronica e Telecomunicazioni (DET), Politecnico di Torino, 10156 Torino, Italy
5   IT Department, College of Computer, Qassim University, Al-Mulida 51431, Saudi Arabia
*   Correspondence: khalil.khan@ajku.edu.pk (K.K.); attique@sejong.ac.kr (M.A.)

Received: 2 June 2019; Accepted: 24 June 2019; Published: 30 June 2019

**Abstract:** Accurate face segmentation strongly benefits the human face image analysis problem. In this paper we propose a unified framework for face image analysis through end-to-end semantic face segmentation. The proposed framework contains a set of stack components for face understanding, which includes head pose estimation, age classification, and gender recognition. A manually labeled face data-set is used for training the Conditional Random Fields (CRFs) based segmentation model. A multi-class face segmentation framework developed through CRFs segments a facial image into six parts. The probabilistic classification strategy is used, and probability maps are generated for each class. The probability maps are used as features descriptors and a Random Decision Forest (RDF) classifier is modeled for each task (head pose, age, and gender). We assess the performance of the proposed framework on several data-sets and report better results as compared to the previously reported results.

**Keywords:** face analysis; face segmentation; head pose estimation; age classification; gender classification

---

## 1. Introduction

The problem of human face image analysis is a fundamental and challenging task in computer vision. It plays a key role in various real world applications such as surveillance, animation and human computer interaction. However, it is still a challenging task due to changes in facial appearance, visual angle, complicated facial expressions and the background. In particular, in the un-constrained conditions it has much more complications.

Each of these face analysis tasks (head pose, age and gender recognition) are approached as individual research problem through various sets of techniques [1–8]. We argue that all these tasks are *very closely* related and essentially can help each other if a prior efficiently segmented face image is given as input. It is also confirmed by psychology literature that face parts such as nose, hair, and mouth helps human visual system in face identity recognition [9,10]. Therefore, performance of all related applications can be improved if a well segmented face image is provided as input to the framework.

The facial attribute information such as head pose estimation, age classification, and gender recognition is already being predicted using facial landmarks information [4,11]. However, the performance of head pose and any other applications in such cases heavily depends on accurate localization of these landmarks [5,7,12]. Locating these face landmarks is itself a *big challenge*. These points localization are greatly affected in certain cases such as occlusion, face rotation and if the quality of the image is very low. Similarly, in far-field imagery conditions, these landmarks extraction are not only difficult but some-times impossible. Lighting conditions and complicated facial expressions also make the localization part challenging. Due to all problems mentioned above, we approach the face analysis task in a complete different way.

In this paper we introduce a unified framework, which addresses all the three face analysis tasks (head pose, age, and gender recognition) through a prior multi-class face segmentation model that was developed through CRFs. We named the newly proposed multitask framework HAG-MSF-CRFs. It is a jointly estimation probability task that tackles it using a very powerful random forest algorithm. Specifically, the proposed framework can be formulated as;

$$(h, a, g) = \arg\max_{h,a,g} p(h, a, g | \mathbf{I}, \mathbf{B}) \tag{1}$$

where head pose, age, and gender recognition are represented by $h$, $a$ and $g$ respectively. Similarly, in Equation (1), $I$ is the input face image and $B$ is the bounding box which is provided by the face detector.

In our previous work we already tackle the problem of multi-class semantic face segmentation (MSF) [13] and its application to head pose estimation [14,15] (MSF-HPE) and gender classification [16]. In most of the previous works, face segmentation is considered as three or some-times four classes face segmentation task. In the MSF, face segmentation is extended to six classes (eyes, nose, mouth, skin, back and hair). However, we were facing some major problems in previously proposed MSF. Firstly, the computation cost of MSF is quite high, as MSF provides a class label to each and every pixel in an image, which ultimately takes a long time. A super-pixel based model is used instead which reduces the processing cost. Secondly, the MSF does not consider any conditional hierarchy between different face parts. For example, it is not possible for the eye region to be near to the mouth region and vice versa. A CRFs based model is introduced in this paper, which couples all labels in a face image in a scaled hierarchy. Going from MSF to the newly proposed MSF-CRFs improves the performance of the segmentation part.

Our proposed multi-task framework is comparable to another approach known as the influence model (IM). This model was first introduced by researchers in the MIT media laboratory [17,18]. The IM estimates how the state of one actor affects another in the system. Our proposed model is somehow similar to the model proposed in [17,18]. In such cases, an outcome in one entity in a system causes outcome in another entity in the same system. In simple words, if one domino is flipped, the next domino will fall automatically and vice versa. In IM it is necessary to know how certain dominoes interact with each other and how one is influenced by another. If the initial state of the dominoes is known with relative location to another, then the outcome of the system is predicted with more accuracy. When the system network structure is already known, the IM enables researchers to infer interaction; however, information about signals from different observations are needed.

To summarize, contributions of the paper are three fold:

- We propose a new multi-class face segmentation algorithm MSF-CRFs. The MSF-CRFs model uses the idea of CRFs between various face parts.
- We develop a new multi-tasks face analysis algorithm HAG-MSF-CRFs. The HAG-MSF-CRFs tackles all the three tasks, which include head pose, age, and gender recognition in a single framework.
- Detailed experiments are conducted on state-of-the-arts (SOA) data-sets, and better results are reported comparatively.

The structure of the remaining paper is as follows: Section 2 describes related works for all the three cases i.e., head pose, age, and gender recognition. Several data-sets are use to evaluate the framework. Details about these databases is given in Section 3. The segmentation model MSF-CRFs is presented in Section 4, whereas the proposed algorithm for face analysis (HAG-MSF-CRFs) is discussed in Section 5. All obtained results are discussed and compared with SOA in Section 6. The paper is summarized with some future directions in Section 7.

## 2. Related work

Our newly proposed model is closely related to IM based built systems. The IM framework is already used in the automatic recognition tasks of social and task-oriented functional roles in group-meetings [17,18]. The classification of social functional roles has been improved as compared to Hidden Markov Models (HMM) and support vector machine (SVM) [18] through IM. The two versions proposed in [18] outperform both HMM and SVM based results in the social functional role problems. The IM methods showed excellent performance, particularly in less populated classes. Media segmentation is performed with IM in cases particularly having rich information [19–21]. The keywords information are exploited in [22] to identify journalists, anchors, and guest speaker if any in a radio program. The maximum entropy algorithm is used for getting the classification accuracy. The IM based algorithms are applied to many audio and visual recognition tasks, for details, more papers can be explored in [23–28].

Before describing the proposed framework, we briefly review related methods for head pose, age, and gender classification. A rich literature and history is already present about all these three topics. However, in this section of the paper we provide a cursory overview of how these tasks were previously approached by researchers.

### 2.1. Head Pose Estimation

Pose of an image can be classified into three broad categories; yaw, pitch, and roll. The yaw angles represents the horizontal orientation and the pitch vertical orientation of a face image. The image plane is represented by the roll angles. We evaluated our proposed algorithm for head pose estimation on four data-sets, which included Pointing'04 [29], Annotated Facial Landmarks in the Wild (AFLW) [30], Boston University (BU) [31], and ICT-3DHPE [32] data-sets.

Two types of information were previously used to approach the head pose estimation i.e., facial landmarks and face image appearance. In the former case, a POSIT algorithm [9] is used to find correspondence between pints in $2D$ shapes and points in $3D$ models. In the latter case, various image appearance features such as SIFT, LBP, HOG etc. are exploited for head pose estimation. Discriminative learning models such as Random Forest and Support Vector Machine (SVM) are trained and tested using the extracted features [4,10]. A more detailed survey on head pose estimation can be explored in [5].

### 2.2. Age Classification

Age classification is a well-researched topic in computer vision society. Previously, age estimation was studied as a classification or regression problem. In the first case, age is associated with a specific range or age group. In the second case, the exact age of a face image is estimated. Recently a survey paper was reported on age estimation in [33]. All data-sets used for age estimation were discussed and a detailed overview was presented about the algorithms proposed thus far. A detailed investigation of age classification between specific ranges or age groups was presented in [34]. Similarly, another algorithm is introduced to classify age from facial images in [35]. Initially, the appearance of face wrinkles is detected and then age categorization is performed based on the extracted wrinkles. The previous idea [35] was further extended in [36] by first localizing the facial features. The modeling of craniofacial growth was performed through psychophysical and anthropometric evidences in [36]. The main drawback of this approach was: accurate localization of facial features is needed in any case.

A subspace method called AGing PatErn subspace is introduced in [37,38]. In these algorithms, aging features from face images were extracted and an adjusted robust regressor was trained to categorize face ages. These methods showed excellent performance compared to SOA methods. However, two serious weaknesses are faced by these algorithms. The input images must be frontal, and the face images must be well-aligned. The approaches proposed in these algorithms are suited for databases collected in indoor environmental conditions. Practical applications of these methods in the un-constrained conditions is almost impossible.

A cost-sensitive hyper-planes ranking method is introduced in [39]. The algorithm proposed in [39] is a multi-stage learning method which is also known as 'a grouping estimation fusion' (DEF) method in the literature. Similarly, a novel features selection method was proposed in [40]. In a nutshell, all these previously mentioned methods showed good performances in indoor lab conditions, but failed when exposed to the real-world conditions.

Recently introduced Deep Convolutional Networks (CNNs) showed excellent performance for different visual recognition problems. A hybrid system for age and gender classification is proposed in [41]. CNNs are used to extract features from the face images, whereas an extreme learning machine (ELM) is used as a classification tool. The authors of the paper named their proposed method as CNNs-ELM. The system is evaluated on two data-sets, MORPH-II [42] and Adience [43]. To the best of our knowledge, this is the best algorithm performing on a joint problem of gender and age recognition thus far. A weakness reported by the authors of the paper is: miss-classification occurs when the system is exposed to younger faces.

### 2.3. Gender Classification

A detailed investigation about gender recognition was conducted by Makinen and Raisamo [44]. The early researchers who worked on gender recognition used neural network [45]. An SVM classifier was used by Moghaddam and Yang [46]. Similarly, an Adaboost classifier was adapted by Baluja and Rowley [47]. In all these methods image was used as one dimensional feature vector and certain features are extracted from it. A joint framework of age and gender recognition was proposed by Toews and Arbel [48]. The model proposed by the authors is a view-point invariant appearance model which is robust to local scale rotations.

Gender classification analysis based on human gait and linear discriminant algorithms was provided by Yu et a. [49]. A new benchmark to study age and gender classification was suggested in [43]. Through the available data, a classification pipeline is presented by the authors of the paper. Khan et al. [50] proposed a semantic pyramid, dealing both gender and action recognition. Annotation for face and upper body was not needed in the proposed method. First part of the name was used as a feature and a modeling mechanism of the name part and face images was performed in the next stage in a method proposed in [51]. Higher accuracy was reported with proposed method as compared to SOA. Recently, a generic algorithm to estimate gender, race, and age in a single framework is proposed in [52].

All the above-mentioned approaches made lots of progress and contribution towards gender recognition. However, most of these methods were aimed either at non-automated estimation methods or only worked well in very constrained imaging environments.

## 3. Databases

In this paper we use six different face databases to perform the three tasks i.e., head pose, age and gender classification. For head pose estimation we use Pointing'04, AFLW, BU, and ICT-3DHPE data-sets. For age classification we use Adience and FERET [53] data-sets. For gender recognition we perform tests with Adience database only.

### 3.1. Head Pose Estimation

- **Pointing'04 database:** The Pointing'04 database is a manually annotated face database. Even though it is a comparatively old head pose data-set, it is still used for research purposes [54–56] due to its challenging nature and large variety with consecutive poses. All the images in the Pointing'04 database are low resolution images captured in low lighting conditions. The Pointing'04 contains 15 sets of face images. Each set is further divided into 2 sets having 93 images for each candidate at various orientations. The age of each subject in the database is kept between the range 20–40 years.

To add more complexity to the database images, five subjects were included with facial hair and seven were wearing glasses. The pan and tilt angle determined the head pose of a subject. Each subject in the database acquisition was asked to look into 93 markers marked on the wall. Each marker represented a specific pose. The given face localization in Pointing'04 may not be accurate due to manual labeling. A sample of the images of a single candidate at 93 different locations is shown in Figure 1. For yaw, the head orientation varied between $-90°$ to $+90°$ with a step size of $15°$ between two adjacent poses. For pitch, the positive values corresponded to the top poses and negative to the bottom poses. The difference between two consecutive poses in the pitch is $30°$.

- **AFLW Data-set:** Images in AFLW exhibited variations in facial expression, lighting conditions, face appearance, and some other environmental factors. All these images were obtained from the internet. The AFLW contained both the frontal and non-frontal images. The frontal images had six facial expressions. More difficulties were added to the images in the form of certain facial accessories. The images were collected from 9 different lighting conditions. In short, AFLW is a very challenging data-set, since the data-set is collected in the real world with un-constrained conditions.

- **BU Data-set:** The BU data-set has two image sequences, i.e., images collected in uniform lighting conditions and images exposed to rather complex scenarios by changing the lighting conditions. We used RGB images only for the experiments. We considered all the three rotations, which included pitch, roll and yaw. A total of 5 subjects participated in the image acquisitions process. A magnetic tracker was attached to each subject's head to obtained the ground truth images.

- **ICT-3DHPE:** A Kinect sensor was used to collect the ICT-3DHPE images. This data-set contains both the depth and RGB images. However, we only used the RGB images in our work. Six male and four females participated in the image collection process. The ground truth images were more accurate in this case as well (like BU data-set), because a magnetic tracker was attached to each participant's head. It must be noted that the ground truth images creation method for Pointing'04 and AFLW is a type of manual labeling method. The chances of error exists while providing labels to the ground truth data.

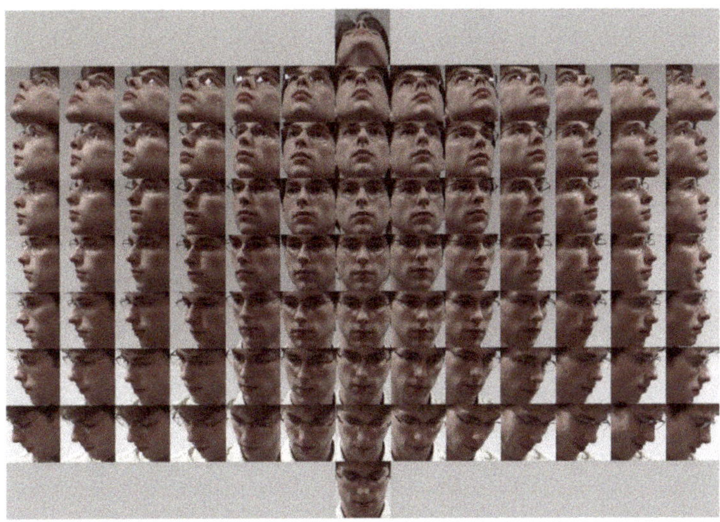

**Figure 1.** Pointing'04 database images of a single subject in all 93 poses.

*3.2. Age and Gender Classification Data-Sets*

- **Adience Benchmark:** It is a recently released un-constrained image database which is used both for age and gender recognition. All these images were created from smart phone devices. These images included variations such as pose, lighting, appearance, noise, and more—meaning the data-set has all conditions of un-constrained image database. The total number of images in Adience are 26,580, whereas the total number of participants are 2284. The exact age of each candidate is not specified, and each subject is assigned to 8 different age groups i.e., [0,2], [4,6], [8,13], [15,20], [25,32], [38,43], [48,53], [60,+]. The data-set can be obtained from the Open University of Israel (computer vision lab).

- **LFW data-set:** The LFW database consists of 13,233 images for 5,749 subjects. The data-set was collected in un-constrained conditions. All these face images were collected from the web. It is an imbalanced database, because the number of male candidates are 10,256 whereas female images are 2,977.

- **FERET data-set:** This is also an old data-set that is widely used to develop and evaluate various facial recognition methods. The database was collected in controlled indoor conditions with gender information for each subject. The data-set is composed of 14,126 images whereas the total number of participants were 1,199. We used the colored version from the FERET database. Some variations of facial expressions, lighting conditions and face pose were kept while image acquisition – made the database a rather challenging one. The database consists of both frontal and non-frontal images. We applied our algorithm to both set of images (between $-45°$ and $+45°$).

### 4. Proposed MSF-CRFs

The overview of the MSF-CRFs model for semantic face segmentation is shown in Figure 2. The labeling problem is modeled efficiently with the proposed MSF-CRFs, which combines the output from the built classifier with image location information. This modeling process helps in maximizing a posteriori. The unary potential models each pixel belonging to each class and the pairwise potential models the relationship between two pixels.

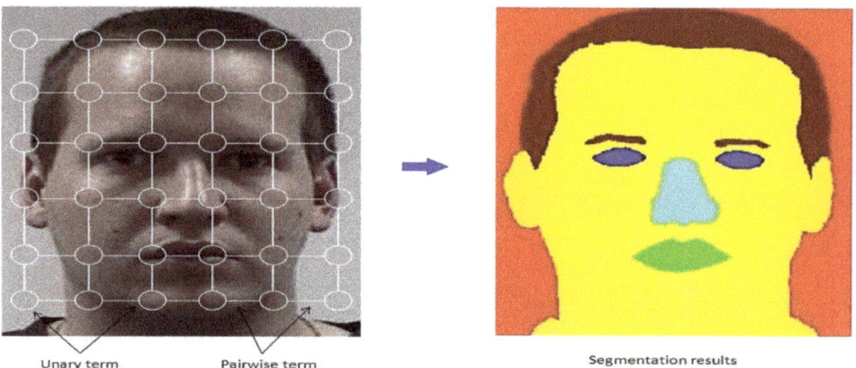


**Figure 2.** The MSF-CRFs graphical model. The input face image in grid cell represents a random variable. The unary potentials are represented by the white circles and the pairwise potential by solid white lines.

As face parts are not localized in most of the images, a face localization algorithm is applied in start. In the literature there are many good methods for face detection, so we use a CNNs based face detector [57]. After localizing the face parts, all face images are re-scaled to a fixed size with a height 256 pixels and the width is adjusted accordingly to keep the original image ratio.

The proposed MSF-CRFs model encodes segmentation probability with features of an image. Initially an image is segmented into super-pixels. The segmentation is represented by $Z$ and this can be represented as $Z = z_1, z_2, ..., z_n$, where n is the total number of super-pixels in the input image. $z_i$ can take the value of any of the six face parts (nose, eyes, mouth, hair, back and skin). For super-pixel segmentation we use SEEDs [58] algorithm.

We also need to develop some conventions about node and edge features. We represent the node features by $Z_m$ and edge features by $Z_e$. We develop a log linear CRFs model which can be written as:

$$\psi(s_i = q, z_i^m) = \sum_{f=1}^{F_m} (X_q^m)_f (z_i^m)_f \tag{2}$$

$$\psi(s_i = q_1, s_j = q_2, = z_{i,j}^e) = \sum_{f=1}^{F_e} (X_{q_1,q_2}^e)_f (z_{i,j}^e)_f \tag{3}$$

In Equations (2) and (3), super-pixel features are represented by $F_m$ whereas $Z_i^m$ represents a vector having length $F_m$. The neighboring super-pixels features are represented by $F_e$. The final resultant feature vector developed is $Z_{i,j}^e$. Similarly, each node and edge weight are adjusted with $X^m$ and $X^e$ respectively. A pair of classification labels in the above Equations is represented by $q_1, q_2$. In the proposed MSF-CRFs model we use symmetric edge potential.

The probability of segmentation conditional on $Z$ can be represented as:

$$P(s|z) = \frac{exp(-\sum_{i=1}^{m} \psi(s_i, z_i^m) - \sum_{i,j} \psi(s_i, s_j, z_{i,j}^e))}{N(Z)} \tag{4}$$

$N(Z)$ represents the partition function in Equation (4). This function acts as a normalization factor for the distribution. We use Bethe Approximation [55] for the partition function in the MSF-CRFs model. Similarly, for marginal approximation we use a loopy belief propagation algorithm. For CRFs optimization, we use the algorithm as in L-BFGS [59]. For weight regulations we also added the Gaussian to the model.

To assess the accuracy of the segmentation estimates, we apply an L1 error to each segmentation estimate. We also penalize each super-pixel as per the difference between the correct label prediction probability and a value 1.0. For example, if a super-pixel has a probability value of 0.7 for being skin (and skin is also the ground truth label of the super-pixel), a penalty value of 0.3 will be incurred as a result.

We compute three types of features for the node listed as; position, HSV color and shape related information (HOG).

For spatial information an $8 \times 8$ grid is considered, and then the relative location of the central pixel is extracted. This location is defined as:

$$f_{loc} = [x/W, y/H] \in R^2 \tag{5}$$

Where $W$ represents the width and $H$ height of the input face image.

For color features, the information from HSV histogram is extracted. The three values (hue, saturation, and variance) are encoded in a single vector constituting a unique feature vector for color information. The dimension of each patch for HSV is kept as $D_{HSV} = 16 \times 16$, whereas the number of bins are set 32. The resulting feature vector for the color information with these values will be $F_{HSV}^{16} \in R^{48}$.

For shape information we use HOG. We keep the dimension of the patch for HOG as $D_{HOG} = 64 \times 64$, which results a feature vector $F_{HOG}^{64 \times 64} \in R^{1764}$

All the three features are concatenated with each other to form a single vector.

## 5. Proposed HAG-MSF-CRFs

Our proposed algorithm is summarized in Algorithm 1. Initially a segmentation model is developed through the CRFs. For face segmentation, the built model MSF-CRFs outputs the most likely class for each super-pixel. The same label is then assigned to each pixel within the super-pixel. For the classification of head pose, age and gender we use the probability maps created during segmentation of each class. Probability maps generated for each class are represented as: $P_{nose}$, $P_{back}$, $P_{eyes}$, $P_{skin}$, $P_{mouth}$, and $P_{hair}$. Figure 3 show some images from Pointing'04 data-set and their probability maps. In the gray-scale images in Figure 3, higher intensity represents higher probability of prediction for a particular class and *vice versa*. For each task (head pose, age, and gender) we train an RDF classifier with a feature vector of the corresponding probability maps. The probability maps are used as feature descriptor.

---

**Algorithm 1** proposed HAG-MSF-CRFs algorithm

---

**Input:** $M_{train} = \{(I_n, T_n)\}_{n=1}^{m}$, $M_{test}$.

where $M_{train}$ is the data used for training model $\mathfrak{A}$, $M_{test}$ is the testing data, $I$ is the input training image and $T(i,j) \in \{1,2,3,4,5,6\}$ is the ground truth data.

**a: Face segmentation part:**

Step a.1: Training a segmentation model $\mathfrak{A}$ through training data (training images and labels)

Step a.2: Finding the center of each super-pixel, extracting patches and passing to the model $\mathfrak{A}$

Step a.3: Using the probabilistic classification method and creating probability maps for each class, represented as:

$p_{skin}$, $p_{mouth}$, $p_{eyes}$, $p_{nose}$, $p_{hair}$, and $p_{back}$

**b. Head pose, age and gender classification part:**

    **if** head pose estimation:

$$f = p_{skin} + p_{mouth} + p_{eyes} + p_{nose} + p_{hair}$$

**Else if** age classification:

$$f = p_{skin} + p_{mouth} + p_{eyes} + p_{nose} + p_{hair}$$

**Else if** gender recognition:

$$f = p_{skin} + p_{eyes} + p_{nose} + p_{hair}$$

where f is the feature vector.

c. Training an RDF classifier for each case (head pose, age and gender)

**Output:** estimated pose, age class and gender.

---

*5.1. Head Pose Estimation*

We manually labeled 10 images from each pose of each data-set. The manually labeled images are used to build an MSF-CRFs model as discussed previously. For all images of every data-set, the probability maps are generated. When a test image is given as input, the MSF-CRFs model creates the probability maps for all classes and all images.

To understand which facial parts help in head pose estimation we conducted a large number of experiments. We use probability maps for the eyes, nose, mouth, skin, and hair. Probability maps in the form of feature descriptors are concatenated to train and test an RDF classifier. We use 10-fold cross validation experiments in our work. Those 10 images, which were previously used to create an MSF-CRFs model were not included in the 10-fold cross validation experiments. The probability maps of a single subject from Pointing'04 data-set are shown in Figure 3. From the Figure 3, it is clear that variation occurs as the pose changes from one position to another. For example taking the skin class (third row), forehead is more exposed to the camera in frontal images. As a result, probability map for brighter part is more concentrated to the center part. Similarly, on extreme left and right profile images, high intensity values are occupied on smaller area. We encoded this information for all classes in the form of feature descriptors and developed a new head pose estimation algorithm.

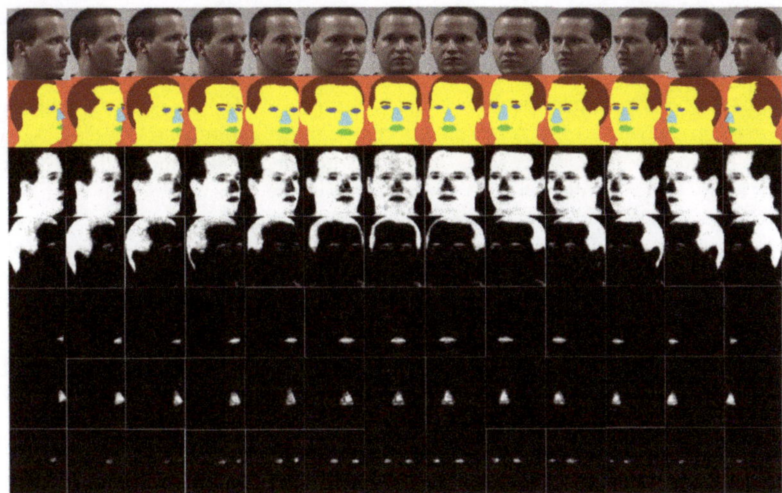

**Figure 3.** Probability maps of a single subject from Pointing'04. Poses vary from −90° to +90° with a step of 15° in the horizontal orientation. Row wise order of the images is as: 1—original images, 2—ground truth images, 3—probability maps for skin, 4—probability maps for hair, 5—probability maps for mouth, 6—probability maps for nose, and 7—probability maps for eyes.

### 5.2. Age Classification

In age classification a face image is assigned to one of the specific age range. From each age group of each data-set, 10 images are manually labeled. The manually labeled images are used to build an MSF-CRFs model. The test face images are passed to the MSF-CRFs model to produce segmentation results and probability maps.

We noted during the experiments that each face part has a contribution towards age classification. Probability maps for each face part differ from one age group to another. Therefore, for age classification we use information about all five face classes, i.e., skin, mouth, hair, and eyes. The probability maps generated are used to train and test an RDF classifier. As in case of head pose, 10-fold cross validation experiments are performed here as well. Manually labeled images which were previously used to create MSF-CRFs model were not included in the 10-fold cross validation experiments.

### 5.3. Gender Recognition

For gender classification, we manually label 30 images for each gender and each data-set. These total 60 images are used to build an MSF-CRFs model for the gender test. A number of qualitative and quantitative experiments are conducted to know which face parts help in gender recognition. After these experiments we train an RDF classifier through probability maps of four classes namely; nose, hair, eyes, and skin.

We perform a detailed study from computer vision and human anatomy literature to know which face parts make a face more feminine or masculine. In the following paragraphs we summarize why we use four classes (skin, nose, hair, and eyes) for gender recognition.

- Usually male forehead is larger compared to female—as the hair line in male lags behind. In male hairline is completely missing in some cases (baldness). This results a larger forehead in male as compared to female. Consequently, brighter part of probability map for the skin is on larger area in case of male.
- Female eyelashes are larger and curly type. Our MSF-CRFs part mis-classified these eyelashes with hairs in females in most of the cases. Even this mis-classification reduces the pixel labeling accuracy of the segmentation part. However, this helps the gender differentiation. In the case of male,

pixel labeling accuracy noted was 79%, resulting better segmentation with brighter probability map. For female the labeling accuracy reduced to 69%, which results a comparatively dimmer probability map.

- A female nose is comparatively smaller with less bridge. On the other hand, male nose is larger and also comparatively longer. A reason reported in the literature for this fact is: as compared to female, the male body is bigger which requires larger lungs and enough passage of air supply towards lungs. Consequently, the male nostrils are larger than female.

- Hairstyle has a very complicated geometry that varies from subject to subject. Our proposed MSF-CRFs reports a pixel labeling accuracy of 97.23%. From the segmentation results (please see Figures 4 and 5), it is clear how efficiently boundary line for hair is detected by our MSF-CRFs model. We encode this information in the form of probability maps and used it in the gender recognition part.

- Sometimes, even eyebrows also help in gender recognition. Male eyebrows are mis-managed and larger, whereas female eyebrows are thinner and curl at the end. In our face segmentation model, we use the same label for eyebrow as hair.

- Literature reports that the mouth must help male and female differentiation. Female lips are clear and visible, whereas in most of the cases upper lip is somehow missing in male images. Unfortunately, we noted no improvement in gender recognition performance with inclusion of the mouth class. Therefore, we did not include mouth class for gender recognition algorithm.

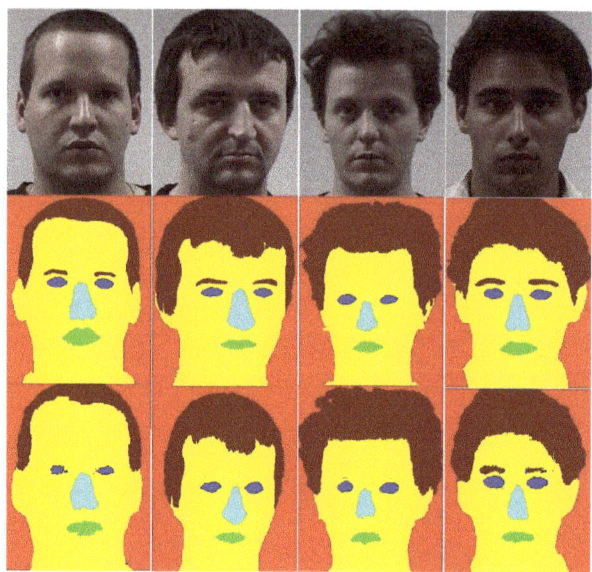

**Figure 4.** Face segmentation results with MSF-CRFs for frontal images on Pointing'04. Images in rows are in order as: row 1—original images, row 2—manually labeled images, row 3—segmentation results produced by MSF-CRFs

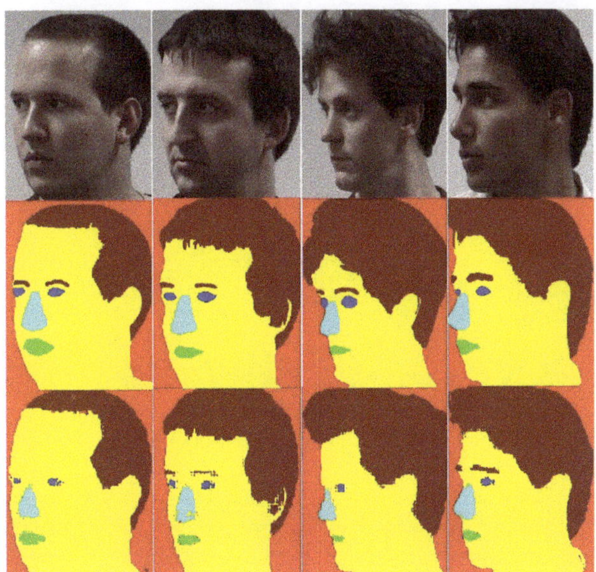

**Figure 5.** Face segmentation results with MSF-CRFs for profile images (+60°) on Pointing'04. Images in rows are in order as: row 1—original images, row 2—manually labeled images, row 3—segmentation results produced by MSF-CRFs

Thus, probability maps for skin, nose, hair, and eyes are concatenated with each other to form a single feature vector. We perform 10-fold cross validation experiments here as well. However, we excluded 60 images which were previously used for training part from each database tests.

## 6. Results and Discussion

### 6.1. Face Segmentation Results

To the best of our knowledge, previously proposed MSF is the first work that considered all six face parts in face segmentation. The main problem with MSF is its computational cost. To remove this deficiency, we used a super-pixel based segmentation in the current model (MSF-CRFs). The processing time of segmentation was improved four times with the MSF-CRFs as compared to the MSF. For example, an image with a $256 \times 240$ pi size took 1.2 min in the MSF model. The same image was segmented with MSF-CRFs in just 18 seconds.

An image is segmented into super-pixels initially. Super-pixel segmentation reduces processing time of segmentation as the number of pixels to be labeled are reduced immensely. In the proposed method we used SEEDs [58] algorithm for super-pixel segmentation. We prefer SEEDS over SLIC and other methods as the speed of the SEEDS is much better than other methods used in SOA [58]. Moreover, SEEDS has much better super-pixel segmentation as reported in standard error metrics.

Face segmentation results for frontal images are much better than profile images. For different super-pixel parameters setting we performed experiments. We noticed better segmentation results with 900 super-pixels. The exact number of super-pixels were less than 900 due to certain segmentation restrictions. The number of super-pixels obtained during the experiments depended on the block levels used and the image size. The super-pixel segmentation was better when the block levels were higher. We used the number of block levels 3, and histogram bins 5. For better accuracy iteration accuracy was kept twice.

Few images from Poinint'04 dataset are shown in Figures 4 and 5. Figure 4 shows some good segmentation results. In Figures 4 and 5, the first row shows the original images, row 2 shows manually labeled images and row 3 shows images segmented with the MSF-CRFs. The frontal images are segmented in Figure 4, whereas the same images rotated at +60° are shown in Figure 5. From these Figs. it is clear that pixel labeling accuracy for frontal images is much better than profile images. It can be noted that as the pose moves to the left or right, labeling accuracy dropped particularly for smaller classes (eyes, nose, and mouth). For extreme profile poses (+90° and −90°) these smaller classes in some images were completely missing.

Performance of the segmentation part highly depends on the quality of the images as well. For example, in the case of AFLW data-set, the images were collected from the internet which included very low quality images. Therefore, poor segmentation results were noticed, ultimately leed to the poor performance of head pose and gender recognition.

*6.2. Head Pose Estimation*

We used two evaluation methods for head pose estimation. The first one is a regression measure i.e., mean absolute error (MAE). MAE is the absolute error between the estimated and ground truth pose. The second one is a classification measure i.e., pose estimation accuracy (PEA). PEA estimates how a particular pose is predicted by a model.

**Pointing'04 data-set:** The results obtained with HAG-MSF-CRFs on the Pointing'04 data-set and its comparison with SOA for both yaw and pitch angles is shown in Table 1. From the Table 1, it is clear that we achieved better results as compared to previously reported results for both the MAEs and PEAs. All possible combination of the six face classes were tried in the experiments. The best results for yaw (average MAE = 2.32° and average PEA = 87%) and pitch (average MAE =1.18° and average PEA = 95%) were obtained with five classes i.e., 'nose', 'mouth' 'skin', 'hair', and 'eyes'. It must be noted that some of the previous methods mentioned in Table 1 may have used a differential experimental setup. For example, 5-fold cross validation experiments were performed in the MLD. We performed our experiments with 10-fold cross validation protocol. Corresponding papers can be explored for the experimental setup and more details for each case.

**Table 1.** Head pose estimation results and its comparison with SOA on Pointing'04 database.

| Method | MAE (Yaw) | Accuracy (Yaw) | MAE (Pitch) | Accuracy (Pitch) |
| --- | --- | --- | --- | --- |
| **HAG-MSF-CRFs** | **2.32°** | **87.7%** | **1.18°** | **95.0%** |
| MSF-HPE [14] | 3.75° | 77.40% | – | – |
| MLD [37] | 4.24° | 73.30% | 6.45° | 86.24% |
| CNN [60] | 5.17° | 69.88% | 5.36° | 77.87% |
| MGD [61] | 6.90° | 64.51% | 8.00° | 62.72% |
| kCovGa [62] | 6.34° | – | 7.14° | – |
| CovGA [62] | 7.27° | – | 8.69° | – |

For a more clear comparison with SOA methods, we also reported the results for each pose both for the MAEs and PEAs. The MAEs results are compared in Figures 6 and 7 for pitch and yaw angles respectively. We had the best results for MAE for all yaw poses (except, 0° and +30°). Similarly, Figures 8 and 9 shows the PEAs results obtained with proposed method and its comparison with SOA for each discrete pose. From the Figure 8, we can see that better results are obtained as compared to SOA for pitch angles. However, CNNs and KCovGA algorithms were performing better at pose −30°.

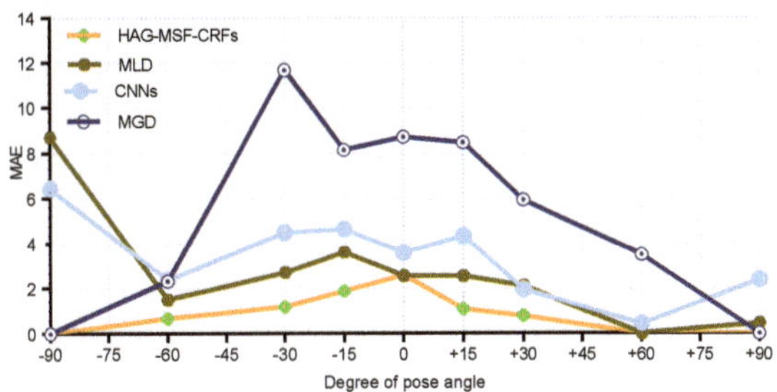

**Figure 6.** MAE comparison with SOA on Pointing'04 (pitch)

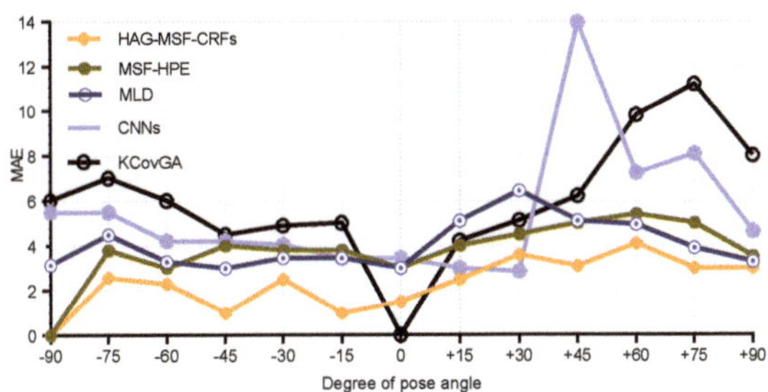

**Figure 7.** MAE comparison with SOA on Pointing'04 (yaw)

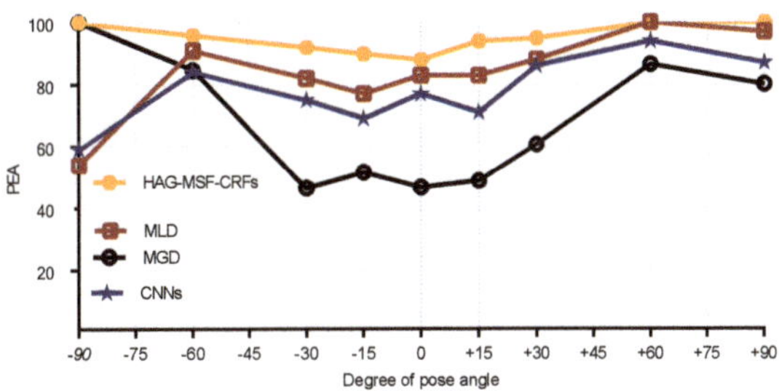

**Figure 8.** PEA comparison with SOA on Pointing'04 (pitch)

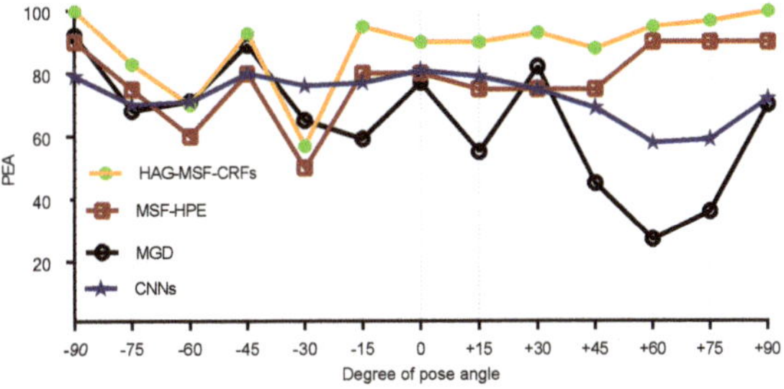

**Figure 9.** PEA comparison with SOA on Pointing'04 (yaw)

For the remaining three data-sets (AFLW, BU and ICT-3DHPE), the results were previously reported in the literature for MAE values only. For a fair comparison, we also compared our results with SOA for MAE only. The summary of the results for all the three cases is reported and compared with SOA in Tables 2–4 for all the three data-sets respectively. From the Tables, it is clear that we had better results in the two cases (BU and ICT-3DHPE) and competitive results for the AFLW database.

AFLW is a database that is collected from the internet. All the images in AFLW are real-world images which are obtained in un-constrained conditions. Importantly, the quality of the images in most of the cases is very poor. Due to this reason, our proposed MSF-CRFs model was not producing promising segmentation results. As a result, we had poor performance as can be seen in the Table 2.

**Table 2.** Head pose estimation results and its comparison with SOA on AFLW database.

| Method | Pitch | Yaw | Roll | Average |
|---|---|---|---|---|
| **QuatNet** [63] | **4.31°** | **3.93°** | **2.59°** | **3.61°** |
| HAG-MSF-CRFs | 4.89° | 4.25° | 3.20° | 4.11° |
| HyperFace [64] | 5.33° | 6.24° | 3.29° | 4.96° |
| Multi-Loss [65] | 5.89° | 6.26° | 3.82° | 5.32° |

The BU and ICT-3DHPE data-sets are also collected in the real-wold conditions. However, in these cases, the quality of the images is much better. We had better results for both the BU and ICT-3DHPE data-sets, as can be seen in the Tables 3 and 4.

**Table 3.** Head pose estimation results and its comparison with SOA on BU database.

| Method | Pitch | Yaw | Roll | Average |
|---|---|---|---|---|
| **HAG-MSF-CRFs** | **2.9°** | **2.1°** | **2.2°** | **2.4°** |
| OpenFace2.0 [66] | 3.2° | 2.4° | 2.4° | 2.6° |
| OpenFace [67] | 3.3° | 2.8° | 2.3° | 2.8° |
| Chehra [68] | 4.6° | 3.8° | 2.8° | 3.8° |
| CLM [6] | 3.5° | 3° | 2.3° | 2.9° |
| FLPD [69] | 5.3° | 4.9° | 3.1° | 4.4° |

**Table 4.** Head pose estimation results and its comparison with SOA on ICT-3DHP database.

| Method | Pitch | Yaw | Roll | Average |
|---|---|---|---|---|
| **HAG-MSF-CRFs** | **3.2°** | **2.6°** | **2.7°** | **3.0°** |
| OpenFace2.0 [66] | 3.5° | 3.1° | 3.1° | 3.2° |
| OpenFace [67] | 3.6° | 3.6° | 3.6° | 3.6° |
| CLM [6] | 4.2° | 4.8° | 4.5° | 4.5° |
| Reg. Forest [70] | 9.4° | 7.2° | 7.5° | 8.0° |
| Chehra [68] | 14.7° | 13.9° | 10.3° | 13.0° |

From the head pose estimation results, it is clear that we had better results in most of the cases, even considering recently proposed CNNs based methods. Through this comparison, we are not disparaging deep learning based methods—rather we believe we need better understanding of the deep learning based methods and their implementation to various tasks.

*6.3. Age Classification*

We reported our age and gender recognition results with term the Classification Rate (CR). We use Adience data-set for age classification. The Adience data-set has eight age categories. We manually labeled 10 images from each age category. A total of 80 images were used to build the MSF-CRFs model for age test. The MSF-CRFs model was used to create segmented images and probability maps. After generating probability maps for all images and all classes, 10-fold cross validation experiments were performed on the remaining images (excluding 80 images which were previously used to build MSF-CRFs model).

For age classification we tried all combination of facial features, as in head pose estimation (excluding background). We noticed that every face part contributed to the age classification. The results reported with HAG-MSF-CRFs and its comparison with SOA are shown in Table 5. From the Table 5, It is clear that we had better results for Adience data-set. Interestingly, for age classification we obtained better results as compared to previous results by a big margin.

**Table 5.** Comparative experiments on age classification using Adience databas.

| Method | Database | CR (%) |
|---|---|---|
| **HAG-MSF-CRFs** | **Adience** | **66.5** |
| Dehghan et al. [71] | Adience | 61.3 ± 3.7 |
| Hou et al. [72] | Adience | 61.1 ± NR |
| CNNs-EML [41] | Adience | 45.1 ± 2.6 |
| Hassner et al. [73] | Adience | 50.7 ± 5.1 |
| CNN-ELM [41] | Adience | 95.00 |

We created Ground truth masks through a commercial image editing software. We did this labeling without any automatic segmentation tool. Such kind of labeling has two main drawbacks. Firstly, this labeling highly depends on subjective perception of a single subject involved in this labeling process. Hence it is very difficult to provide an accurate label to all pixels in an image—particularly on the boundary region of the different face parts. For example, differentiating the nose region from the skin and drawing a boundary between the two is very difficult. Secondly, creating manually labeled images is very time consuming and tedious work. Due to this reason, our age part is limited to age classification only. We did not perform tests on the regression part of the age task. For that case, we would need a large number of manually labeled face images for each age number.

*6.4. Gender Recognition*

We performed gender recognition tests with three data-sets, which included Adience, LFW and FERET. The CR values for all three data-sets are shown in Table 6. We also compared our reported results with SOA methods in Table 6.

As in head pose estimation, the possible combinations for all facial features were tried. We obtained the best results with skin, hair, eyes, and nose. After localizing face parts, each image was re-scaled to a height 256 and width was varied accordingly. We manually labeled 30 images from each gender and each data-set. A total of 60 images were used to train an MSF-CRFs (gender) model for each database individually. We performed no cross tests, same database images were used to train an MSF-CRFs model and then some other images of the same data-set were used to evaluate the model.

A fair and exact comparison is very hard to achieve, as different authors use different image settings and different validation protocols. For evaluation of gender recognition, we performed 10-fold cross validation experiments. We manually labeled 60 images, performed 10-fold cross validation experiments, while excluding 60 images which were previously used to build MSF-CRFs model for gender.

Gender classification results with proposed HAG-MSF-CRFs and its comparison with SOA are reported in Table 6. In general, classification accuracy was better than previously reported results. Again, we had poor results as compared to other results for LFW data-set.

As a whole, performance of the newly proposed HAG-MSF-CRFs was very interesting. We introduced a new idea of face image analysis which is using pixel level labeling information for a face image. In a nutshell, we derived an important observation from the reported results *"a strong correlation exists between face parts segmentation and its pose, age and gender. An accurate face segmentation leads to exact head pose, age and gender recognition and vice versa."*

**Table 6.** Comparative experiments on gender recognition using Adience, LFW and FERET data-sets.

| Method | Database | CR (%) |
| --- | --- | --- |
| **HAG-MSF-CRFs** | **Adience** | **89.7** |
| Levi et al. [74] | Adience | 86.8 |
| Lapuschkin et al. [75] | Adience | 85.9 |
| CNNs-EML [41] | Adience | **77.8** |
| Hassner et al. [73] | Adience | 79.3 |
| **Van et al. [76]** | **LFW** | **94.4** |
| HyperFace [64] | LFW | 94.0 |
| LNets+ANet [77] | LFW | 94.0 |
| HAG-MSF-CRFs | LFW | 93.9 |
| Moeini et al. [78] | LFW | 93.6 |
| PANDA-1 [79] | LFW | 92.0 |
| ANet [40] | LFW | 91.0 |
| Rai and Khanna [80] | LFW | 89.1 |
| **HAG-MSF-CRFs** | **FERET** | **100** |
| Moeini et al. [78] | FERET | 99.5 |
| Tapia and Perez [81] | FERET | 99.1 |
| Rai and Khanna [80] | FERET | 98.4 |
| Afifi and Abdelrahman [82] | FERET | 99.4 |
| A priori-driven PCA [83] | FERET | 84.0 |

## 7. Conclusions

In this paper we propose an end-to-end semantic face segmentation algorithm (MSF-CRFs) which tries to solve the challenging problems of head pose, age, and gender recognition. The segmentation model is built using the idea of CRFs between various face parts. Three kinds of features are extracted to build the segmentation model. The MSF-CRFs model classify each pixel in the face image to one of the six classes (hair, eyes, skin, nose, mouth, and background). A probabilistic classification strategy is used to generate probability maps for each face class. Random Decision Forest classifier is trained for each task (head pose, age and gender) through different probability maps combination. A large number of experiments are conducted to know which face parts help in head pose, age and gender recognition. Experimental results are validated on six different face data-sets obtaining better or competitive results compared to SOA.

The segmentation results provide sufficient information for different hidden variable in a face image. A route towards different more classification problems in a face image is provided. For example, we are planing to add some more tasks to the framework such as complicated facial expression recognition, ethnicity classification and many more. We are also planing to improve performance of the segmentation part for example using recently introduced CNNs based methods.

**Author Contributions:** Conceptualization, K.K. and I.S.; methodology, K.K.; software, R.U.K.; validation, K.K. and A.G.; formal analysis, A.I.; investigation, R.U.K.; resources, M.A.; data curation, K.K.; writing—original draft preparation, K.K.; writing—review and editing, K.K., A.I., I.S. and M.A.; visualization, K.K.; supervision, K.K.; project administration, K.K.

**Funding:** This research received no external funding.

**Acknowledgments:** We are immensely grateful to the anonymous reviewers and editor for their comments on an earlier version of the manuscript.

**Conflicts of Interest:** The authors declare no conflict of interest.

## References

1. Asthana, A.; Zafeiriou, S.; Cheng, S.; Pantic. M. Robust discriminative response map fitting with constrained local models. In Proceedings of the IEEE Conference on Computer Vision and Pattern Recognition (CVPR), Portland, OR, USA, 23–28 June 2013; pp. 3444–3451.

2.   Belhumeur, P. N.; Jacobs, D.W.; Kriegman, D.J.; Kumar. N. Localizing parts of faces using a consensus of exemplars. *IEEE Trans. Pattern Anal. Mach. Intell.* **2013**, *35*, 2930– 2940. [CrossRef] [PubMed]

3.   Cao, X.; Wei, Y.; Wen, F.; Sun. J. Face alignment by explicit shape regression. *Int. J. Comput. Vis.* **2014**, *107*, 177–190. [CrossRef]

4.   Dantone, M.; Gall, J.; Fanelli, G.; Van Gool, L. Real-time facial feature detection using conditional regression forests. In Proceedings of the IEEE Conference on Computer Vision and Pattern Recognition, Providence, RI, USA, 16–21 June 2012.

5.   Murphy-Chutorian, E.; Trivedi, M.M. Head pose estimation in computer vision: A survey. *IEEE Trans. Pattern Anal. Mach. Intell.* **2009**, *31*, 607–626. [CrossRef] [PubMed]

6.   Saragih, J.M.; Lucey, S.; Cohn. J.F. Deformable model fitting by regularized landmark mean-shift. *Int. J. Comput. Vis.* **2011**, *91*, 200–215. [CrossRef]

7.   Shan, C.; Gong, S.; McOwan, P.W. Facial expression recognition based on local binary patterns: A comprehensive study. *Image Vis. Comput.* **2009**, *27*, 803–816. [CrossRef]

8.   Xiong, X.; De la Torre, F. Supervised descent method and its applications to face alignment. In Proceedings of the IEEE Conference on Computer Vision and Pattern Recognition (CVPR), Portland, OR, USA, 23–28 June 2013; pp. 532–539.

9.   Davies, G.; Ellis, H.; Shepherd, J. Perceiving and remembering faces. *Am. J. Psychol.* **1983**, *96*, 151–154.

10.   Sinha, P.; Balas, B.; Ostrovsky, Y.; Russell, R. Face recognition by humans: Nineteen results all computer vision researchers should know about. *Proc. IEEE* **2006**, *94*, 1948–1962. [CrossRef]

11.   Gross R.; Baker, S. Generic vs. person specific active appearance models. *Vis. Comput.* **2005**, *23*, 1080–1093. [CrossRef]

12.   Haj,M.A.; Gonzalez, J.; Davis, L.S.. On partial least squares in head pose estimation: How to simultaneously deal with misalignment. In Proceedings of the IEEE Conference on Computer Vision and Pattern Recognition, Providence, RI, USA, 16–21 June 2012.

13.   Khan, K.; Mauro, M.; Leonardi, R. Multi-class semantic segmentation of faces. In Proceedings of the IEEE International Conference on Image Processing (ICIP), Quebec City, QC, Canada, 27–30 September 2015; pp. 827–831.

14.   Khan, K.; Mauro, M.; Migliorati, P.; Leonardi, R. Head pose estimation through multi-class face segmentation. In Proceedings of the IEEE International Conference on Multimedia and Expo (ICME), Hong Kong, China, 10–14 July 2017; pp. 253–258.

15.   Khan, K.; Mauro, M.; Migliorati, P.; Leonardi, R. Gender and expression analysis based on semantic face segmentation. In Proceedings of the International Conference on Image Analysis and Processing (ICIAP), Catania, Italy, 11–15 September 2017; pp. 37–47.

16.   Khan, K.; Attique, M.; Syed, I.; Gul, A. Automatic Gender Classification through Face Segmentation. *Symmetry* **2019**, *11*, 770. [CrossRef]

17.   Pan, W.; Dong, W.; Cebrian, M.; Kim, T.; Fowler, J.H.; Pentland, A.S. Modeling dynamical influence in human interaction: Using data to make better inferences about influence within social systems. *IEEE Signal Process. Mag.* **2012**, *29*, 77–86. [CrossRef]

18.   Dong, W.; Lepri, B.; Pianesi, F.; Alex, P. Modeling functional roles dynamics in small group interactions. *IEEE Trans. Multimedia* **2012**, *15*, 83–95. [CrossRef]

19.   Maskey, S.; Hirschberg, J. Automatic summarization of broadcast news using structural features. In Proceedings of the 8th European Conference on Speech Communication and Technology, Geneva, Switzerland, 1–4 September 2003; pp. 1173–1176.

20.   Vinciarelli, A. Sociometry based multiparty audio recordings segmentation. In Proceedings of the IEEE International Conference on Multimedia and Expo, Toronto, ON, Canada, 9–12 July 2006; pp. 1801–1804.

21.   Weng, C.Y.; Chu, W.T.; Wu, J.L. Movie analysis based on roles' social network. In Proceddings of the IEEE International Conference on Multimedia and Expo, Beijing, China, 2–5 July 2007; pp. 1403–1406.

22.   Barzilay, R.; Collins, M.; Hirschberg, J.; Whittaker, S. The rules behind roles: Identifying speaker role in radio broadcasts. In Proceddings of the Seventeenth. National Conference on Artificial Intelligence, Austin, TX, USA, 30 July–1 August 2000, pp. 679–684.

23.   Banerjee, S.; Rudnicky, A.I. Using simple speech-based features to detect the state of a meeting and the roles of the meeting participants. In Proceedings of the 8th International Conference on Spoken Language, Jeju Island, Korea, 4–8 October 2004; pp. 2189–2192.

24.  Vinciarelli, A. Role recognition in broadcast news using Bernoulli distributions. In Proceedings of the IEEE International Conference on Multimedia and Expo, Beijing, China, 2–5 July 2007; pp. 1551–1554.

25.  Favre, S.; Salamin, H.; Dines, J.; Vinciarelli, A. Role recognition in multiparty recordings using social affiliation networks and discrete distributions. In Proceedings of the 10th international conference on Multimodal interfaces, Crete, Greece, 20–22 October 2008; pp. 29–36.

26.  Salamin, H.; Vinciarelli, A.; Truong, K.; Mohammadi, G.; Bimbo, A.D.; Chang, S.-F.; Smeulders, A.W.M. Automatic role recognition based on conversational and prosodic behaviour. In Proceedings of the 18th ACM international conference on Multimedia, Firenze, Italy, 25–29 October 2010; pp. 847–850.

27.  Garg, N.P.; Favre, S.; Salamin, H.; Hakkani-Tür, D.Z.; Vinciarelli, A. Role recognition for meeting participants: An approach based on lexical information and social network analysis. In Proceedings of the 16th ACM international conference on Multimedia, Vancouver, BC, Canada, 26–31 October 2008; pp. 693–696.

28.  Jayagopi, D.B.; Ba, S.O.; Odobez, J.-M.; Gatica-Perez, D. Predicting two facets of social verticality in meetings from five-minute time slices and nonverbal cues. In Proceedings of the 10th international conference on Multimodal interfaces, Crete, Greece, 20–22 October 2008; pp. 45–52.

29.  Gourier, N.; Hall, D.; Crowley, J.L. Estimating face orientation from robust detection of salient facial features. In Proceedings of the International Workshop on Visual Observation of Deictic Gestures, Cambridge, UK, 22 August 2004.

30.  Kostinger, M.; Wohlhart, P.; Roth, P.; Bischof, H. Annotated facial landmarks in the wild: A large-scale, real-world database for facial landmark localization. In Proceedings of the IEEE International Conference on Computer Vision Workshops, Barcelona, Spain, 6–13 November 2011; pp. 2144–2151.

31.  Cascia, M. L.; Sclaroff, S.; Athitsos, V. Fast, Reliable Head Tracking under Varying Illumination: An Approach Based on Registration of Texture-Mapped 3D Models. Technical Report BUCS-1999-005, Computer Science Department, Boston University, Boston, MA, USA, 23 April 1999.

32.  Baltrusaitis, T.; Robinson, P.; Morency, L.-P. 3D Constrained Local Model for Rigid and Non-Rigid Facial Tracking. In Prceedings of the IEEE Conference on Computer Vision and Pattern Recognition, Providence, RI, USA, 16–21 June 2012.

33.  Atallah, R.R.; Kamsin, A.; Ismail, M.A.; Abdelrahman, S.A.; Zerdoumi, S. Face recognition and age estimation implications of changes in facial features: A critical review study. *IEEE Access* **2018**, *6*, 28290–28304. [CrossRef]

34.  Fu, Y.; Guo, G.; Huang, T.S. Age synthesis and estimation via faces: A survey. *IEEE Trans. Pattern Anal. Mach. Intell.* **2010**, *32*, 1955–1976. [PubMed]

35.  Kwon, Y.H.; da Vitoria Lobo, N. Age classification from facial images. *Comput. Vis. Image Underst.* **1999**, *74*, 1–21. [CrossRef]

36.  Ramanathan, N.; Chellappa, R. Modeling age progression in young faces. In Proceedings of the IEEE Computer Society Conference on Computer Vision and Pattern Recognition, New York, NY, USA, 17–22 June 2006; pp. 387–394.

37.  Geng, X.; Xia, Y. Head pose estimation based on multivariate label distribution. In Proceedings of the Computer Vision and Pattern Recognition (CVPR), Columbus, OH, USA, 24–27 June 2014; pp. 1837–1842.

38.  Guo, G.; Fu, Y.; Dyer, C.R.; Huang, T.S. Image-based human age estimation by manifold learning and locally adjusted robust regression. *IEEE Trans. Image Process.* **2008**, *17*, 1178–1188. [PubMed]

39.  Chang, K.Y.; Chen, C.S. A learning framework for age rank estimation based on face images with scattering transform. *IEEE Trans. Image Process.* **2015**, *24*, 785–798. [CrossRef]

40.  Li, C.; Liu, Q.; Dong, W.; Zhu, X.; Liu, J.; Lu, H. Human age estimation based on locality and ordinal information. *IEEE Trans. Cybern.* **2014**, *45*, 2522–2534. [CrossRef]

41.  Duan, M.; Li, K.; Yang, C.; Li, K. A hybrid deep learning CNN–ELM for age and gender classification. *Neurocomputing* **2018**, *275*, 448–461. [CrossRef]

42.  Ricanek, K.; Tesafaye, T.; Morph: A longitudinal image database of normal adult age-progression. In Proceedings of the Seventh International Conference on Automatic Face and Gesture Recognition, Southampton, UK, 10–12 April 2006; pp. 341–345.

43.  Eidinger, E.; Enbar, R.; Hassner, T. Age and gender estimation of unfiltered faces. *IEEE Trans. Inf. Forensics Secur.* **2014**, *9*, 2170–2179. [CrossRef]

44.  Makinen, E.; Raisamo, R. Evaluation of gender classification methods with automatically detected and aligned faces. *IEEE Trans. Pattern Anal. Mach. Intell.* **2008**, *30*, 541–547. [CrossRef]

45.	Golomb, B.A.; Lawrence, D.T.; Sejnowski, T.J. Sexnet: A neural network identifies sex from human faces. In Proceedings of the 1990 Conference on Advances in neural information processing systems, Denver, CO, USA, 26–29 November 1990; pp. 572–577.

46.	Moghaddam, B.; Yang, M.-H. Learning gender with support faces. *IEEE Trans. Pattern Anal. Mach. Intell.* **2002**, *24*, 707–711. [CrossRef]

47.	Baluja, S.; Rowley, H.A. Boosting sex identification performance. *Int. J. Comput. Vis.* **2006**, *71*, 111–119. [CrossRef]

48.	Toews, M.; Arbel, T. Detection, localization, and sex classification of faces from arbitrary viewpoints and under occlusion. *IEEE Trans. Pattern Anal. Mach. Intell.* **2009**, *31*, 1567–1581. [CrossRef] [PubMed]

49.	Yu, S.; Tan, T.; Huang, K.; Jia, K.; Wu, X. A study on gait-based gender classification. *IEEE Trans. Image Process.* **2009**, *18*, 1905–1910. [PubMed]

50.	Khan, F.S.; van de Weijer, J.; Anwer, R.M.; Felsberg, M.; Gatta, C. Semantic pyramids for gender and action recognition. *IEEE Trans. Image Process.* **2014**, *23*, 3633–3645. [CrossRef] [PubMed]

51.	Chen, H.; Gallagher, A.; Girod, B. Face modeling with first name attributes. *IEEE Trans. Pattern Anal. Mach. Intell.* **2014**, *36*, 1860–1873. [CrossRef] [PubMed]

52.	Han, H.; Otto, C.; Liu, X.; Jain, A.K. Demographic estimation from face images: Human vs. machine performance. *IEEE Trans. Pattern Anal. Mach. Intell.* **2015**, *37*, 1148–1161. [CrossRef] [PubMed]

53.	Phillips, P.J.; Wechsler, H.; Huang, J.; Rauss, P.J. The FERET database and evaluation procedure for face-recognition algorithms. *Image Vis. Comput.* **1998** *16*, 295–306. [CrossRef]

54.	Liu, Y.; Xie, Z.; Yuan, X.; Chen, J.; Song, W. Multi-level structured hybrid forest for joint head detection and pose estimation. *Neurocomputing* **2017**, *266*, 206–215. [CrossRef]

55.	Chang, X.; Nie, F.; Wang, S.; Yang, Y.; Zhou, X.; Zhang, C. Compound rank-k projections for bilinear analysis. *IEEE Trans. Neural Netw. Learn. Syst.* **2016**, *27*, 1502–1513. [CrossRef]

56.	Schwarz, A.; Haurilet, M.; Martinez, M.; Stiefelhagen, R. DriveAHead-a large-scale driver head pose dataset. In Proceedings of the IEEE Conference on Computer Vision and Pattern Recognition Workshops, Honolulu, HI, USA, 22–25 July 2017; pp. 1–10.

57.	Liu, W.; Anguelov, D.; Erhan, D.; Szegedy, C.; Reed, S.; Fu, C.-Y.; Berg, A.C. SSD: Single shot multibox detector. In Proceedings of the European Conference on Computer Vision, Amsterdam, The Netherlands, 11–14 October 2016; pp. 21–37.

58.	Van den Bergh, M.; Boix, X.; Roig, G.; Van Gool, L. Seeds: Superpixels extracted via energy-driven sampling. *Int. J. Comput. Vis.* **2015**, *111*, 298–314. [CrossRef]

59.	Huang, G.B.; Narayana, M.; Learned-Miller, E. Towards unconstrained face recognition. In Proceeings of the Computer Vision and Pattern Recognition Workshops, Anchorage, AL, USA, 24–26 June 2008; pp. 1–8.

60.	Lee, S.; Saitoh, T. Head Pose Estimation Using Convolutional Neural Network. In Proceedings of the IT Convergence and Security, Seoul, Korea, 25–28 September 2017; pp. 164–171.

61.	Jain, V.; Crowley, J.L. Head pose estimation using multi-scale Gaussian derivatives. In Prceedings of the Scandinavian Conference on Image Analysis, Providence, Berlin, Germany, 17 June 2013; pp. 319–328.

62.	Ma, B.; Li, A.; Chai, X.; Shan, S. CovGa: A novel descriptor based on symmetry of regions for head pose estimation. *Neurocomputing* **2014**, *143*, 97–108. [CrossRef]

63.	Hsu, H.-W.; Wu, T.-Y.; Wan, S.; Wong, W.H.; Lee, C.-Y. Quatnet: Quaternion-based head pose estimation with multiregression loss. *IEEE Trans. Multimedia* **2019**, *21*, 1035–1046. [CrossRef]

64.	Ranjan, R.; Patel, V.M.; Chellappa, R. Hyperface: A deep multi-task learning framework for face detection, landmark localization, pose estimation, and gender recognition. *IEEE Trans. Pattern Anal. Mach. Intell.* **2019**, *41*, 121–135. [CrossRef]

65.	Ruiz, N.; Chong, E.; Rehg, J.M. Fine-grained head pose estimation without keypoints. In Proceedings of the IEEE Conference on Computer Vision and Pattern Recognition Workshops, Salt Lake City, UT, USA, 19–21 June 2018; pp. 2074–2083.

66.	Baltrusaitis, T.; Zadeh, A.; Chong Lim, Y.C.; Morency, L.P. Openface 2.0: Facial behavior analysis toolkit. In Proceedings of the 13th IEEE International Conference on Automatic Face & Gesture Recognition (FG 2018), Xi'an, China, 15–19 May 2018; pp. 59–66.

67.	Baltrušaitis, T.; Robinson, P.; Morency, L.-P. Openface: An open source facial behavior analysis toolkit. In Proceedings IEEE Winter Conference on Applications of Computer Vision (WACV), Lake Placid, NY, USA, 7–10 March 2016; pp. 1–10.

68. Asthana, A.; Zafeiriou, S.; Cheng, S.; Pantic. M. Incremental Face Alignment in the Wild. In Proceedings of the IEEE Conference on Computer Vision and Pattern Recognition (CVPR), Columbus, OH, USA, 24–27 June 2014; pp. 1859–1866.

69. Wu, Y.; Gou, C.; Ji, Q. Simultaneous facial landmark detection, pose and deformation estimation under facial occlusion. In Proceedings of the IEEE Conference on Computer Vision and Pattern Recognition, Honolulu, HI, USA, 22–25 July 2017; pp. 3471–3480.

70. Fanelli, G.; Weise, T.; Gall, J.; Van Gool, L. Real Time Head Pose Estimation from Consumer Depth Cameras. In *Pattern Recognition*; Mester, R., Felsberg, M., Eds.; Springer: Berlin, Germany, 2011; Volume 6835.

71. Dehghan, A.; Ortiz, E.G.; Shu, G.; Masood, S.Z. Dager: Deep age, gender and emotion recognition using convolutional neural network. *arXiv* **2017**, arXiv:1702.04280.

72. Hou, L.; Yu, C.P.; Samaras, D. Squared earth mover's distance-based loss for training deep neural networks. *arXiv preprint* **2016**, arXiv:1611.05916.

73. Hassner, T.; Harel, S.; Paz, E.; Enbar, R. Effective face frontalization in unconstrained images. In Proceedings of IEEE Conference on Computer Vision and Pattern Recognition (CVPR), Boston, MA, USA, 7–12 June 2015; pp. 4295–4304.

74. Levi, G.; Hassner, T. Age and gender classification using convolutional neural networks. In Proceedings of IEEE Conference on Computer Vision and Pattern Recognition (CVPR), Boston, MA, USA, 7–12 June 2015; pp. 34–42.

75. Lapuschkin, S.; Binder, A.; Muller, K.-R.; Samek, W. Understanding and comparing deep neural networks for age and gender classification. In Proceedings of the IEEE International Conference on Computer Vision, Venice, Italy, 22–29 October 2017; 1629–1638.

76. van de Wolfshaar, J.; Karaaba, M.F.; Wiering, M.A. Deep convolutional neural networks and support vector machines for gender recognition. In Proceedings of the IEEE Symposium Series on Computational Intelligence, Cape Town, South Africa, 7–10 December 2015; pp. 188–195.

77. Kumar, N.; Belhumeur, P.N.; Nayar. S.K. FaceTracer: A Search Engine for Large Collections of Images with Faces. In Proceedings of the European Conference on Computer Vision (ECCV), Marseille, France, 12–18 October 2008; pp. 340–353.

78. Moeini, H.; Mozaffari, S. Gender dictionary learning for gender classification. *J. Vis. Commun. Image Represent.* **2017**, *42*, 1–13. [CrossRef]

79. Zhang, N.; Paluri, M.; Ranzato, M.; Darrell, T.; Bourdev. L. Panda: Pose aligned networks for deep attribute modeling. In Proceedings of the IEEE Conference on Computer Vision and Pattern Recognition (CVPR), Columbus, OH, USA, 24–27 Jun 2014; pp. 1637–1644.

80. Rai, P.; Khanna, P. An illumination, expression, and noise invariant gender classifier using two-directional 2DPCA on real Gabor space. *J. Vis. Lang. Comput.* **2015** *26*, 15–28. [CrossRef]

81. Tapia, J.E.; P´erez, C.A. Gender classification based on fusion of different spatial scale features selected by mutual information from histogram of LBP, intensity, and shape. *IEEE Trans. Inf. Forensics Secur.* **2013**, *8*, 488–499. [CrossRef]

82. Afifi, M.; Abdelhamed, A. AFIF4: Deep gender classification based on AdaBoost-based fusion of isolated facial features and foggy faces. *J. Vis. Commun. Image Represent.* **2019**, *62*, 77–86 [CrossRef]

83. Thomaz, C.; Giraldi, G.; Costa, J.; Gillies, D. A priori-driven PCA. In Proceedings of the 11th Asian Conference on Computer Vision, Daejeon, Korea, 5–9 November 2012, pp. 236–247.

Article

# Emotion Recognition from Skeletal Movements

**Tomasz Sapiński [1,†], Dorota Kamińska [1,*,†], Adam Pelikant [1] and Gholamreza Anbarjafari [2,3,4]**

[1]  Institute of Mechatronics and Information Systems Lodz University of Technology, 90-924 Lodz, Poland
[2]  iCV Lab, Institute of Technology, University of Tartu, 51014 Tartu, Estonia
[3]  Faculty of Engineering, Hasan Kalyoncu University, 27000 Sahinbey, Gaziantep, Turkey
[4]  Institute of Digital Technologies, Loughborough University London, London E15 2GZ, UK
*  Correspondence: dorota.kaminska@p.lodz.pl; Tel.: +48-631-25-78
†  These authors contributed equally to this work.

Received: 1 March 2019; Accepted: 26 June 2019; Published: 29 June 2019

**Abstract:** Automatic emotion recognition has become an important trend in many artificial intelligence (AI) based applications and has been widely explored in recent years. Most research in the area of automated emotion recognition is based on facial expressions or speech signals. Although the influence of the emotional state on body movements is undeniable, this source of expression is still underestimated in automatic analysis. In this paper, we propose a novel method to recognise seven basic emotional states—namely, happy, sad, surprise, fear, anger, disgust and neutral—utilising body movement. We analyse motion capture data under seven basic emotional states recorded by professional actor/actresses using Microsoft Kinect v2 sensor. We propose a new representation of affective movements, based on sequences of body joints. The proposed algorithm creates a sequential model of affective movement based on low level features inferred from the spacial location and the orientation of joints within the tracked skeleton. In the experimental results, different deep neural networks were employed and compared to recognise the emotional state of the acquired motion sequences. The experimental results conducted in this work show the feasibility of automatic emotion recognition from sequences of body gestures, which can serve as an additional source of information in multimodal emotion recognition.

**Keywords:** emotion recognition; gestures; body movements; Kinect sensor; neural networks; deep learning

---

## 1. Introduction

People express their feelings through different modalities. There is evidence that the affective state of individuals is strongly correlated with facial expressions [1], body language [2] voice [3] and different types of physiological changes [4]. On the basis of external behaviour one can easily determine the internal state of the interlocutor. For example, burst of laughter generally signals amusement, frowning signals nervousness or irritation, crying is closely related to sadness and weakness [5–7]. Mehrabian formulated the principle 7-38-55, according to which the percentage distribution of the message is as follows: 7% verbal signals and words, 38% strength, height, and rhythm and 55% body movements and facial expressions [8]. This suggests that words serve in particular to convey the information and the body language to form conversation or even to substitute the verbal communication. However, it has to be emphasised that this relation is applicable only if a communicator is talking about their feelings or attitudes [9].

Currently, human-computer interaction (HCI) is one of the most rapidly growing fields of research. The main goal of HCI is to facilitate the interaction using several parallel channels of communication between the user and the machine. Although computers are now a part of human life, the relation between human and computer is not natural. Knowledge of the emotional state of the user would allow the machine to boost the effectiveness of cooperation. That is why affect detection became an important trend in pattern recognition and has been widely explored, especially in the case of facial expressions and speech signals [10]. Body gestures and posture receive considerably less focus. With recent developments and the increasing reliability of motion capture technologies, the literature about automatic recognition of expressive movements has been increasing in quantity and quality. Despite the rising interest in this topic, affective body movements in automatic analysis are still underestimated [11].

The most natural and intuitive method for body movement projection is based on the skeleton, which represents hierarchically arranged joint kinematics along with body segments [12]. In the past, research on body tracking was based on video data, which made it extremely challenging and usually amounted to single frame analysis [13–15]. However, the definition of motion is a change in position over time, thus it should be described as a set of consecutive frame sequences. Skeleton tracking has become much easier with the appearance of motion capture systems, which automatically generate the human skeleton represented by 3-dimensional (3D) coordinates. Additionally, it brought up an increase of research on body movement, such as unusual event detection and crime prevention [16–20].

Affective movement may be described by displacement, distance, velocity, acceleration, time and speed by extracting dynamic features from analysed model. For example in Reference [21], the authors were tracking trajectories of head and hands from a frontal and a lateral view. They combined shape and dynamic expressive gesture features, creating a 4D model of emotion expression that effectively classified emotions according to their valence and arousal. Dynamic features were also considered in Reference [22], where the authors suggested that the timing of the motion is an accurate representation of the properties of emotional expressions.

Very promising results are presented in Reference [23]. The authors analysed Microsoft Kinect v2. recordings of body movements expressing five basic emotions, namely, anger, happiness, fear, sadness and surprise. They used a deep neural network consisting of stacked RBMs , which outperformed all other classifiers, achieving an overall recognition rate of 93%. However, it must be emphasised that the superior performance is associated with the type of analysed data. In Reference [23] emotions are represented as predetermined gestures (each emotion is assigned to particular type of gesture, for example, power pose to happiness). The actors/actresses are instructed how to present particular emotional state prior to recording. Such an approach narrows the research down to the posture recognition problem, which may not be as effective with more complex gestures, despite such promising results.

More viable research is presented by Kleinsmith et al. in Reference [24], where the Gypsy 5 motion capture system was used to record the spontaneous body gestures of Nintendo Wii sports games players. The authors used low-level posture configuration features to create affective movement models for states of concentration, defeat and triumph. An overall accuracy of 66.7% was obtained using a multilayer perceptron. The emotional behaviour of Nintendo Wii tennis players was also analysed in Reference [25]. The authors based their experiment on time-related features such as body segment rotation, angular velocity, angular frequency, orientation, angular acceleration, body directionality and amount of movement. Results obtained using recurrent neural network (RNN), whose average recognition rate is 58.4%, are comparable to human observers' benchmarks.

More recent research [26] presents analysis of human gait recordings performed by professional actors/actresses, captured by Vicon system. The motion data is encoded with HMMs , which are subsequently used to derive a Fisher Score (FS) representation. SVM classification is performed in the HMM-based FS space. The authors obtained a total average recognition rate of 78% for the same subject and 69% for interpersonal recognition. Classification was performed for four emotional states: neutral, joy, anger and sadness. In Reference [27], Vicon was used to collect a full body dataset of emotion including anger, happiness, fear and sadness, expressed by 13 subjects. The authors proposed a stochastic model of the affective movement dynamics using hidden Markov models, performance of which was tested with SVM classifier and resulted in 74% recognition rate.

Despite much lower accuracy compared to affective speech or facial expressions, gesture analysis can serve as a complement to a multimodal system. For example in Reference [28], the authors expanded their studies on emotional facial expressions by analysing sequences of images presenting the motion of arms and upper body. They used a deep neural networks model to recognise dynamic gestures with minimal image pre-processing. By summing up all the absolute differences of each pair of images of particular sequence they created a shape representation of the motion. The experiment demonstrated a significant increase of recognition accuracy achieved by using multimodal information. Their model improves the accuracy of state-of-the-art approaches from 82.5% reported in the literature to 91.3%, using the bi-modal face and body benchmark database (FABO) [29].

Considering all these works, one can observe that there is still a lack of comprehensive affective human analysis from body language [30] mainly because there is no clear consensus about the input and output space. The contributions of this paper are summarised as follows:

(a) We propose a different representation of affective movements, based on sequence of joints positions and orientations. Together with classification using selected neural networks and a comparison of classification performance with methods used in action recognition, for seven affective states: neutral, sadness, surprise, fear, disgust, anger and happiness.

(b) The presented algorithms utilise a sequential model of affective movement based on low level features, which are positions and orientation of joints within the skeleton provided by Kinect v2. By using such intuitive and easily interpretable representation, we created an emotional gestures recognition system independent of skeleton specifications and with minimum preprocessing requirements (eliminating features extraction from the process).

(c) Research is carried out on a new, comprehensive database that comprises a large variety of emotion expressions [31]. Although the recordings are performed by professional actors/actresses, the movements were freely portrayed not imposed by the authors. Thus, it may be treated as quasi-natural.

(d) By comparing results with action/posture recognition approaches, we have shown that emotion recognition is a more complex problem. The analysis should focus on dependencies in the sequence of frames rather than describing whole movement by general features.

This paper adopts the following outline. First, in Section 2, we describe our pipeline for automatic recognition of emotional body gestures and discuss technical aspects of each component. In Section 3, we present results obtained using proposed algorithm, which are thoroughly discussed. Finally, the paper concludes with a summary, followed by suggestions for potential future studies in Section 4.

## 2. The Proposed Method

In this section, we present the main components of the proposed system, starting with data acquisition, followed by its pre-processing and ending with classification methods. The structure of proposed emotional gestures expression recognition approach is presented in Figure 1.

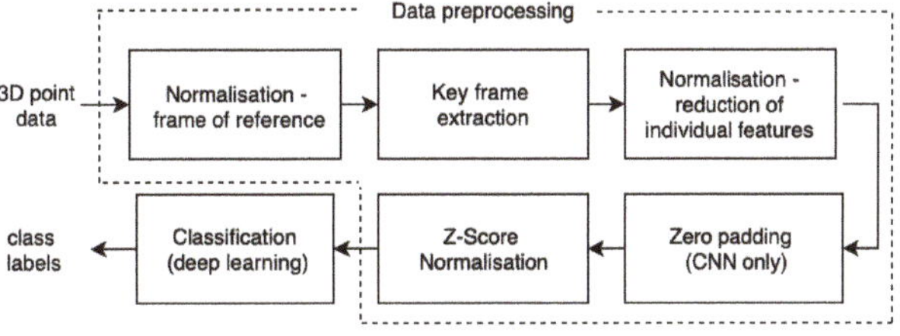


**Figure 1.** The structure of proposed emotional gestures expression recognition approach.

### 2.1. 3D Point Data—Emotional Gestures and Body Movements Corpora

Motion capture data used for the purpose of this research is a subset of the multimodal database of emotional speech, video and gestures. In this work, we used our recently gathered database [31]. This section is dedicated to recordings of human skeleton. The recordings were conducted in the rehearsal room of *Teatr Nowy im. Kazimierza Dejmka w Łodzi*. Each recorded person was a professional actor/actress from the aforementioned theatre. A total of 16 people were recorded: 8 male and 8 female, aged from 25 to 64. Each person was recorded separately. Before the recording, all actors/actresses were asked to perform the emotional states in the following order: neutral, sadness, surprise, fear, disgust, anger and happiness (this set of discreet emotions was based on examination conducted by Ekman in Reference [32]). In addition, they were asked to utter a short sentence in Polish, with the same emotional state as their corresponding gesture. The sentence was *Każdy z nas odczuwa emocje na swój sposób* (English translation: *Each of us perceives emotions in a different manner*). No additional instructions were given on how a particular state should be expressed. All emotions were acted out 5 times, without any guidelines or prompts from the researchers. The total number of gathered samples amounted to 560, which includes 80 samples per each emotional state. Recordings took place in a quiet environment with no lighting issue, against a green background. Cloud point and skeletal data feeds were captured using a Kinect v2 sensor. The full body was in frame, including the legs, as shown in Figure 2. The data were gathered in the form of XEF files.

We are fully aware that there are many disadvantages of an acted emotional database. However, in order to obtain three different modalities simultaneously and gather clean and high quality samples in a controlled, undisturbed environment the decision was made to create a set of acted out emotions. This approach provides crucial fundamentals for creating a corpus with a reasonable number of recorded samples, diversity of gender and age of the actor/actress and the same verbal content. What is more, the actor/actress had complete freedom during recording: movements were not imposed and previously defined, there were no additional restrictions, every repetition is different and simulated by the actor/actress themselves. Thus, presented database may be treated as a quasi-natural one. The database is available for research upon request.

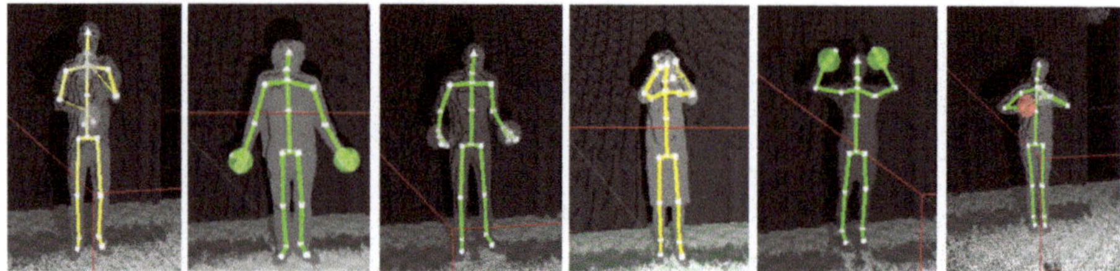

**Figure 2.** Selected frames of actor/actress' poses in six basic emotions: fear, surprise, anger, sadness, happiness, disgust.

For the purpose of this research some of the samples were rejected due to technical reasons, for example, inaccurate position recognition of upper or lower extremities. The final database of affective recordings selected for this study contains 474 samples. The exact number of recordings as well as their average length for each emotional state is presented in Table 1.

**Table 1.** The amount of samples used in the research and the average length of recordings per emotion (in seconds).

| Emotional State | Neutral | Sadness | Surprise | Fear | Anger | Disgust | Happiness |
|---|---|---|---|---|---|---|---|
| No. of samples | 64 | 63 | 70 | 72 | 70 | 65 | 70 |
| Average recordings length in second | 3.7 | 4.16 | 4.59 | 3.79 | 4.15 | 4.76 | 4.03 |

Data acquired from the Kinect v2 determines the 3D position and orientation of 25 individual joints, as shown in Figure 3a. The position of each joint is defined by the vector $[x, y, z]$, where the basic unit is $1m$ and the origin of the coordinate system is Kinect v2 sensor itself. The orientation is also determined with three values expressed in degrees. The device does not return orientation values of head, hands, knees and feet.

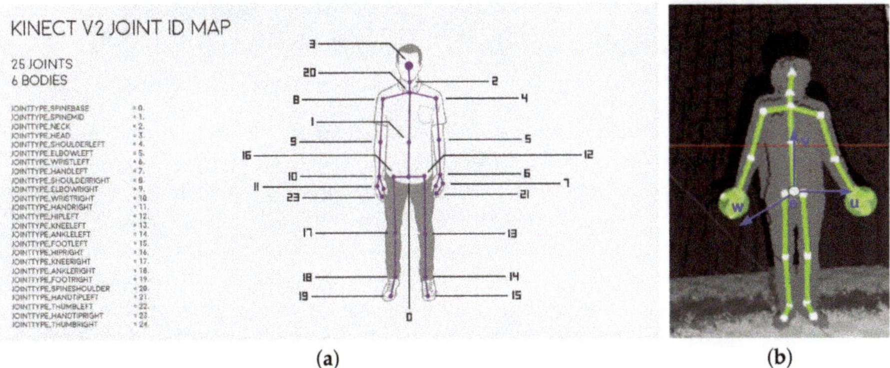


**Figure 3.** (**a**) Skeleton mapping in relation to the human body [33]. (**b**) An example frame of Kinect recording showing the skeleton.

## 2.2. Preprocessing

Raw Kinect v2 data output needs to be subjected to several steps of processing before it can be used in classification—each step is described in following section. The assumption of this research was to reduce data preprocessing to minimum in order to make the path between data acquisition and classification as short as possible, maintaining effective emotion recognition at the same time.

### 2.2.1. Normalisation—Frame of Reference

Kinect v2 provides data of 3D joints position and orientation, in the space relative to the sensor itself $[x, y, z]$ (where $x$ is pointing left from the sensor, $y$ is pointing upwards, $z$ is the forward axis of the sensor). This kind of data is influenced by the distance between the actor/actress and the sensor during recording. Thus, skeleton coordinates had to be projected from the sensor space $[x, y, z]$ onto a local space of the body $[u, v, w]$ with the center of this space in the *SpineBase* joint of the Kinect skeleton (presented in Figure 3a, called the main joint or root joint), where $u$ is pointing left, $v$ is pointing up, $w$ is pointing forward in relation to the *SpineBase* joint, all $[u, v, w]$ coordinates were calculated in respect to the main joint rotation, as shown in Figure 3b. As a result, a vector containing the positions and orientation of all joints in relation to the main one was obtained. This operation is performed for each frame in every sample. Positions and orientations of the main joint in the first frame are treated as the initial state, while the changes in the displacement or rotation of the main joint in subsequent frames are calculated in relation to the first frame.

### 2.2.2. Key Frame Extraction

Gestures and body movements can be analyzed as a set of key frames. The key frame should contain crucial information about a particular pose for a given motion sequence. For this purpose, body movement should be divided into separate frames as can be seen in Figure 4.

**Figure 4.** Sequence of three key frames extracted from point cloud data representing happiness.

There are many methods for key frame extraction. Most of them fall into three categories, namely, curve simplification (CS), clustering and matrix factorisation [34]. For the purpose of this research, CS method was used. In this method, the motion sequence is represented as a trajectory curve in 3D space of features and CS algorithms are applied to these trajectory curves. CS utilises Lowe's algorithm [35] for curve simplification, which represents the values of a single joint in a sequence of motion. Starting with the line connecting the beginning and the end of the trajectory, the algorithm divides it into two sublines (intervals), if the maximum deviation of any point on the curve is greater than a certain level of error. The algorithm performs the same process recursively for each subline, until the error rate is small enough for each subline. In this study, we examined the following values of error rate: 1 cm, 2 cm, 3 cm, 5 cm, 10 cm and 15 cm. For the error rate of 1 cm and 2 cm, the obtained number of key frames is almost identical to the number of frames of the recording, even for neutral state in which the actor/actress stay almost still. Thus, this level of error rate is considered as a Kinect v2 measurement error (especially in the case of hand movement, which is described in Section 2.3). For the error rate of 10 cm and 15 cm, the obtained number of key frames is not sufficient to adequately describe emotional movement. The average number of key frames oscillates around 2, which means that only a few frames between the first and the last one were selected. Thus, error rate values of 1 cm, 2 cm, 10 cm and 15 cm were excluded from further analysis.

### 2.2.3. Normalisation—Reduction of Individual Features

It is assumed that every human is built in proportion to his or her height and the length of legs and arms is proportional to the overall body structure [36]. To unify the value of the position of the joints between the higher and lower individuals, we propose normalisation based on the distance between two joints with the lowest noise value of their position on all recordings: *SpineBase* and *SpineShoulder*. The distance used for normalisation is measured for each frame of the actor/actress's neutral recordings. Normalisation of all joints within a given sequence of frames follows Equation (1), where skeleton consists of 25 joints, $d_i$ is the distance vector between the $i$ and $J_0$ joints normalised to the median of distances between the joints $J_0$ and $J_{20}$ (*SpineBase* and *SpineShoulder*) of all neutral recordings for each individual.

$$d_i = \frac{J_i}{\widetilde{J_{20} - J_0}} \tag{1}$$

where $i = 1, \ldots, 25$ is the number of joints. This process is performed for all joints, relative to the skeleton in the neutral position of particular individual. Neutral state is used to preserve information about special movements such as jump or squat occurring in emotional recordings (e.g., joyful hop). Considering the same degree of freedom of each body part for all recorded individuals, values of joint orientation did not require any additional processing.

The output of the key frames extraction is a set of sequences of varying lengths, which can not be considered as an input for all types of classifiers, in our case CNN. In order to unify the length of the sequences, we applied zero padding algorithm to prepare the data for CNN.

Next, all sequences are subjected to z-score normalisation, which is a widely used step to accelerate the process of neural networks learning [37–39]. For the purpose of this research we apply *sequence-wise normalisation* [38] for each key frame sequence. In this method, mean and standard deviation is calculated among data from all sequences excluding zero frames added during the previous step.

## 2.3. Datasets Division

During data preparation, a relative average quantity of motion (distance covered by a specific joint) was measured for each emotional state. Calculations were made according to the formula (2).

$$avg_{je} = \sum_{ne=1}^{N_e} \frac{|p_{je}(f_{ne}) - p_{je}(f_{ne}-1)|}{F_{ne}N_e} \tag{2}$$

where: $j = 0,\ldots,25$—the number of the joint; $e$—emotional state (Ne—neutral, Sa—sadness, Su—surprise, Fe—fear, An—anger, Di—disgust, Ha—happiness); $N_e$—is a number of recordings per emotion $e$; $ne = 1,\ldots,N_e$—the index of the emotional state $e$ recording; $F_{ne}$—is a number of frames per recording $n$ of emotional state $e$; $f_{ne} = 2,\ldots,F_{ne}$—frame index in recording $n$ of emotional state $e$ (excluding first frame); $p_{je}(f_{ne})$—position of joint $j$ in frame $f_{ne}$ in hierarchical local coordinates.

For each joint, the calculated values are relative, based on changes in the local coordinate system of the given joint, the centre of which is located in a superior joint in hierarchical skeleton construction (e.g., for the *WristRight* joint corresponding to the position of right hand wrist, the origin of the local coordinate system is the *ElbowRight* joint corresponding to the position of the right hand elbow. These calculations were made separately for each emotion. The results are shown in Figure 5.

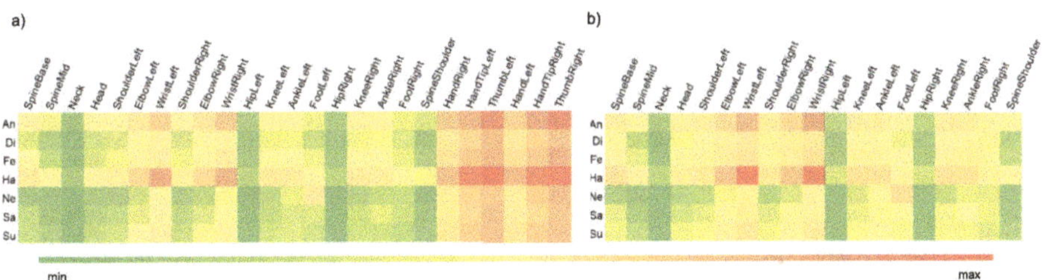

**Figure 5.** Heat-map presenting distribution of joints involvement for particular emotional state (**a**) for all joints (**b**) excluding hands.

One can observe in Figure 5a that the largest involvement in emotional expression is observed for hands and thumbs (*HandTipLeft*, *HandTipRight*, *HandTipLeft*, *HandTipRight*, *HandLeft*, *HandRight*). However, the intensity of movement of these particular joints is caused by the measurement error of Kinect v2. Thus, in further analysis it is assumed that the hand position is determined by position of the wrists (*WristLeft* i *WristRight*) and all hand related joints were excluded from the datasets. According to Figure 5b, the largest involvement is observed for wrists and arm related joints, which is common for emotion expression. It is worth emphasising that the involvement of legs is visible, especially for the knees (*KneeLeft*, *KneeRight*) and ankles (*AnkleLeft*, *AnkleRight*).

Most state-of-the-art research focuses only on the upper body, thus in this study, the influence of leg movement on affective gestures was examined. In addition, we investigated which type of data (joint orientation, position or mixture of both) is best suited for the classification of emotional sates from gestures. In order to conduct such research we examined the datasets presented in Table 2.

**Table 2.** Input datasets for classification

| Dataset | Dataset Content | Dataset Features Count |
| --- | --- | --- |
| PO | Positions and orientation, upper and lower body | 115 |
| POU | Positions and orientation, upper body | 67 |
| P | Positions, upper and lower body | 58 |
| O | Orientation, upper and lower body | 58 |
| PU | Positions, upper body | 34 |
| OU | Orientation, upper body | 34 |

### 2.4. Classification—Models of Neural Networks

The final step of the proposed method is classification, which aims to assign input data to a specific category $k$ (in this case: neutral state, sadness, surprise, fear, anger, disgust, happiness). In this work, we apply different deep learning Neural Networks (NN) to the proposed combination of datasets Table 2 in order to compare their performances, based on the recognition rates. We use a Convolutional Neural Network (CNN), a Recurrent Neural Network (RNN) and a Recurrent Neural Network with Long Short-Term Memory Network (RNN-LSTM) with low level features (positions and orientation of joints within the skeleton), in terms of motion emotion recognition efficiency. The proposed approach of adjusting the abovementioned neural networks to motion sequence analysis is presented in the following section.

#### 2.4.1. Convolutional Neural Network

The scope of use of CNNs has expanded greatly to different application domains, including the classification of signals representing emotional states [40,41]. Due to its well configured structures consisting of multiple layers, this kind of network is able to determine the most distinctive features based on enormous collections of data. The possibility of reducing the number of parameters required for images over a regular network makes CNN the most commonly used classifier for image processing. CNN considers an image as a matrix and uses the convolution operation [42] to implement a filter, which is sliding through the input matrix. In a multi-layered CNN, the input of each convolution layer is comprised of the filtered output matrix of the previous layer. The convolutional filter values are adjusted during the training phase. The process of using a CNN for gestures-based emotion recognition from sequence of movement is presented in Figure 6a.

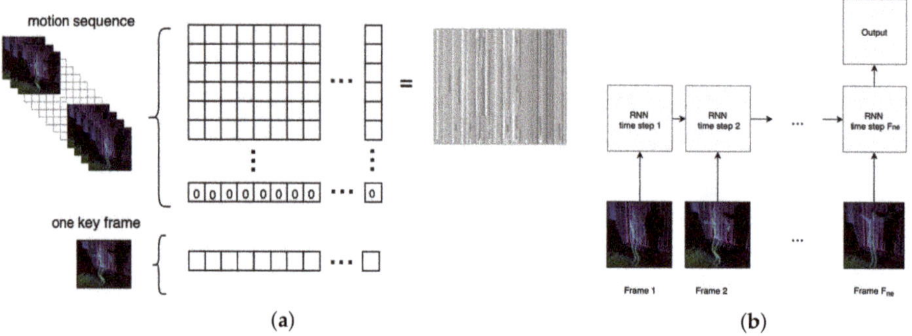


**Figure 6.** (a) The process of using a Convolutional Neural Network (CNN) for gestures-based emotion recognition shows the process of creating an matrices based on motion sequence. (b) The process of using a Recurrent Neural Network (RNN) for motion sequence analysis—each time step of the motion sequence is evaluated by a RNN.

### 2.4.2. Recurrent Neural Network

RNNs allow operation directly on time sequences. They are successfully applied to tasks involving temporal data such as speech recognition, language modelling, translation, image captioning or gestures analysis. In RNN, the output of the previous sequence time step is taken into consideration when calculating the result of the next one. However, standard RNN does not handle long term dependencies well, due to the vanishing gradient problem [43].

The Long Short Term Memory network (RNN-LSTM) is an extension for RNN, which works much better than the standard version. In RNN-LSTM architecture, RNN uses gateway units in addition to the common activation function, which extend its memory [44]. Such an architecture allows the network to learn and "remember" dependencies over more time steps, linking causes and effects remotely [45]. The process of using a RNN and RNN-LSTM for gestures-based emotion recognition sequence of movement is presented in Figure 6b.

## 3. Results and Discussion

*Selection of the Optimal Classification Model*

For each of the neural network types mentioned in Section 2.4, the following architectures were tested:

- CNN networks containing from 2 to 3 convolution layers (each convolution layer was followed by a max pooling layer) followed by 1 to 2 dense layers, from 50 to 400 neurons for convolution and 50 to 200 for dense neurons;
- RNN networks containing from 2 to 4 layers, built from 50 to 400 neurons;
- RNN-LSTM networks containing from 2 to 4 layers, built from 50 to 400 neurons;

For all NN types, separate models were built increasing the neuron count on each layer by 25 for each new model (i.e., for RNN starting with a network containing 2 layers of 50 recurrent neurons and finishing with 4 layers containing 400 neurons). Table 3 shows the results obtained using three types of neural network for the above mentioned datasets. For CNN, the best results were obtained for a network of 4 layers, 3 layers of convolution neurons 250, 250, 100 for each layer respectively and a dense layer of 100 neurons. For RNN best results were obtained for a 3 layer model with 3 recurrent layers of 300, 150, and 100 neurons. RNN-LSTM achieved best results for a 3 layer architecture of 250, 300, 300 neurons. In addition, all NNs had a single dense layer of 7 neurons as the output layer. We used 10-fold leave-one-subject-out cross-validation and repeat the process for 10 iterations, averaging the final score. All NNs were trained using ADAM [46] for gradient descent optimisation and cross–entropy as the cost function, as it is a robust method based on well known classical probabilistic and statistical principles and is therefore problem-independent. It is capable of efficiently solving difficult deterministic and stochastic optimisation problems [47]. Training was set to 500 epochs with an early stop condition if no loss decrease was detected for more than 30 epochs.

**Table 3.** Classification performances of different feature representations in for the set of 7 basic emotions. Numbers in bold highlight the maximum classification rates achieved in each column. PO—Positions and orientation, upper and lower body, POU—Positions and orientation, upper body, P—Positions, upper and lower body, O—Orientation, upper and lower body, PU—Positions, upper body, OU—Orientation, upper body.

| Features Set | PO | | POU | | P | | O | | PU | | OU | |
|---|---|---|---|---|---|---|---|---|---|---|---|---|
| Error Rate | 3 | 5 | 3 | 5 | 3 | 5 | 3 | 5 | 3 | 5 | 3 | 5 |
| CNN | 56.6 | 54.8 | 38.8 | 38.8 | **58.1** | 56.8 | 33.6 | 33.0 | 41.6 | 38 | 50.2 | 49.0 |
| RNN | 55.4 | 55.2 | 49.2 | 49.0 | **59.4** | 59.4 | 36.4 | 33.8 | 54.6 | 54.2 | 34.4 | 31.8 |
| RNN-LSTM | 65.2 | 59.6 | 64.6 | 61.25 | **69.0** | 67.0 | 55.0 | 54.6 | 65.8 | 64.2 | 54.2 | 53.8 |
| ResNet20 | - | - | - | - | 27.8 | 27.5 | - | - | 25 | 23.7 | - | - |

One can easily observe that the best results (69%) were obtained using RNN-LSTM on the $P$ set containing position of all skeletal joins (upper and lower body). In general, this set of features gives the best results for all types of networks (58.1% for CNN, 59.4% for RNN). This suggests that this kind of features provide the best description for emotional expressions from all considered feature types. In case of the $PU$ set, results for all networks are lower than 5%, which indicates the effect of the lower part of the body on recognition. Using orientation $O$ as a features set, even if complimenting the position ($PO$ or $POU$), results in much lower recognition. According to Table 3 results indicate a slight impact of error rate—better results were achieved using the 3 error rate almost for every dataset and NN, in few cases the results were equal. This may suggest that even a small movement or displacement can affect the recognition of emotions and the error rate of 5 cm might not be precise enough to represent all relevant movement data.

In addition, the experiment was conducted on sequences without the keyframing step in the pre-processing (containing all the recorded frames) for all NN models and all the datasets. The results of classification were 5–10% lower (depending on the model and set) than those acquired by key frames with error rate of 3 cm. Moreover, the time of NN training rose significantly due to a large increase in the data volume. Lower recognition results for sets without keyframing might have been caused by the Kinect v2 sensor noise, as the device output is not very precise and produces small variations in returned positions and orientations from frame to frame. This can be mitigated by applying filtering on the signal, however it is a time and computational consuming process, which does not fall into the assumption of reducing data pre-processing to a minimum. In our approach, the keyframing process allowed us to avoid the sensor precision related issues.

Performance of the proposed NN models was compared with the state-of-the-art NN architecture, ResNet. It has won several classification competitions, achieving promising results on tasks related to detection, localisation and segmentation [48]. The core idea of this model is to use a so-called identity shortcut connection to jump over one or more layers [49]. ResNets use the convolutional layers with $3 \times 3$ filters, which are followed by batch normalisation and rectified linear unit ReLU. Plenty of experiments showed that the use of the shortcut connections makes the model more accurate and faster compared to their equivalent models. We recreated the exact process as described in Reference [48], as the results obtained for action recognition in Reference [48] look very promising (accuracy over 99%) and as initially assumed, the method might be applicable for emotional gestures classification. The 3D coordinates of the Kinect skeleton (from our $P$ and $PU$ datasets) were transformed into RGB images. The sets were also augmented according to the description in the source paper. For our experiment, we prepared the testing and training set following the 10-fold leave-one-subject-out cross-validation method, meaning that the testing set did not contain the training samples and samples obtained from training set samples augmentation. Accuracy achieved using ResNet is significantly lower than that of the other NN types. This might be caused by the size of the original dataset, which contains only 474 unique samples and the process of argumentation presented in Reference [48] does not produce a diverse enough set to train such a deep NN.

For each type of neural network, the best results are presented in a form of confusion matrix (see Figure 7). One can observe that the best results were obtained for the neutral state as it differs greatly from other expressions (the actor/actress stood still, while there was a relatively bigger amount of movement while expressing other states).

Happiness, sadness and anger have a high rate of recognition and are sporadically classified as other emotions, as gestures in those three states are highly distinctive and differ from other emotional states (in terms of dynamics, body and limb positions and movement), even when the gestures are not exaggerated. Disgust and fear were confused with one another most frequently, this might be caused by the way they were performed by the actor/actress, as this confusion pattern is analogous for all three NN types. It is clearly visible on the recordings that those two emotions were acted out very similarly in terms of gestures (usually backing out movement with hands placed near head or neck for both states).

**Figure 7 (a)** — CNN on $P$ set, 7 classes:

| | Fe | Ne | Ha | Sa | Su | An | Di |
|---|---|---|---|---|---|---|---|
| Fe | 0.49 | 0.04 | 0.03 | 0.2 | 0.09 | 0.03 | 0.16 |
| Ne | 0 | 0.79 | 0.06 | 0.09 | 0.06 | 0 | 0.03 |
| Ha | 0.05 | 0 | 0.63 | 0 | 0.09 | 0.19 | 0.03 |
| Sa | 0.19 | 0.14 | 0 | 0.54 | 0.06 | 0.03 | 0 |
| Su | 0.11 | 0.04 | 0.06 | 0.06 | 0.45 | 0.08 | 0.25 |
| An | 0.03 | 0 | 0.17 | 0.03 | 0.12 | 0.64 | 0 |
| Di | 0.14 | 0 | 0.06 | 0.09 | 0.12 | 0.03 | 0.53 |

**Figure 7 (b)** — RNN on $P$ set, 7 classes:

| | Fe | Ne | Ha | Sa | Su | An | Di |
|---|---|---|---|---|---|---|---|
| Fe | 0.6 | 0.02 | 0.05 | 0.04 | 0.05 | 0.03 | 0.25 |
| Ne | 0.01 | 0.8 | 0.04 | 0.09 | 0.02 | 0 | 0 |
| Ha | 0.04 | 0.03 | 0.61 | 0.06 | 0.12 | 0.1 | 0.03 |
| Sa | 0 | 0.09 | 0.01 | 0.57 | 0.16 | 0.03 | 0.1 |
| Su | 0.1 | 0.05 | 0.12 | 0.12 | 0.47 | 0.11 | 0.13 |
| An | 0 | 0 | 0.16 | 0.06 | 0.07 | 0.64 | 0.05 |
| Di | 0.25 | 0.02 | 0 | 0.07 | 0.11 | 0.1 | 0.44 |

**Figure 7 (c)** — RNN-LSTM on $P$ set, 7 classes:

| | Fe | Ne | Ha | Sa | Su | An | Di |
|---|---|---|---|---|---|---|---|
| Fe | 0.69 | 0.03 | 0 | 0.03 | 0.03 | 0.02 | 0.23 |
| Ne | 0 | 0.78 | 0.03 | 0.09 | 0 | 0 | 0.03 |
| Ha | 0.06 | 0.03 | 0.69 | 0.03 | 0.1 | 0.15 | 0 |
| Sa | 0.03 | 0.09 | 0.03 | 0.74 | 0.07 | 0 | 0 |
| Su | 0.06 | 0.03 | 0.06 | 0.03 | 0.67 | 0.1 | 0.17 |
| An | 0 | 0 | 0.06 | 0 | 0.1 | 0.71 | 0.03 |
| Di | 0.17 | 0.03 | 0.12 | 0.09 | 0.03 | 0.02 | 0.53 |

**Figure 7.** Confusion matrix for (**a**) CNN on $P$ set with 3 cm error rate (**b**) RNN on $P$ set with 3 cm error rate (**c**) RNN-LSTM on $P$ set with 3 cm error rate. Seven emotional states: Ne—neutral, Sa—sadness, Su—surprise, Fe—fear, An—anger, Di—disgust, Ha—happiness.

Since the recognition accuracy of the neutral class far exceeds other emotional states, as the samples for this state contain the least amount of motion and it differs from all the other states greatly, in the next step we analyse two sets without this class. From the first one we merely exclude neutral state, thus it consists of sadness, surprise, fear, anger, disgust and happiness. The second set contains emotional states, which are most commonly used in the literature: sadness, fear, anger, and happiness. Experimental results of the above-mentioned datasets are presented in Table 4. As in the case of seven classes, the best results were obtained using $P$ set. Similarly, RNN-LSTM proved to be the most effective, providing 72% in case of 6 classes and 82.7% in the case of 4. Confusion matrices for the above-mentioedn sets are presented in Figures 8 and 9.

**Table 4.** Classification performances of different feature representations for the set of basic emotions. PO—Positions and orientation, upper and lower body, POU—Positions and orientation, upper body, P—Positions, upper and lower body, O—Orientation, upper and lower body, PU—Positions, upper body, OU—Orientation, upper body.

| Features Set | PO | | POU | | P | | O | | PU | | OU | |
|---|---|---|---|---|---|---|---|---|---|---|---|---|
| #Emotions / #Classes | 6 | 4 | 6 | 4 | 6 | 4 | 6 | 4 | 6 | 4 | 6 | 4 |
| CNN | 50.5 | 55.2 | 51.5 | 55.5 | **54.2** | **63.6** | 47.8 | 50.5 | 53.7 | 60.7 | 47.4 | 49.2 |
| RNN | 54.4 | 66.8 | 58.6 | 70.8 | **59.2** | **80.8** | 39 | 55.2 | 54.4 | 66.8 | 40 | 57.2 |
| RNN-LSTM | 66.2 | 80 | 59.6 | 74.2 | **72** | **82.7** | 51.8 | 62.4 | 64.6 | 75.8 | 47.4 | 58.9 |
| ResNet20 | - | - | - | - | 30.6 | 40.2 | - | - | 30.1 | 39.7 | - | - |

**Figure 8 (a)** — CNN on $P$ set, 6 classes:

| | Fe | Ha | Sa | Su | An | Di |
|---|---|---|---|---|---|---|
| Fe | 0.46 | 0.05 | 0.35 | 0.05 | 0.03 | 0.05 |
| Ha | 0.03 | 0.61 | 0.00 | 0.18 | 0.12 | 0.06 |
| Sa | 0.07 | 0.00 | 0.87 | 0.00 | 0.00 | 0.07 |
| Su | 0.08 | 0.14 | 0.25 | 0.33 | 0.17 | 0.03 |
| An | 0.09 | 0.26 | 0.06 | 0.00 | 0.57 | 0.03 |
| Di | 0.13 | 0.03 | 0.30 | 0.00 | 0.07 | 0.47 |

**Figure 8 (b)** — RNN on $P$ set, 6 classes:

| | Fe | Ha | Sa | Su | An | Di |
|---|---|---|---|---|---|---|
| Fe | 0.70 | 0.05 | 0.05 | 0.05 | 0.00 | 0.14 |
| Ha | 0.03 | 0.70 | 0.00 | 0.06 | 0.12 | 0.09 |
| Sa | 0.07 | 0.03 | 0.67 | 0.07 | 0.10 | 0.07 |
| Su | 0.08 | 0.17 | 0.08 | 0.36 | 0.14 | 0.17 |
| An | 0.06 | 0.11 | 0.03 | 0.03 | 0.60 | 0.17 |
| Di | 0.13 | 0.10 | 0.07 | 0.13 | 0.03 | 0.53 |

**Figure 8 (c)** — RNN-LSTM on $P$ set, 6 classes:

| | Fe | Ha | Sa | Su | An | Di |
|---|---|---|---|---|---|---|
| Fe | 0.77 | 0.04 | 0.03 | 0.03 | 0.04 | 0.09 |
| Ha | 0.02 | 0.76 | 0.00 | 0.03 | 0.11 | 0.09 |
| Sa | 0.03 | 0.02 | 0.87 | 0.05 | 0.00 | 0.03 |
| Su | 0.08 | 0.04 | 0.07 | 0.57 | 0.08 | 0.15 |
| An | 0.07 | 0.07 | 0.07 | 0.03 | 0.74 | 0.01 |
| Di | 0.10 | 0.10 | 0.08 | 0.05 | 0.03 | 0.63 |

**Figure 8.** Confusion matrices for (**a**) CNN on $P$ set with 3 cm error rate (**b**) RNN on $P$ set with 3 cm error rate (**c**) RNN-LSTM on $P$ set with 3 cm error rate. Six emotional states: Fe—fear, Ha—happiness, Sa—sadness, Su—surprise, An—anger, Di—disgust.

**(a)**

|      | Fe   | Ha   | Sa   | An   |
|------|------|------|------|------|
| Fe   | 0.51 | 0.08 | 0.35 | 0.05 |
| Ha   | 0.20 | 0.61 | 0.01 | 0.17 |
| Sa   | 0.17 | 0.08 | 0.72 | 0.03 |
| An   | 0.09 | 0.11 | 0.10 | 0.70 |

**(b)**

|      | Fe   | Ha   | Sa   | An   |
|------|------|------|------|------|
| Fe   | 0.78 | 0.05 | 0.11 | 0.05 |
| Ha   | 0.09 | 0.80 | 0.03 | 0.09 |
| Sa   | 0.08 | 0.00 | 0.69 | 0.06 |
| An   | 0.00 | 0.06 | 0.06 | 0.54 |

**(c)**

|      | Fe   | Ha   | Sa   | An   |
|------|------|------|------|------|
| Fe   | 0.80 | 0.04 | 0.09 | 0.04 |
| Ha   | 0.07 | 0.81 | 0.04 | 0.07 |
| Sa   | 0.10 | 0.04 | 0.81 | 0.06 |
| An   | 0.03 | 0.06 | 0.04 | 0.85 |

**Figure 9.** Confusion matrices for (**a**) CNN on $P$ set with 3 cm error rate (**b**) RNN on $P$ set with 3 cm error rate (**c**) RNN-LSTM on $P$ set with 3 cm error rate. Four emotional states: Fe—fear, Ha—happiness, Sa—sadness, An—anger.

In order to compare the proposed method with other classification methods, we calculated the most commonly used features, such as kinematic related features (velocity, acceleration, kinetic energy), spatial extent related features (bounding box volume, contraction index, density), smoothness related features, leaning related features and distance related features. During features extraction we strictly followed approach presented in Reference [23], since the authors obtained very promising results on a database derived from Kinect recordings. We juxtaposed several well known classification methods to verify the above-mentioned features and their effectiveness in gestures-based emotion recognition. The obtained results are presented in Table 5.

**Table 5.** The performance of some well-known classifiers.

| Classifier | #Emotions/#Classes | | |
|------------|------|------|------|
|            | 7    | 6    | 4    |
| J48            | 45.36     | 37.07     | 56.36 |
| Random Forests | **52.95** | **50.73** | **64** |
| k-NN           | 43.46     | 42.92     | 61.09 |
| S-PCA + k-NN   | 35.86     | 37.33     | 51.27 |
| SVM            | 41.98     | 42.93     | 59.27 |
| MLP            | 42.19     | 46.09     | 61.45 |

To determine the performance of the above-mentioned classifiers we used the WEKA [50] environment. All parameters of the classifiers were set empirically in order to achieve the highest efficiency. As one can easily observe, the best results were obtained in the case of Random Forests. However, it should be emphasised that none of the methods listed above achieve better results than the proposed approach. This is a result of the generalisation of features from the whole recording, an approach which might be appropriate for simple gestures recognition; however, it becomes inaccurate for more complex and non-repeatable expressions.

## 4. Conclusions

In this paper, we presented a sequential model of affective movement as well as how different sets of low level features (positions and orientation of joints) performed on CNN, RNN and RNN-LSTM. The training and testing data contained samples representing seven basic emotions. The database consisted of recordings of constant affective movements, in contrast with other research, which is mostly reduced to specific single gesture recognition. Thus, we did not analyse solely separated selected frames but the whole movement as a unit. This experiment highlighted how challenging the task of recognising an emotional state based merely on gestures might be. The performance was much lower than in the case of particular gesture recognition; however, it was still higher than a human's performance (63%) [31].

The obtained results showed that body movements can serve as an additional source of information in a more comprehensive study. Thus, for future work we plan to combine all the three modalities, namely audio, facial expressions and gestures, which are signals perceived by a healthy human during a typical conversation. We believe that additional patterns extracted from affective movement may have a significant impact on the quality of recognition, especially in the case of emotion recognition in the wild [41]. In addition, we plan to extend our analysis using the Denspose [51] method and fuse and juxtapose with features provided by Kinect v2.

What is more, we will explore and compare methods used for action recognition, such as those presented in References [52–54], as they provide interesting expansion of the models used in this paper. For example, in Reference [52] the authors use a similar RNN-LSTM network architecture, instead of raw skeletal data, geometrical features extracted from the skeleton are fed to the NN. Also an interesting approach for RNN-LSTM is presented in Reference [53], where spatial attention joint-selection gates and temporal attention with frame-selection gates are added to RNN-LSTM. In Reference [54], the authors used F2C CNN -based network architecture for action recognition, with superior results compared to other classification modes. We plan to incorporate methods used in action recognition for the purpose of gesture based emotion classification, as the problem poses similar challenges in both areas.

**Author Contributions:** conceptualisation, T.S. and D.K.; methodology, T.S. and D.K.; software, T.S. and D.K.; validation, T.S. and D.K.; formal analysis, T.S. and D.K.; investigation, T.S. and D.K.; resources, T.S. and D.K.; data collection, T.S. and D.K.; writing—original draft preparation, T.S. and D.K.; writing—review and editing, G.A.; visualisation, T.S. and D.K.; supervision, G.A. and A.P.; project administration, G.A.; funding acquisition, D.K.

**Funding:** This research received no external funding.

**Acknowledgments:** The Estonian Centre of Excellence in IT (EXCITE) funded by the European Regional Development Fund and the Scientific and Technological Research Council of Turkey (TÜBITAK) (Project 1001 - 116E097).

**Conflicts of Interest:** The authors declare no conflict of interest.

## Abbreviations

The following abbreviations are used in this manuscript:

| | |
|---|---|
| ADAM | Adaptive Moment Estimation |
| CNN | Convolutional Neural Network |
| CS | Curve Simplification |
| HMM | Hidden Markov Model |
| k-NN | k-nearest neighbors |
| LSTM | Long short-term memory |
| NN | Neural Network |
| RBM | Restricted Boltzmann Machine |
| RNN | Recurrent Neural Network |
| ResNet | Residual Network |
| SVM | Support Vector Machine |
| MLP | Multilayer perceptron |

## References

1. Ekman, P. Facial action coding system (FACS). *A Human Face* **2002**. Available online: https://www.cs.cmu.edu/~face/facs.htm (accessed on 28 June 2019).
2. Pease, A.; McIntosh, J.; Cullen, P. *Body Language*; Camel; Malor Books: Los Altos, CA, USA, 1981.
3. Izdebski, K. *Emotions in the Human Voice, Volume 3: Culture and Perception*; Plural Publishing: San Diego, CA, USA, 2008; Volume 3.
4. Kim, J.; André, E. Emotion recognition based on physiological changes in music listening. *IEEE Trans. Pattern Anal. Mach. Intell.* **2008**, *30*, 2067–2083. [CrossRef] [PubMed]
5. Ekman, P. *Emotions Revealed: Understanding Faces and Feelings*; Hachette: London, UK, 2012.
6. Hess, U.; Fischer, A. Emotional mimicry: Why and when we mimic emotions. *Soc. Personal. Psychol. Compass* **2014**, *8*, 45–57. [CrossRef]

7.  Kulkarni, K.; Corneanu, C.; Ofodile, I.; Escalera, S.; Baro, X.; Hyniewska, S.; Allik, J.; Anbarjafari, G. Automatic recognition of facial displays of unfelt emotions. *IEEE Trans. Affect. Comput.* **2018**. [CrossRef]
8.  Mehrabian, A. *Nonverbal Communication*; Routledge: London, UK, 2017.
9.  Mehrabian, A. *Silent Messages*; Wadsworth: Belmont, CA, USA, 1971; Volume 8.
10. Poria, S.; Cambria, E.; Bajpai, R.; Hussain, A. A review of affective computing: From unimodal analysis to multimodal fusion. *Inf. Fusion* **2017**, *37*, 98–125. [CrossRef]
11. Corneanu, C.; Noroozi, F.; Kaminska, D.; Sapinski, T.; Escalera, S.; Anbarjafari, G. Survey on emotional body gesture recognition. *IEEE Trans. Affect. Comput.* **2018**. [CrossRef]
12. Ofli, F.; Chaudhry, R.; Kurillo, G.; Vidal, R.; Bajcsy, R. Sequence of the most informative joints (smij): A new representation for human skeletal action recognition. *J. Vis. Commun. Image Represent.* **2014**, *25*, 24–38. [CrossRef]
13. Gunes, H.; Piccardi, M. Affect recognition from face and body: Early fusion vs. late fusion. In Proceedings of the 2005 IEEE International Conference on Systems, Man and Cybernetics, Waikoloa, HI, USA, 12 October 2005; Volume 4, pp. 3437–3443.
14. Ofodile, I.; Helmi, A.; Clapés, A.; Avots, E.; Peensoo, K.M.; Valdma, S.M.; Valdmann, A.; Valtna-Lukner, H.; Omelkov, S.; Escalera, S.; et al. Action Recognition Using Single-Pixel Time-of-Flight Detection. *Entropy* **2019**, *21*, 414. [CrossRef]
15. Kipp, M.; Martin, J.C. Gesture and emotion: Can basic gestural form features discriminate emotions? In Proceedings of the 3rd International Conference on Affective Computing and Intelligent Interaction and Workshops (ACII 2009), Amsterdam, The Netherlands, 10–12 September 2009; pp. 1–8.
16. Bernhardt, D.; Robinson, P. Detecting emotions from connected action sequences. In *Visual Informatics: Bridging Research and Practice, Proceedings of the International Visual Informatics Conference (IVIC 2009), Kuala Lumpur, Malaysia, 11–13 November 2009*; Springer: Berlin/Heidelberg, Germany, 2009; pp. 1–11.
17. Rasti, P.; Uiboupin, T.; Escalera, S.; Anbarjafari, G. Convolutional neural network super resolution for face recognition in surveillance monitoring. In *Articulated Motion and Deformable Objects (AMDO 2016)*; Springer: Cham, Switzerland, 2016; pp. 175–184.
18. Demirel, H.; Anbarjafari, G. Data fusion boosted face recognition based on probability distribution functions in different colour channels. *Eurasip J. Adv. Signal Process.* **2009**, *2009*, 25. [CrossRef]
19. Litvin, A.; Nasrollahi, K.; Ozcinar, C.; Guerrero, S.E.; Moeslund, T.B.; Anbarjafari, G. A Novel Deep Network Architecture for Reconstructing RGB Facial Images from Thermal for Face Recognition. *Multimed. Tools Appl.* **2019**. [CrossRef]
20. Nasrollahi, K.; Escalera, S.; Rasti, P.; Anbarjafari, G.; Baro, X.; Escalante, H.J.; Moeslund, T.B. Deep learning based super-resolution for improved action recognition. In Proceedings of the IEEE 2015 International Conference on Image Processing Theory, Tools and Applications (IPTA), Orleans, France, 10–13 November 2015; pp. 67–72.
21. Glowinski, D.; Mortillaro, M.; Scherer, K.; Dael, N.; Volpe, G.; Camurri, A. Towards a minimal representation of affective gestures. In Proceedings of the 2015 International Conference on Affective Computing and Intelligent Interaction (ACII), Xi'an, China, 21–24 September 2015; pp. 498–504.
22. Castellano, G. Movement Expressivity Analysis in Affective Computers: From Recognition to Expression of Emotion. Ph.D. Thesis, Department of Communication, Computer and System Sciences, University of Genoa, Genoa, Italy, 2008. (Unpublished).
23. Kaza, K.; Psaltis, A.; Stefanidis, K.; Apostolakis, K.C.; Thermos, S.; Dimitropoulos, K.; Daras, P. Body motion analysis for emotion recognition in serious games. In *Universal Access in Human-Computer Interaction, Proceedings of the International Conference on Universal Access in Human-Computer Interaction, Toronto, ON, Canada, 17–22 July 2016*; Springer: Cham, Switzerland, 2016; pp. 33–42.
24. Kleinsmith, A.; Bianchi-Berthouze, N.; Steed, A. Automatic recognition of non-acted affective postures. *IEEE Trans. Syst. Man, Cybern. Part B (Cybern.)* **2011**, *41*, 1027–1038. [CrossRef]
25. Savva, N.; Scarinzi, A.; Bianchi-Berthouze, N. Continuous recognition of player's affective body expression as dynamic quality of aesthetic experience. *IEEE Trans. Comput. Intell. Games* **2012**, *4*, 199–212. [CrossRef]
26. Venture, G.; Kadone, H.; Zhang, T.; Grèzes, J.; Berthoz, A.; Hicheur, H. Recognizing emotions conveyed by human gait. *Int. J. Soc. Robot.* **2014**, *6*, 621–632. [CrossRef]
27. Samadani, A.A.; Gorbet, R.; Kulić, D. Affective movement recognition based on generative and discriminative stochastic dynamic models. *IEEE Trans. Hum. Mach. Syst.* **2014**, *44*, 454–467. [CrossRef]

28. Barros, P.; Jirak, D.; Weber, C.; Wermter, S. Multimodal emotional state recognition using sequence-dependent deep hierarchical features. *Neural Netw.* **2015**, *72*, 140–151. [CrossRef] [PubMed]
29. Gunes, H.; Piccardi, M. A bimodal face and body gesture database for automatic analysis of human nonverbal affective behavior. In Proceedings of the IEEE 18th International Conference on Pattern Recognition (ICPR 2006), Hong Kong, China, 20–24 August 2006; Volume 1, pp. 1148–1153.
30. Li, B.; Bai, B.; Han, C. Upper body motion recognition based on key frame and random forest regression. *Multimed. Tools Appl.* **2018**, 1–16. [CrossRef]
31. Sapiński, T.; Kamińska, D.; Pelikant, A.; Ozcinar, C.; Avots, E.; Anbarjafari, G. Multimodal Database of Emotional Speech, Video and Gestures. In *Pattern Recognition and Information Forensics, Proceedings of the International Conference on Pattern Recognitionm, Beijing, China, 20–24 August 2018*; Springer: Cham, Switzerland, 2018, pp. 153–163.
32. Ekman, P.; Friesen, W.V. Constants across cultures in the face and emotion. *J. Personal. Soc. Psychol.* **1971**, *17*, 124. [CrossRef]
33. Microsoft Kinect. Available online: https://msdn.microsoft.com/ (accessed on 11 January 2018).
34. Bulut, E.; Capin, T. Key frame extraction from motion capture data by curve saliency. *Comput. Animat. Soc. Agents* **2007**, 119. Available online: https://s3.amazonaws.com/academia.edu.documents/42103016/casa.pdf?response-content-disposition=inline%3B%20filename%3DKey_frame_extraction_from_motion_capture.pdf&X-Amz-Algorithm=AWS4-HMAC-SHA256&X-Amz-Credential=AKIAIWOWYYGZ2Y53UL3A%2F20190629%2Fus-east-1%2Fs3%2Faws4_request&X-Amz-Date=20190629T015324Z&X-Amz-Expires=3600&X-Amz-SignedHeaders=host&X-Amz-Signature=7c38895c4f79ebe3faf97dc8839ec237a2851828bd91bc26c8518cabfce692d6 (accessed on 29 June 2019).
35. Lowe, D.G. Three-dimensional object recognition from single two-dimensional images. *Artif. Intell.* **1987**, *31*, 355–395. [CrossRef]
36. Bogin, B.; Varela-Silva, M.I. Leg length, body proportion, and health: a review with a note on beauty. *Int. J. Environ. Res. Public Health* **2010**, *7*, 1047–1075. [CrossRef]
37. Ioffe, S.; Szegedy, C. Batch normalization: Accelerating deep network training by reducing internal covariate shift. *arXiv* **2015**, arXiv:1502.03167 .
38. Laurent, C.; Pereyra, G.; Brakel, P.; Zhang, Y.; Bengio, Y. Batch normalized recurrent neural networks. In Proceedings of the IEEE 2016 IEEE International Conference on Acoustics, Speech and Signal Processing (ICASSP), Shanghai, China, 20–25 March 2016; pp. 2657–2661.
39. Sola, J.; Sevilla, J. Importance of input data normalization for the application of neural networks to complex industrial problems. *IEEE Trans. Nucl. Sci.* **1997**, *44*, 1464–1468. [CrossRef]
40. Noroozi, F.; Marjanovic, M.; Njegus, A.; Escalera, S.; Anbarjafari, G. A Study of Language and Classifier-independent Feature Analysis for Vocal Emotion Recognition. *arXiv* **2018**, arXiv:1811.08935.
41. Avots, E.; Sapiński, T.; Bachmann, M.; Kamińska, D. Audiovisual emotion recognition in wild. *Mach. Vis. Appl.* **2018**, 1–11. [CrossRef]
42. Krizhevsky, A.; Sutskever, I.; Hinton, G.E. Imagenet classification with deep convolutional neural networks. *Adv. Neural Inf. Process. Syst.* **2012**, 1097–1105. [CrossRef]
43. Hochreiter, S. The vanishing gradient problem during learning recurrent neural nets and problem solutions. *Int. J. Uncertain. Fuzziness Knowl. Based Syst.* **1998**, *6*, 107–116. [CrossRef]
44. Avola, D.; Bernardi, M.; Cinque, L.; Foresti, G.L.; Massaroni, C. Exploiting recurrent neural networks and leap motion controller for the recognition of sign language and semaphoric hand gestures. *IEEE Trans. Multimed.* **2018**, *21*, 234–245. [CrossRef]
45. Hermans, M.; Schrauwen, B. Training and analysing deep recurrent neural networks. *Adv. Neural Inf. Process. Syst.* **2013**, *1*, 190–198.
46. Kingma, D.P.; Ba, J. Adam: A Method for Stochastic Optimization. *CoRR* **2014**, *abs/1412.6980*. Available online: https://arxiv.org/abs/1412.6980 (accessed on 28 June 2019).
47. De Boer, P.T.; Kroese, D.P.; Mannor, S.; Rubinstein, R.Y. A tutorial on the cross-entropy method. *Ann. Oper. Res.* **2005**, *134*, 19–67. [CrossRef]
48. Pham, H.H.; Khoudour, L.; Crouzil, A.; Zegers, P.; Velastin, S.A. Learning and recognizing human action from skeleton movement with deep residual neural networks. 2017. Available online: https://arxiv.org/abs/1803.07780 (accessed on 28 June 2019).

49. He, K.; Zhang, X.; Ren, S.; Sun, J. Deep residual learning for image recognition. In Proceedings of the IEEE Conference On Computer Vision And Pattern Recognition, Las Vegas, NV, USA, 26 June–1 July 2016; pp. 770–778.

50. Holmes, G.; Donkin, A.; Witten, I.H. Weka: A machine learning workbench. In Proceedings of the ANZIIS '94—Australian New Zealnd Intelligent Information Systems Conference, Brisbane, Australia, 29 November–2 December 1994.

51. Güler, R.A.; Neverova, N.; Kokkinos, I. Densepose: Dense human pose estimation in the wild. *arXiv* **2018**, arXiv:1802.00434 .

52. Zhang, S.; Liu, X.; Xiao, J. On geometric features for skeleton-based action recognition using multilayer lstm networks. In Proceedings of the 2017 IEEE Winter Conference on Applications of Computer Vision (WACV), Santa Rosa, CA, USA, 24–31 March 2017; pp. 148–157.

53. Song, S.; Lan, C.; Xing, J.; Zeng, W.; Liu, J. An end-to-end spatio-temporal attention model for human action recognition from skeleton data. In Proceedings of the Thirty-First AAAI Conference on Artificial Intelligence, San Francisco, CA, USA, 4–9 February 2017.

54. Minh, T.L.; Inoue, N.; Shinoda, K. A fine-to-coarse convolutional neural network for 3d human action recognition. *arXiv* **2018**, arXiv:1805.11790.

*Article*

# Enhanced Approach Using Reduced SBTFD Features and Modified Individual Behavior Estimation for Crowd Condition Prediction

**Fatai Idowu Sadiq [1,2,*], Ali Selamat [1,3,4,*], Roliana Ibrahim [1] and Ondrej Krejcar [3]**

[1] Faculty of Engineering, School of Computing, UTM & Media and Games Center of Excellence (MagicX), Universiti Teknologi Malaysia, 81310 Johor Bahru, Malaysia; roliana@utm.my

[2] Faculty of Physical Sciences, Ambrose Alli University, P.M.B 14, 310101 Ekpoma, Edo State, Nigeria

[3] Center for Basic and Applied Research, Faculty of Informatics and Management, University of Hradec Kralove, Rokitanskeho 62, 500 03 Hradec Kralove, Czech Republic; ondrej.krejcar@uhk.cz

[4] Malaysia Japan International Institute of Technology (MJIIT), Universiti Teknologi Malaysia, 54100 Kuala Lumpur, Malaysia

* Correspondence: sfatai2011@gmail.com (F.I.S.); aselamat@utm.my (A.S.); Tel.: +234-8037441378 or +60-187649242 (F.I.S.); +60-197363500 (A.S.)

Received: 28 February 2019; Accepted: 7 May 2019; Published: 13 May 2019

**Abstract:** Sensor technology provides the real-time monitoring of data in several scenarios that contribute to the improved security of life and property. Crowd condition monitoring is an area that has benefited from this. The basic context-aware framework (BCF) uses activity recognition based on emerging intelligent technology and is among the best that has been proposed for this purpose. However, accuracy is low, and the false negative rate (FNR) remains high. Thus, the need for an enhanced framework that offers reduced FNR and higher accuracy becomes necessary. This article reports our work on the development of an enhanced context-aware framework (EHCAF) using smartphone participatory sensing for crowd monitoring, dimensionality reduction of statistical-based time-frequency domain (SBTFD) features, and enhanced individual behavior estimation (IBE$_{enhcaf}$). The experimental results achieved 99.1% accuracy and an FNR of 2.8%, showing a clear improvement over the 92.0% accuracy, and an FNR of 31.3% of the BCF.

**Keywords:** context-aware framework; accuracy; false negative rate; individual behavior estimation; statistical-based time-frequency domain and crowd condition

---

## 1. Introduction

Crowd abnormality monitor (CAM) is a process of determining individual behavior in a crowd to prevent accidents in crowd-prone areas. Crowd monitoring using activity recognition (AR) to analyze individual behavior is maturing rapidly due to the current advancement in sensor technologies [1]. Increased research focus on human activity recognition (HAR) in diverse application domains highlights the significance of human–computer interaction (HCI) [2]. Two conventional methods are employed in the analysis of abnormal behavior in crowds. According to Zhang et al. [3], the "object-based" method identifies a crowd as a collection of individuals, while segmentation methods are used for analyses of crowd behaviors. In crowd behavior analysis, the performance of segmentation or detection of objects is usually faced with the complexity in the detection of objects [3]. Previous studies have demonstrated the object-based method with individual activity recognition. Issues in ongoing research have been extensively discussed, with initial solutions suggested in [4]. Context-aware approaches have been proposed previously; for example, [5]. However, only one [6] focused on crowd abnormality monitor and mitigation with the use of individual AR. However, the threshold used for crowd density in terms of the prediction of crowd condition is unclear [6]. An efficient approach should be able to accurately determine the number of persons within a square meter in order to prevent accidents during an emergency in a crowd scenario [7]. In the study by [6], the simulation was done inside a university building and conducted with a system of CAM [6], thus reducing the practical applicability of the system. Therefore, an alternative with high accuracy performance and a low false negative rate (FNR), which measures the false alarm to promote the efficient and reliable prediction of crowd conditions based on individual behavior [6], is needed. This will be based on an extension of the proposed basic context-aware framework (BCF) proposed [6]. A potential solution is to advance the previous BCF using the reduction of relevant statistical-based time-frequency domain (SBTFD) features with improved accuracy, reduced the FNR, and $IBE_{enhcaf}$ for individual and crowd condition prediction.

The motivation of this article proposes an enhanced context-aware framework using $IBE_{enhcaf}$ to improve the safety of human lives in a crowd-prone environment. The proposed approach utilized reduced features, with high-accuracy performance previously reported [4,8]. This study reports the result of an ongoing study on other sensor data validation, which included the effect of low FNR, and a clear definition of crowd density threshold for individuals per square meter ($m^2$) for crowd monitoring. The proposed approach employs the crowd density definition suggested in [7] and utilizes individual contexts from sensor signals in real time. In addition, the detection of five or more persons per $m^2$ is considered an extremely high density [9] to minimize the risk of accident in a moving crowd. The suggested solution promises accurate and reliable feedback to likely accident victims in an unforeseen situation. In this article, the context-aware framework is defined as a BCF that utilizes contexts such as individual user activities, location, and time [6]. The contexts are hidden information derived from smartphone sensor data [6]. The contributions of this article are:

(1) To present the validation result of other sensors used for individual behavior estimation (IBE) to extend the BCF.

(2) To suggest a clear crowd density threshold (CDT) per $m^2$ using a low FNR from reduced features to extend BCF.

(3) To propose an enhanced approach with reduced SBTFD features and modified IBE for crowd condition prediction with CDT to improve on BCF.

The proposed solution has the potential to minimize incessant death occurrences in social gatherings through a viable technology concept. The rest of the article is organized as follows: Section 2 discusses the current approaches to crowd monitoring, Section 3 presents the materials and methodology used in the study, Section 4 presents experimental results for the investigated issue to achieve the contributions in the article. The results are discussed in Section 5, while Section 6 addresses the conclusion and future work.

## 2. Current Approaches in Crowd Monitoring System

The crowd monitoring system (CMS) currently has three approaches, namely: (i) computer vision-based methods, (ii) sensor data analysis, and (iii) social media data analysis [10]. The most commonly used is sensor data analysis, which is also employed in this study [11] for several reasons. These include (i) a tendency for the provision of accurate and real-time information, (ii) nowadays, the new sensors on smartphones having the potential to revolutionize how we manage information, (iii) offering safety and enhancing security if well utilized in crowded places, (iv) wider coverage, as smartphones are used by almost everyone, and (v) feedback to potential victims in case of accidents [12]. Besides, sensor data analysis is widely used in AR with promising results [1,2,5]. Several feature extraction methods (FEM) have been employed in recent studies [13,14]. Table 1 presents the strengths and limitations of existing feature extraction methods.

The following section presents an analysis of FEM, including time domain (TD), frequency domain (FD), and feature reduction, and highlights those that can potentially be used for individual and crowd condition monitoring. Then, feature reduction based on feature selection methods (FSM) is examined for CMS for the minimization of time, classification, and accurate prediction. Related studies in context-aware frameworks are also discussed.

## 2.1. Time Domain (TD)

TD features include mean, median, range, variance, maximum, minimum, skewness, and kurtosis, to name a few. The features are widely used in HAR [15–17]. According to [17], the integral method has been applied to extract energy expenditure information from raw sensor signal data, where the total integral of the modulus of acceleration (IMA) was employed. The method is referred to as the time integral of the module of accelerometer signals, and is expressed in Equation (1):

$$IMA_{tot} = \int_{t=1}^{N} |a_x| dt + \int_{t=0}^{N} |a_y| dt + \int_{t=0}^{N} |a_z| dt \tag{1}$$

where $a_x, a_y, a_z$ represent the orthogonal components of acceleration, t denotes time, and $N$ is the window length. Some of the methods of extracting features rely on the ability to transform input signals to and from different domains [14]. To apply feature computations on a smartphone, one needs to be careful due to computational complexity as a result of limited memory, processing time, and battery lifetime. According to [18], almost all TD features are suitable for mobile devices, because their correlation operations have higher computational cost. A feature extracted from the raw sensor signal's data from individual activity recognition is such a piece of information, and can be used when classifying activity recognition to determine the characteristics of the individual in a crowd scenario in this thesis. In order to create features from the AR sensor raw dataset, different methods and mathematical calculations are applied to the raw dataset, and new features are extracted. Other time domain features such as zero crossing, signal vector magnitude, the signal magnitude area, and angular velocity have also been used in AR [19,20].

## 2.2. Frequency Domain (FD)

Features in this domain are important because the Fourier domain in AR sensor data has a much greater range than the AR in the spatial domain. To be sufficiently accurate, its values are usually calculated and in float values. Fast Fourier transform (FFT) also preserves information from the original raw signal and ensures that important features are not lost as a result of FFT [21]. FD splits the signal into sinusoidal waves with various frequencies using Equation (2):

$$f = \int_{1}^{w} x(t)e^{-j2\pi ft} dt; \; x(t) = \int_{1}^{w} X(f)e^{j2\pi ft} dt \tag{2}$$

where $t$ = time; $f$ = frequency; $X(f)$ = inverse Fourier transform; and $x(t)$ depicts Fourier transformation [22].

The proper selection of FD feature and sampling frequency is a key factor for extracting the frequency components; an inability to realize this may result in a false prediction of an individual in a crowd [3]. Zheng [3] transforms $x(t)$ to overcomes the drawback of inaccurate detection by introducing a frequency domain component and obtaining relevant information for AR [3,23]. Other important domains include the wavelet domain (WD), which are better noted in the analysis if irregular data patterns are used; that is, impulses exist at different time intervals [12], and therefore, require the selection of a proper mother wavelet. The heuristic domain (HD) works by using the assignment of the correct value to suggest the best corrective measure of sensor signals [16]. Therefore, HD requires input from multiple experts aggregates the result. The time domain–frequency domain (TDFD) produces an efficient performance for individual's representation in the crowd [14]; however, the use of FFT_RMS as the only FD may not assume the performance of other TD features.

**Table 1.** Strength and limitations of existing feature extraction methods.

| Feature Domain | Feature Extraction Methods | Merits | Demerits |
|---|---|---|---|
| TD | Mean | Is a good discriminator of individual characteristics calculated with small computational cost and a small memory requirement, is commonly used a feature in activity recognition (AR) research [12,16,22] | Does not produce a good result when isolated from other measures. |
| | Standard deviation | Derived through the use of mean to reveal any deviation in AR sensor data [6] | Frequency domain absence hinders its performance |
| | Correlation | Help to determine the correlation between one individual's characteristic feature and the other to express [6]. | Failure to produce the FD along the corresponding axis affects the performance of AR accuracy. |
| | Root Mean Square | Quality of sensor's data may dictate its tendency to reveal the actual location for individual in the prediction of crowd disaster [6]. | Could not work in isolation from other measures. |
| FD | FFT_RMS | Good tool for stationary signal processing [6,18]. | Weakness in analysing non-stationary signals from sensor data. |
| TDFD | Time domain -frequency domain | Produce an efficient performance for individual's representation in the crowd [6,14]. | The use of FFT_RMS as the only FD may not assume the performance of other TD features. |

[1] Note: TD = Time domain feature; FD = Frequency domain feature; TDFD = Time domain–frequency domain feature; FFT_RMS = Fast Fourier Transform of Root Mean Square.

Table 2 presents a synthesis of existing FEMs and their names in AR. It shows the features used in a crowd condition, the application domain, and the researcher, and those that have not been used in crowd conditions are also indicated. Table 2 shows that only conventional FEMs have been used in previous crowd-related research with Mean, Std, along x, y, and z [16,18,22], and variance along x, y, and z [14,18]. This could be responsible for the observed inaccuracy of 92% reported for CAM, which has also been noted by [24] to be generally low. It can also be noted that some salient TDFD features that are capable of accurate prediction were overlooked in the BCF, thus strengthening the need for further studies.

**Table 2.** Summary feature extraction methods (FEM) methods used and those that have not been used in crowd-related studies.

| Feature Extracted Methods in Activity Recognition | Application Domain | Features That Have Been Used in a Crowd | Reference |
|---|---|---|---|
| DD: Discrete cosine transform (DCT) 48 coefficients DCT features extracted | Daily activity | N/A | [25] |
| Variance (Var.) ax, ay, and az; number is not specified | Crowd behavior | Var. along x, y, and z | [26] |
| TD: Mean; std.; mad; max; min; sma; interquartile range (Iqr); entropy; arCoeff; cor.; maxfreq.; meanfreq.; FD: Max; min; sma; interquartile (iqr); skewness; kurtosis, energy band; angle; TDFD: 561 features | Daily living activity | Mean, Std, along x, y, and z | [18] |
| TD: mean, std., correlation (corr.), rms $a_x$ $a_y$ $a_z$. FD: FFT_rms $a_x$ $a_y$ $a_z$; TDFD: 15 features | Crowd abnormality monitor (CAM) | Features in the baseline study Known as BCF | [6] |
| TD: all time domain features in Table 1; FD: spectral coefficient; max. frequency; entropy of coefficient; dominating frequency; discrete coefficient; empirical cumulative distribution function (ECDF): with the setting of parameter value based on bin used for inverse computation; number is not specified | Motion sensing in daily life | Mean, Std, along x, y, and z | [27] |
| TD: mean, max, min, std., zero cross, median, range, sum of square, rms and var. TD: 30 features | Individual activity contexts | Mean, Std, along x, y, and z | [11] |
| TD: Mean; std.; max.; min.; corr.; Iqr.; DD: Dynamic time warping distance discrete time wavelet (DTW). FD: FFT coefficients as frequency domain features; except the first FFT coefficient. WD: wavelet energy TDFD and WD: 89 features | Motion sensor for daily activity | Mean, Std, along x, y, and z | [28] |
| TD: min, max, mean, STD, signal magnitude area (SMA), signal vector magnitude (SVM), tilt angle, FD: power spectral density (PSD), signal entropy, special energy: 60 features | User's daily detection of abnormality | Mean, Std, along x, y, and z | [29] |
| Improved SBTFD features presented in our previous work | Individual and crowd condition prediction | 15 features are newly suggested as improved TD for SBTD, and 24 features as improved FD for SBFD | [4] |

### 2.3. Related Works on Feature Reduction, Context-Aware Framework (CAF), and Activity Recognition (AR)

Feature reduction methods are important approaches that help avoid the cause of dimensionality [30], that is, the number of feature spaces in a feature vector. It targets a reduction in the number of previously used features on a mobile device in AR. High dimensionality on the accuracy of classification performance has been an important domain of research in HAR [31,32]. Feature reduction can facilitate the early detection of an emergency in an unforeseen circumstance [29]. Thus, the risk associated with individual activity recognition (IAR) in a crowd condition can be minimized by the reduction of FNR. The issue of high false alarm with FNR was not addressed in BCF. The solution proposed in our previous work as Phase 2 was reported [4].

The review of AR recognition works on individuals and crowds explains the potential of features dimensionality reduction for accurate and efficient crowd conditions; however, a feature reduction-based feature selection method has never been applied for this purpose. The work of [33] on early recognition supports this objective; it predicts a one-shot learning-based pattern transition for early detection recognition. A great benefit of the approach proposed by [34] utilized a smaller number of features for the prediction of ovarian cancer survival, and achieved very limited computational efforts. The use of a smart selection of a lesser number of relevant features compared with the number of features used with FEM in BCF diminishes the computational effort greatly, and reduced the false negative alarm. Moreover, an unclear definition of CDT has been noted by [7,9] as a major challenge in BCF. An inappropriate threshold of high density used for individual behavior estimation by [6], and a lack of feedback to victims resulting to a high false alarm in an emergency led to an unreliable prediction of crowd conditions, such as for example crowd abnormality behavior. Chang et al. [35] introduced a context-aware mobile platform for an intellectual disaster alerts system (IDAS); it focused on how environmental changes can result in accidents and disasters. According to the authors, a quick and accurate alert delivered to victims is essential in a disaster situation. However, their work focuses on addressing disaster issues, rather than crowd monitoring for safety.

Context-aware computing, an application concept that can sense the physical environment and reacts accordingly, was proposed by [36]. It is aimed at facilitating the quick and efficient development of a framework that combines context-aware service and machine learning [36]. The study led to the development of context-aware and pattern oriented machine-learning framework (CAPOMF). It focused on how commuters can avoid potholes to save vehicle repair costs. In previous context-awareness research, machine learning is rarely used [36–38] for the realization of context-aware framework. The studies of [6,39] also emphasized that context-aware application and its services remain open research issues. Prior to [6], no context-aware research with activity recognition have been applied or proposed for crowd abnormality mitigation in the literature. The outstanding problems that constitute a challenge in context-aware research regarding their affects on crowd disaster mitigation are itemized as follows:

(1)   Context acquisition, modeling, inference, and sensing.
(2)   Determination of appropriate sensors and the nature of contexts to be acquired.
(3)   Real-time management of sensors and context-based action generation.

As of June 2018, context-aware computing was worth US$120 billion [40]. Its research finds application in many domains with only few in disaster management. The extant literature highlights three methods used in context-aware framework: (i) scenario-based with a hypothetical example using a develop application, (ii) comparative analysis using a side-by-side comparison of components [41], and metric evaluation with accuracy, precision, recall, and f-score with an experiment on related activities [35]. Table 3 presents related works and highlights gaps in previous research.

**Table 3.** Related context-aware frameworks and activity recognition methods with the research gaps for individual and crowd condition prediction.

| Context-Aware Framework/AR | ARAC | FSM | CCP | Features Used | Why the Features Are Not Enough |
|---|---|---|---|---|---|
| CAM-BCF [6,42] | 92% based on TDFD | N/A | A high false negative rate | TD: mean x, y, z, std. x, y, z; cor. xy, yz, xz; rms. x, y, z; FD: FFT rms along x, y, z-axes as TDFD features | Salient TD and FD features with better result commonly used in literature were overlooked |
| IDAS [36] | N/A | N/A | N/A | N/A | N/A |
| Context recognition [11] | 55–98% based on TD | N/A | N/A | TD: Mean, STD.; Med. Min., Max., Zero Crossing, (ZC), Sum of Squares (SOS), rms, Range, Var | Attention paid to the only TD without giving consideration to FD that compliments TD features |
| Feature analysis [42] | 86–93% based on FSM | CFS, CHI, MRMR | N/A | 75th Percentile (PE): PE_y, min-max: mm_x, mm_y, PE_x, mm_z, PE_z | Negligence of FD features in selected features and 86.6% reported for MRMR |
| Coupling HAR [43] | 86–91% based on TDFD | N/A | N/A | Not specified | The detail was not given |

[3] Note: ARAC = Activity recognition accuracy, AR = Activity recognition, FSM = Feature selection method adopted to reduce features and CCP = Crowd condition prediction. CFS = Correlation-based feature selection, CHI = Chi-square feature selection and MRMR = Minimum redundancy–maximum relevance feature selection.

## 3. Materials and Methods

This section presents the methodology employed in this study. It provides a description of the development of the context-aware activity recognition application used for data collection, data validation outcome, adopted and modified algorithm implementation, and results in analysis approaches.

We developed an Android application called Context Activity Data Collector (CADC) based on Java programming as a client, and the crowd controller station (CCS) as a server to store the CADC in real-time for offline data analysis. The CADC runs on an Android 3.0.2 version of a Samsung Galaxy SM-G530H. Figure 1 shows the CADC data collection interface. An example of the sensor signals collected at a Malaysian public institution between March and April (2015) is shown in Figure 1. The eight (8) classes considered in the experiment conducted are selected from multiple possible conditions of an individual in the considered scenario. The scenarios considered are: climb down (V1), climb up (V2), fall (V3), jogging (V4), peak shake while standing (V5), standing (V6), still (V7), and walking (V8).

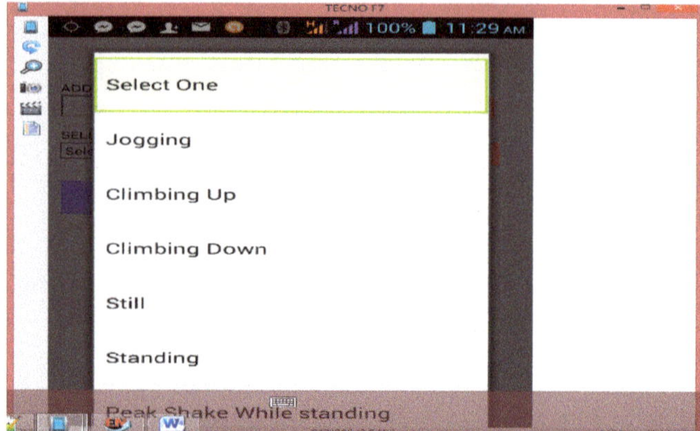


**Figure 1.** Sensor signals dataset collection interface used by volunteers during the experiment.

Several instances were captured for each scenario performed by volunteers (node S), yielding 22,350 class instances. In this case, S is referred to as the volunteers that make use of Figure 1 in the experiment conducted. The class instances obtained from S during the experiment include V1: 1975, V2: 2410, V3: 3159, V4: 2952, V5: 2937, V6: 2757, V7:3230, and V8: 3470 for dataset D1. The validated results of other sensor signals (captured as six additional classes, V12 to V18) for D1, which include a digital compass, longitude, latitude, and timestamps used for individual behavior estimation, were reported for dataset D1 based on IAR. Table 4 summarized the D1 dataset used for this research.

**Table 4.** Summary of sensor signals for the D1 raw dataset based on experiment conducted.

| Attribute | Dataset 1 (D1) [4] | Class | Activity/Sensors Name |
|---|---|---|---|
| Age | 25–51 years | V1 | Climb down |
| Activity count | 8 | V2 | Climb up |
| No of instances | 22,350 | V3 | Fall |
| No of participants | 20 | V4 | Jogging |
| Sensor type | Accelerometer x, y, and z digital compass (DC), longitude, latitude, timestamp | V5 | Peak shake while standing |
| Position placement | Hand | V6 | Standing |
| No. of devices | 20 smartphone | V7 | Still |
| Dataset gathering | Crowd controller as a server | V8<br>V12 | Walking<br>Latitude |
| | | V13 | Longitude |
| | | V14 | Speed |
| | | V15 | Altitude |
| | | V16 | Timestamp |
| | | V17 | Digital compass |
| | | V18 | Accuracy |

### 3.1. Methodology for the Proposed Enhanced Approach

The methodology in this article focuses on Phase 4 of Figure 2, while phases 1–3 were activities presented in the previous work [4,8]. They are important to achieve Phase 4 focused in this article as stated in the objective highlighted in Section 1, and the need for the reflection of these parts in Figure 2 for clear flow and understanding of this article.

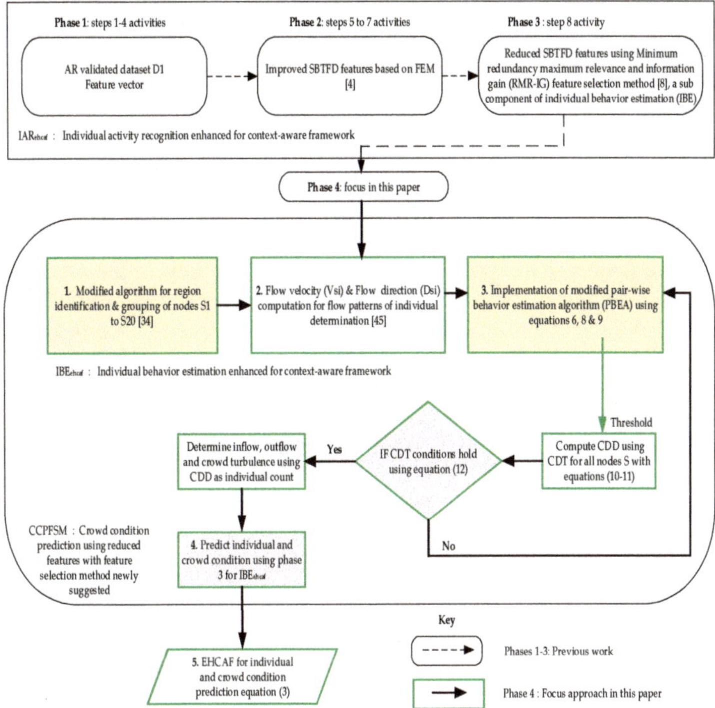

**Figure 2.** The process flow of the methodology used for the enhanced context-aware framework approach (EHCAF).

A high accuracy and reduction of a negative false alarm are highly desirable and central to crowd condition prediction; however, the approach cannot be adopted without adequate changes to the algorithm using the same data collection with the activity recognition method as shown in Figure 1 using Table 4. This was done by adopting the suitable threshold, which is called the crowd density threshold (CDT) (Figure 2) in Equation (4), while modifying the algorithms presented in BCF with a clear threshold definition of crowd density estimation to accurately detect individual per $m^2$ in crowd scenarios experimented. The crowd density in this study is defined as >2 persons/$m^2$. In order to achieve the stated objectives, the following tasks were carried out as summarized in Figure 2:

Step 1: Design: experimental; data type: sensor-based real-time IAR; Sample: 20 volunteers; provided: 22350 instances for D1 dataset.

Step 2: Procedure: development of CADC application (Figure 1) with algorithm implemented based on CDT using Java installed on volunteers' phones; sensors (digital compass, longitude, latitude as Global Positioning System (GPS) data for location etc., as presented in Table 4.

Step 3: Functioning of CADC: internet-enabled with hotspots; 50 to 100 $m^2$ coverage.

Step 4: Server setup: crowd controller station (CCS); volunteers (node S) launch the CADC app by pressing the start button; select activity scenario; perform each for 10 min while maintaining a range of 1 $m^2$ to each other, which was done collectively until all activity is reached; CCS store the sensor signals' collected data in text format; each volunteer stops the app as specified to end the data collection; duration was 5 h for each round of data collection. The guideline in the previous AR data set is employed [11,13,20]. The D1 collection became necessary because the sensors required were not available in the public domain [11,13,20] at the time of this study.

Step 5: Validation: The validation of raw sensor signals [44] was performed using an analysis of variance (ANOVA). This helps for the significant test of the dataset used in this study.

Step 6: Data analysis: Missing data was handled by employing moving average; noise removal from D1 was achieved using segmentation with 50% overlapping based on 256 sliding windows; for detail, see [4].

Step 7: Improved SBTFD features with newly suggested 39 features based on FEM (total 54 features) yields 7.1% accuracy improvement; this was implemented in Python; and reported in [4].

Step 8: Feature reduction using a feature selection method newly introduced to this domain produced seven (7) effective features; this again yields 99.1% accuracy, which is also an enhancement in AR and crowd monitoring studies; details are provided in [8].

This section described the procedure for enhanced IBE. Following the AR in steps 7 and 8; it is necessary to obtain other necessary features that can identify and estimate the behavior of an individual [6]. It begins with the implementation of a modified algorithm for the identification and grouping of individual participants (smartphone) as node S by the crowd controller station (CCS) using GPS as sensor data [5]. This is followed by the implementation of adopted algorithms, which determines abnormal movement behavior among individuals using the flow velocity Vsi estimation and flow direction Dsi identification [44]. The Vsi and Dsi were computed using the sensor fusion method based on Kalman filter as reported in [44].

The next stage picks the Vsi and Dsi, and combines them with the seven best (reduced) features previously achieved in step 8 from each class of activity scenario e.g., V2; for detail, see [33]. Thereafter, the combined Vsi, Dsi, and reduced features were used as input to modify the pairwise behavior estimation algorithm (PBEA). The PBEA was implemented to identify and determine the behavior of the individual in a crowd with a disparity value computed using the disparity matrix. The final stage employs the IBE using the reduced features based on CDT to evaluate the individual crowd density determination (CDD) per $m^2$. The CDD help to appraise the inflow and outflow of moving individuals to ascertain crowd turbulence. This was realized using the CCS, which triggers up a context-aware alert to predict the abnormal behavior of an individual and crowd condition. It also determines the participation of the individual in a crowd scenario based on disparity values to develop the proposed approach, an enhanced context-aware framework (EHCAF), which is an improvement on the BCF.

The following sections present details of the steps in the research methodology after the IAR using the reduced features in Phase 3 to achieve an IAR flow pattern. The flow pattern differentiates the behavior of one node from the other nodes in the experiment [5]. In the following section, a brief description of these sensors' validation is presented.

### 3.2. D1 Validation of Sensor Signals apart from Accelerometer Data

The result of the accelerometer signals of D1 was earlier reported [4]. D1 validation was carried out to validate the processed raw sensor signals for other sensors used for $IBE_{ehcaf}$ in this article. The validation task was carried out to ascertain the quality of the D1 dataset displayed in Figure 1. We have applied the statistical validation technique (SVT) commonly used in the literature [3,22] based on the parametric nature of the dataset. For the validation, two hypotheses were formulated and tested using IBM SPSS 22.0. The hypotheses are as follows:

(1) Null hypothesis $H_0$: $\mu_1 = \mu_2 = \mu_3 \dots, \mu_{11}$; there is no significant difference between the means of the variables V12, V13, ... , V18 used for the analysis of D1 for prediction in this study.

(2) Alternative hypothesis $H_A$: $\mu_1 \neq \mu_2 \neq \mu_3 \neq \dots$ ; there is a significant difference in at least one of the means of the variables V12, V13, ... , V18 used for the analysis of D1 for prediction in this study.

### 3.2.1. Reduced Features from Improved Statistical-Based Time-Frequency Domain (SBTFD)

This section discusses the reduced features from SBTFD employed for an enhanced context-aware framework for individual activity recognition ($IAR_{ehcaf}$) in (Phase 2 of Figure 2) based on improved SBTFD features reported in our previous works [4]. In this article, we focus on the individual behavior estimation enhancement ($IBE_{ehcaf}$) while utilizing the reduced features (Phase 3 of Figure 2) for crowd condition prediction using the feature selection method ($CCPFSM$) to enhance the proposed approach shown in Equation (5) in Phase 4 of Figure 2 using Equation (3). The $EHCAF$ is discussed as follows:

$$EHCAF = IAR_{ehcaf} + IBE_{ehcaf} + CCPFSM \tag{3}$$

where $EHCAF$ comprises the improved $SBTFD$ and reduced features from the FSM in our previous work [8]. $IBE_{ehcaf}$ represents the newly reduced features achieved using the employed FSM combined with Vsi and Dsi performed for IBE implementation with the modified and adopted algorithms (1) and (2). This serves as input to the modified Algorithm (3) in Figure 2, and are employed in this article. Note that the detail about improved SBTFD features and dimensionality reduction based on FSM (phases 1–3) are out of the scope of this article.

$CCPFSM$ denotes the prediction achieved by the reduced features and other parameters known as flow velocity $V_{si}$ and flow direction $D_{si}$ in Equation (2) (Phase 4), which were used to perform a task for the prediction of crowd condition in Equation (3). It employs an enhanced context-aware framework through the use of context-sensing from node S and crowd density determination (CDD) in Phase 4 for the inflow and outflow movement of individual behavior to evaluate the possible causes of abnormality in a crowd using the proposed approach as a solution. This helps to realize the development of $EHCAF$ shown in Equation (3).

### 3.2.2. Modified Algorithm for Region Identification and Grouping of Nodes S

Crowd behavior monitoring was done with the use of sensor signals for identifying each participant with a smartphone as node S, based on an individual followed up by a grouping of the nodes (S) (see Algorithm 1 in Appendix A). It was conducted using the individual sensor analyses in Step 4 (Section 3.1) with context recognition performed on the activity recognition of an individual, in order to estimate participants' behavior. The mapping between the program sensors and activities considered were utilized as input to algorithm 1 (Appendix A) implementation. In Algorithm 1, S is the participant node used as input in Step 4 (Section 3.1).

The crowd formation distribution is divided into sets of sub-regions using the crowd controller station (CCS). When a new participant node S is detected, the context-aware application notifies the crowd controller station, which automatically adds the new node to the specific sub-region of the present location in line 19 (Algorithm 1 in Appendix A). The region identification of participant is actualized with the smartphone of the participant as a node S, line 1, with the GPS data in lines 2–3 with respect to time (line 4 of Algorithm 1 in Appendix A) using the data displayed in Figure 1.

The grouping of participants into the sub-region list $SA_1$, $SA_2$, and $SA_n$ is achieved using line 20 of Algorithm 1 in Appendix A. It takes care of the movement of the participant from one place to another for the scenario used in the experiment. Node S was equipped with the context-aware mobile application prototype during the experiment, whenever the distance moved by the participant is greater than a threshold value in (line 18 of Algorithm 1 in Appendix A), as adopted in the work of [6]. The threshold value is about 20 m from the hotspot for effective monitoring via communication within the coverage area. Once the node is outside the hotspot range, it is exempted. The algorithm also determines the neighbouring nodes in a sub-area by estimating the distance between two participant nodes and other nodes monitored by the CCS. Based on the work of [6], if the distance between nodes is less than 10 m, the new participant node will be added to the same area using line 19 of Algorithm 1 in Appendix A. The distance of 10 m was selected for the hotspot to allow for ease of assessments in case of an emergency. The distance estimation is based on Vincenty's formula and is adopted for computing latitude and longitude coordinate points [5,44].

### 3.2.3. Flow Velocity Estimation and Flow Direction Identification Based on Activity Recognition

The implementation of this algorithm takes the contexts from sensor signals—specifically latitude, longitude (GPS data), accelerometer x, accelerometer y, accelerometer z, and timestamp—as input to Equation (3) of Figure 1. The input data were used to compute the flow velocity estimation and also used to determine the flow direction of individual movement behavior. The output from the implementation of the algorithm is flow Velocity ($V_{si}$) and flow Direction ($D_{si}$) [44]. The $V_{si}$ and $D_{si}$ are important informative features used to obtain hidden context information from individual behaviors in a crowd scenario that is considered to determine flow patterns of individual movement.

### 3.2.4. Implementation of Modified PBEA Algorithm

The disparity matrix is the difference between a node and any other nodes used in (Algorithm 2 of Appendix B). For example, u and v; $s_i$ or $s_j$. The diagonal elements of the disparity matrix are usually defined as zero, which implies that zero is the measure of disparity between an element and itself [44,45].

Given two R-dimensional $x_i = (x_i^1, x_i^2, \ldots x_i^R)$ and $x_j = (x_j^1, x_j^2, \ldots x_j^R)$, the Euclidean distance (EUD) $d\ (i, j)$ as observed in [45] is expressed in Equation (4):

$$d_{i,j} \sqrt{(x_i^1 - x_j^1)^2 + (x_i^2 - x_j^2)^2 + \ldots + (x_i^R - x_j^R)^2} \tag{4}$$

where $d\ i,j$ denotes the Euclidean distance in Equation (4).

The computation was performed to calculate the distance between nodes for the input data from S1 to S20. This is to determine the disparity value for individual estimation in each region where node S is located. The variables $x_i^1, x_j^1$ correspond to the features and their instances in pairs; based on SBTFD, a reduced feature set (fft_corxz, y_fft_mean, z_fft_mean, z_fft_min, y_fft_min, z_fft_std, y_fft_std) is then combined with Vsi and Dsi contexts from the sensor signals of D1. These serve as input to the PBEA. Euclidean distance (EUD) is commonly used in research across different domains. It has been used to compute the distance between two points with reliable results; hence, the choice of using it to generate distance from each participant to every other participant based on nodes [45,46]. In addition, the investigation revealed that EUD is suitable for the modified PBEA adopted from the BCF implemented in this research.

The algorithm caters for n numbers of nodes, but the location used for an experiment does not vary for all the activities performed. This was due to the aforementioned communication range stated in (Algorithm 1 of Appendix A). Thereafter, the clustered results obtained were similar beyond three sub-areas, since the location considered is uniform for the experiment. This was noticed from the GPS data for longitude and latitude obtained in the experiment used with D1. It was observed that there is a variation between nodes whose monitor's device is represented by S for identification. The cluster of nodes was performed using Equation (5):

$$EUD\ (d_{i,j}) = \sum_{i=1}^{n} \sum_{p\ \varepsilon\ K_i} dist(p, k_i)^2 \tag{5}$$

In Equation (5), EUD represents the Sum of the Square Error (SSE). SSE is determined by using the node of the participant that is nearest to each pair of the participant node, which helps for S identification in the monitoring group and subsequent ones in the group. The advantages of K-means that were adopted and used in Algorithm 1 in Appendix A were discussed in [44,46]. Equation (6) was applied to perform the IBE*ehcaf* in Equation (3) (of Phase 4).

For the IBE$_{\text{ehcaf}}$ task, let $\delta$ be a matrix of pairwise between $n$ attributes in Equation (6) [26]:

$$\delta_{i,j} = \begin{pmatrix} \delta_{1,1} & \delta_{1,2} & \delta_{1,3} & \delta_{1,4} \ldots \delta_{1,n} \\ \delta_{2,1} & \delta_{2,2} & \delta_{2,3} & \delta_{2,4} \ldots \delta_{2,n} \\ \delta_{3,1} & \delta_{3,2} & \delta_{3,3} & \delta_{3,4} \ldots \delta_{3,n} \\ \delta_{n,1} & \delta_{n,2} & \delta_{n,3} & \delta_{n,4} \ldots \delta_{n,n} \end{pmatrix} \tag{6}$$

where $\delta_{i,j}$ represents the disparity between the aforementioned features $i$ and $j$. Also, let $f(\delta_{i,j})$ be a monotonically increasing function that transforms differences into disparities using Equation (6).

The equation produces an R-dimensional matrix (where R $\leq$ n) configuration of points.

$x_i = (x_1, x_2, \ldots, x_i, \ldots x_j, \ldots x_n)$; likewise, $x_i = (x_i^1, x_i^2, \ldots x_i^R)$ and $x_j = (x_j^1, x_j^2, \ldots x_j^R)$, for $(1 \leq i, j \leq n)$. The *EUD* between any two nodes, S of $x_i$ and $x_j$ in this configuration, equals the disparities between features $i$ and $j$ expressed using Equation (7):

$$d_{i,j} \approx f(\delta_{i,j}) \tag{7}$$

The $d_{i,j}$ is defined by Equation (6). The measure has been applied to find the pairwise (Euclidean distance) between two cities with minimum possible distortion by [47], as reported in [46]. In this case, we represent the $n$ nodes of the matrix D (N, A) where u =N and v =A for B$^{(s)}$ with the positive integers 1, 2, 3,... n. Then, a distance matrix, B$^{(s+1)}$, is set up with elements, and is expressed using Equation (8) [46]:

$$d_0(i, j) = \begin{cases} l(i, j) & if\ participant(node)(i, j)\ exist \\ d_{i,j} = 0 & if\ i = j \\ d_{i,j} > 0 & if\ i \neq j \end{cases} \tag{8}$$

The length, $d(i, j)$, of the path from node $i$ to node $j$ is given by element D $(u, v)$ of the final matrix D$^{(n)}$ B$^{(n)}$, which makes it possible for the tracing back of each one of the node paths. An example of disparity matrix computation can be computed using Equation (9) as employed for the participant estimation algorithm noted in [5,24]:

$$D_{(u;vT)} = g(Corr(f(B_{si}, T),\ f(B_{si + 1}, T))) \tag{9}$$

where D is the disparity based on function $f$, and g is a variable that provides the mapping to a disparity value $f$. The disparity value is computed based on the input data, specifically fft_corxz, y_fft_mean, z_fft_mean, z_fft_min, y_fft_min, z_fft_std, y_fft_std, $V_{si}$ and $D_{si}$. While $f$ depicts correlation (*Corr*) performed on a matrix containing the input data in pairs; $B_{si}$ is an individual participant node; $u$ is the number of nodes of the participant along the column of the matrix; $v$ is nodes of the participant along a row of the matrix, and $T$ denotes time. The function $f$, Corr, and $g$ depend on the specific crowd that is considered. Typically, $f$ is a pre-processing function. Corr computes a measure of differences between the input data for every $(i, j)$ pair of nodes to determine an individual in a crowd scenario. Finally, g maps to a disparity value. The disparity value is defined to be zero if the two participants are likely resulting from their participation in the same crowd. Conversely, the disparity tends to one or more if the node $s$ is not likely to be the result of participation in the same crowd. The outcome generates a disparity matrix $D_T = [D_{((u;vT))}]_{m\ x\ n}$ at time $T$. The reduced features set achieved and other parameters derived as features previously reported in [33]—namely, $V_{si}$ and $D_{si}$ [44], are fed into the PBEA, as shown in Equation (6) of (Phase 4) as input to generate the output for individual and crowd condition prediction illustrated in the next section.

### 3.2.5. Crowd Density Threshold Condition

This study adopted the conditions that trigger abnormality to set a threshold for crowd density determination within the coverage area as established in previous studies [48] and employed in other studies [3,4,6,49]. The threshold adopted in this study was first suggested by [6], who defined a crowd as made up of three or more persons. This study employs two persons per m$^2$ for the experiment based on [6]. However, the monitoring of participants occurs within the coverage areas and range of distance for the hotspot, and can be assessed using the device of a participant smartphone, which is referred to as node S. It is generally acknowledged that five persons/m$^2$ is an extremely high density, four persons/m$^2$ is high density, three persons/m$^2$ is medium density, two persons/m$^2$ is low density, while one or no persons/m$^2$ is considered very low density [7]. In addition, six or more persons/m$^2$ is considered extremely dangerous, with the potential to cause abnormality [7]. Crowd density determination (CDD) was employed to compute the density of the monitored crowd of moving nodes based on a crowd density threshold (CDT) condition shown in Equations (10)–(12) of (Phase 4). Node S is recognized by the crowd controller station (CCS) based on node count using Equations (10) and (11) [50].

$$Density \; = \; LN \; < \; area\ in\ m^2 \; * \; 5 \tag{10}$$

$$\mathrm{CDD} = 1 + 4 * \left[ \frac{Density - \lambda}{\psi - \lambda} \right] \tag{11}$$

where $LN$ represents the number of participants monitored, $\lambda$ denotes the minimum density level, and $\psi$ is the maximum density observed in the experiment at a particular time. The maximum capacity has also been proposed to be calculated using the number of participants < area in m$^2$ × 10; where 10 is regarded as extreme crowd density, as noted in the work of [50]. More than two participants per m$^2$ exceed the threshold. In order to explain the disparity matrix (a low value and high value) employed by [5], which is used to explain the type of crowd observed in the analyses of the result for this article, Equation (12) shows the crowd density threshold condition (CDT) used for the CDD evaluation.

$$\left\{ \begin{array}{c} 1.\ If\ CDT\ for\ d_{i,j}\ per\ sqm^2 \leq 2\ then \\ low\ crowd\ density\ occur \\ 2.\ else\ If\ CDT\ for\ d_{i,j}\ per\ sqm^2 = 3\ then \\ medium\ crowd\ density\ occur \\ 3.\ else\ If\ CDT\ for\ d_{i,j}\ per\ sqm^2 = 4\ then \\ high\ crowd\ density\ occur \\ 4.\ else \\ extremely\ high\ crowd\ density\ occur \end{array} \right\} \tag{12}$$

## 4. Experimental Results

This section presents results based on the highlighted objectives as follows: the raw sensor data validation, and the descriptive analysis for the validation summarized for all classes N: 22,350, which consists of V1 to V8. V12 provided a mean of 4.735, the standard deviation of 2.519, and a standard error of 0.2216. V13 provided a mean of 47.762, the standard deviation of 47.501, and a standard error of 0.4179. V15 produced a mean of 21.629, the standard deviation of 82.162, and a standard error of 0.7228. Meanwhile, V18 provided a mean of 48.891, the standard deviation of 106.286, and a standard error of 2.255. Inferential statistics for the ANOVA test conducted at $p = 0.05$ shows V12, V13, V15, and V18 having F-values of 46644.20, 4653.71, 196.41, and 967.01, respectively. The $p$-value = 0.000 is statistically significant. Hence, we reject H$_0$, and accept H$_{A,}$ and conclude that there is a significant difference in at least one of the means of the variables V12, V13, … , V18 used for the analysis of D1. This conclusion implies that the D1 dataset is valid, consistent, and adequate for the analysis conducted in this study.

*4.1. Result on the Classification of Raw Dataset D1*

The results of classification after validation is as follows. In Table 5, out of the 22,350 instances (last row); about 10,692 (bold in diagonal) of the confusion matrix were correctly predicted, while the remaining 11,658 instances were wrongly predicted. In Figure 3, the summary of classification results for baseline, a raw dataset D1, an improved SBTFD with 54 features, and seven reduced SBTFD features newly introduced to extend the BCF to produce an enhanced approach (EHCAF) is presented in Equation (3). The best ARAC, FNR, and RMSE are achieved with EHCAF-7 features having 99.1%, 2.8%, and 7.9%, respectively. This is against 92.0%, 31.3%, and 21.6%, respectively.

**Table 5.** Confusion matrix from the classification result of individual activity recognition (IAR) using the sensor signals of the D1 raw dataset.

| Class Label | Predicted Class | | | | | | | | Actual Class |
|---|---|---|---|---|---|---|---|---|---|
| | V1 | V2 | V3 | V4 | V5 | V6 | V7 | V8 | TP + FN |
| Climb down: V1 | 591 | 425 | 228 | 147 | 106 | 137 | 41 | 300 | 1975 |
| Climb up: V2 | 405 | 705 | 292 | 178 | 161 | 186 | 57 | 426 | 2410 |
| Fall: V3 | 188 | 273 | 778 | 325 | 858 | 254 | 99 | 384 | 3159 |
| Jogging: V4 | 147 | 163 | 269 | 1698 | 190 | 131 | 42 | 312 | 2952 |
| Peak shake_wst: V5 | 113 | 161 | 854 | 233 | 767 | 101 | 24 | 144 | 2397 |
| Standing: V6 | 106 | 142 | 210 | 110 | 70 | 1813 | 85 | 221 | 2757 |
| Still: V7 | 40 | 67 | 112 | 49 | 47 | 110 | 2733 | 72 | 3230 |
| Walking: V8 | 273 | 380 | 418 | 312 | 159 | 255 | 66 | 1607 | 3470 |
| Total | | | | | | | | | 22350 |

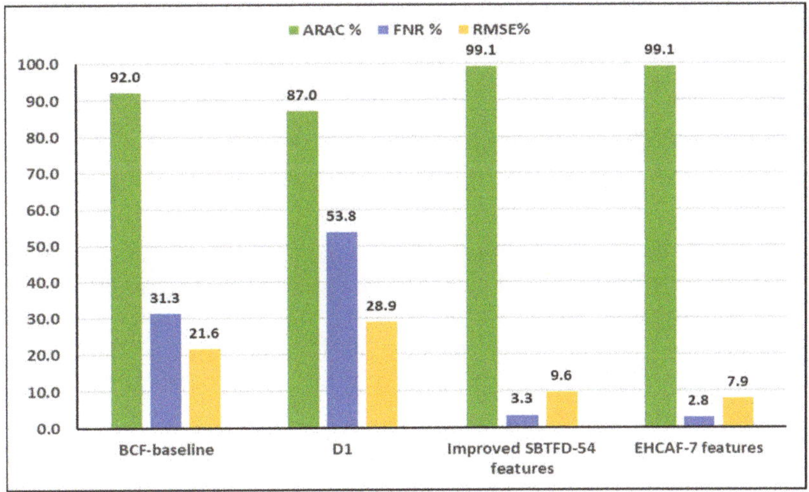


**Figure 3.** Comparison of BCF—baseline classification results, raw dataset—D1, improved statistical-based time-frequency domain (SBTFD), and reduced features for the enhanced approach.

*4.2. Results of Region Identification and Grouping of Nodes Using Clusters*

Figure 4 provided a higher number of clusters, which shows that more participant nodes gathered in subarea $SA_1$ than subareas $SA_2$ and $SA_3$ in the experiment. Thus, $SA_1$ is more prone to risk than $SA_2$ and $SA_3$.

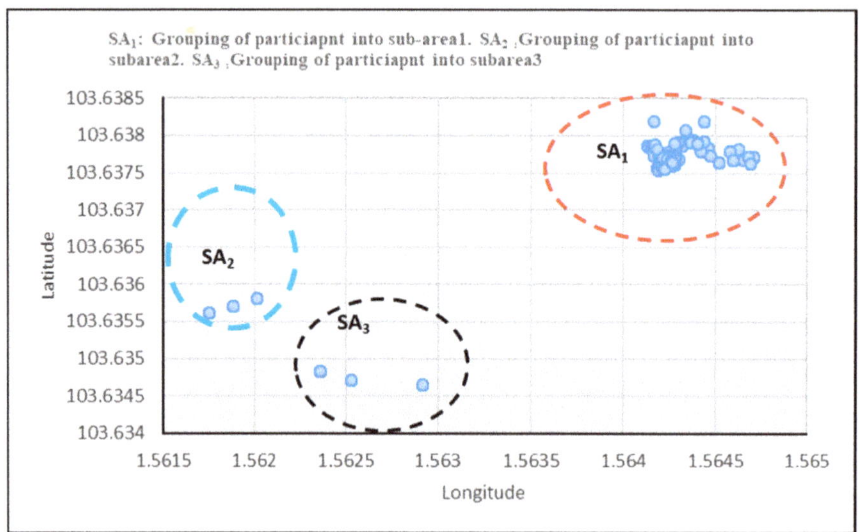



**Figure 4.** Results of clusters for identifying and grouping participant into subareas with GPS data.

*4.3. Results on the Algorithm Implemented for Flow Velocity and Flow Direction*

For details of the algorithm implemented for flow velocity and flow direction, please refer to [44]. This article focuses on the individual behavior estimation method combined with reduced features, which were not considered in the BCF.

*4.4. Modified PBEA Using Reduced Features and Enhanced Individual Behavior Estimation*

The output serves as input to the modified PBEA as shown in Figure 2 to produce an enhanced context-aware framework for individual and crowd conditions. The analysis is based on pairs of the node; for example, 1 and 2, 1 and 3, 1 and 4... up to 20 for individual behavior estimations. A disparity matrix was computed for the estimation of an individual based on the 20 nodes used as input for S1 to S20 for different nodes in the experiment. The experimental result revealed the interaction of participating (nodes) and their behavioral patterns in a crowd scenario based on the CDT employed and crowd density estimate. It shows two, three, three, and 12 nodes of a different number of individuals per m$^2$ (Appendix C).

4.4.1. Crowd Condition Prediction Using Individual Behaviour Estimation

For crowd estimation, it is necessary to estimate individual activity recognition and behavior initially. This had been addressed in our earlier works [4,8]. The crowd condition prediction using seven reduced features with Vsi and Dsi is newly introduced. This achieved higher accuracy by 99.1% against 92.0%. Also, a marginal reduction of the false negative rate by 28.5% from 2.8% against 31.3%, which is an improvement over the BCF [5], was obtained to achieved EHCAF see Figure A2 of Appendix D. The individual behavior estimation with suggested CDT and crowd density determination computation for crowd count serve as a means to extend the BCF [5]. This could help identify early danger by using context sensing through a smartphone with a context-awareness alert, thus minimizing the level of abnormality behavior in a crowd-prone area.

### 4.4.2. Implication of Low False Negative Alarm on the Enhanced Approach Based on PBEA Experiment

Figure 5 shows that the experimental results based on the proposed approach using reduced features and enhanced IBE in this article for crowd condition prediction has a low false negative rate (FNR), achieving an FNR of 2.8% and an ARAC of 99.1%, compared with an FNR of 31.3% based on an ARAC of 92% in the baseline. The results suggest that the higher the false negative rate (FNR) of AR, the higher the number of participants that may be at risk. Figure 5 also shows the comparative risk situation for EHCAF in blue color and BCF in red color, showing one (1) participant (node) in 20 and 28 participants in 1000 for the EHCAF, and six in 20 and 313 participants for 1000 in the BCF. The value was computed using a FNR of 2.8/100 * Number of the participants (NOPs) based on a crowd of people considered which will be varied in a real-life scenario when the proposed is applied.

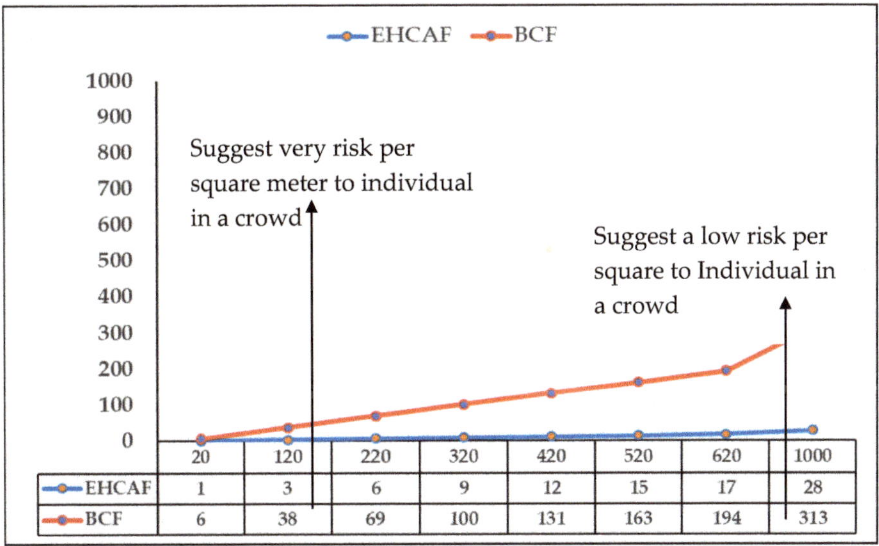


| | 20 | 120 | 220 | 320 | 420 | 520 | 620 | 1000 |
|---|---|---|---|---|---|---|---|---|
| EHCAF | 1 | 3 | 6 | 9 | 12 | 15 | 17 | 28 |
| BCF | 6 | 38 | 69 | 100 | 131 | 163 | 194 | 313 |

**Figure 5.** Effects of the false negative rate on the proposed approach when applying to human behavior monitoring in real life in a crowd condition.

This section presents the details of benchmarking with related works in the literature [5,51,52]. To confirm that the achieved higher results for the proposed approach is significantly better on the evaluation measurements used, Statistical t-tests were carried out using SPSS version 22.0 on dataset D1 and the BCF. The results of the seven reduced features based on FSM from method A, with $p$-values of 0.003 for the improved SBTFD and 0.021 against BCF, indicates $p < 0.05$, implying that the performance of the proposed approach is statistically significant at an 0.05 alpha level.

This supports the objective presented in this article. Based on the analysis of results, the enhanced context-aware framework (EHCAF) depicted in Figure A2 (Appendix D) is an improvement on the basic context-aware framework (BCF) benchmark, as shown in Table 6. However, Table 6 shows the components for EHCAF; likewise, the justification for improved parameters to establish the validity of our findings in the entire study.

**Table 6.** Comparison between BCF [6] and proposed approach (EHCAF).

| Components | EHCAF<br>IAR$_{ehcaf}$ | Justification |
|---|---|---|
| AR dataset | Validation of D1 performed with ANOVA is significant | Explain the suitability of the D1 in line with the literature. Quality of data is very important for crowd monitoring and accurate prediction |
| Accuracy | 99.1%, 98.0%, and 99.0% were achieved | An improvement over BCF with enhanced accuracy performance is achieved |
| Feature selection method (FSM) | Minimum Redundancy Maximum Relevance with Information Gain (MRMR-IG) with SBTFD provided seven reduced features (Corr_xz-fft, y_fft_mean, z_fft_mean, z_fft_min, y_fft_min, z_fft_std, and y_fft_std) | Reduces the dimensionality of features space on the monitoring devices. Lower computational task. Facilitates early recognition and utilizes less time for classification |
| Classifier | J48, Random forest (RF) | Compatible with an Android device and widely used in AR |
| Accuracy & FNR | 99.1%; 2.8% | Improvement of 7.1% accuracy and 28.5% FNR over BCF |
| Individual Behavior Estimation | IBE$_{ehcaf}$ | Provide accurate prediction to enhanced the safety of human lives |
| Region identification<br>Grouping of node S into Sub-area | Modified algorithm using k-means to implement Algorithms 1 and 2 with D1 to identify the region, cluster nodes S, and group into sub-areas | Potential to reveal susceptible clusters nodes in sub-areas that are prone to danger. Ascertain threshold with the specify coverage of nodes |
| Flow velocity and flow direction | Adopted and implemented using D1 | Serve as informative features to extract individual context behavior not possible for IAR in phases 1 to 3 |
| IBE | Modified PBEA using flow velocity (Vsi), flow direction (Dsi), and seven reduced features for IBE | Estimation of nodes per m$^2$ and analysis within coverage areas experimented with volunteers |
| Threshold | Threshold > two per m$^2$ | An efficient method should measure accurately the number of volunteers (node) within per m$^2$ to prevent abnormality occurrence in a crowd. |
| Inflow, outflow & crowd turbulence | Compute and evaluated using CDD based on individual count | Potential to identify person prone to danger early using context-awareness alert |
| Crowd condition | Crowd abnormality behavior | To enhanced the safety of human lives in a crowded area |
| Prediction | Crowd condition prediction using modified PBEA with reduced features (CCPFSM) | Enhanced approach with improved accuracy and FNR performance |
| Validation | Inferential statistics and paired sample statistics test was used to validate all the three methods employed for the enhanced approach | Improved SBTFD with 0.002; reduced features with 0.003 and 0.021 of $p < 0.05$ are statistically significant |

## 5. Discussion of Results

The result achieved an improvement of 7.1% and a false negative rate of 28.5% with an error reduction of 13.7% in terms of root mean square errors. This suggests safety to human lives in a crowd-prone situation when applying to real-life applications against the BCF by [5] as analysed in Table 7. In Figure 4, the susceptible area where crowd abnormality is likely to occur suggests sub-area list $SA_1$; this was obvious from the plot as more clustered nodes were observed in the area, which is an indication of more participants interacting together at a very close range to one another, as shown in Figure A1 (of Appendix C).

Based on the flow velocity Vsi and flow direction Dsi from accelerometer sensor signals analyzed, the V3 fall scenario revealed that only 778 were correctly recognized as TP, out of the 3159 expected among the instances of 22,350. Meanwhile, the rest consists of FP: 2383, FN: 2831, and TN: 16808 in Table 5. In Table 5, the unrecognized individual activity from 2381 which accounted for the abnormal behavior of individuals could be responsible for disaster manifestation. In a nutshell, the incorrect recognition demands effective features such as those suggested with the statistical-based time-domain in [10–16] and statistical-based frequency domain in [27,52], which informed the solution adopted in our previous work [4,33].

**Table 7.** Comparison of the proposed approach (EHCAF), activity recognition, and basic context-aware framework (BCF).

| Context-Aware Frameworks | SCI | ARAC | FEM | FSM | CCP | RMSE |
|---|---|---|---|---|---|---|
| BCF-baseline [5] | ✓ | 92.0% | TDFD-15 | N/A | High FNR (31.3%) | 21.6% |
| [11] | ✓ | 55% to 98.0% | TD-30 | N/A | N/A | N/A |
| [40] | N/A | N/A | TDFD Wavelet | MRMR 86.6% | High FNR (56.5%) | 31.0% |
| Proposed approach (EHCAF) | ✓ | 99.1% | Improved SBTFD-54 | 7 reduced features using MRMR-IG (method A)-99.1% | Low FNR (2.8%) | 7.9% |

Note: SCI: Context-aware issues. ARAC: Activity recognition accuracy. FEM: Feature extraction method. FSM: Reduced features achieved using Feature Selection Method. CCP: Crowd Condition Prediction. RMSE: Root mean square error. N/A: Not applicable.

Figure A1 (Appendix C) showed four distinct groups with the highest and lowest number of participants with 12, three, three, and two nodes, respectively. It shows the interactions and range at which those nodes interconnected for the scenario used as an example. Another plot from the data using a different set of 20 nodes to compute a different set of disparity values based on the disparity matrix with implemented algorithm three gave a similar result. The 12 nodes suggested a dangerous situation in terms of crowd scenario according to [6,7]. This implies a high inflow and outflow, which could bring about high crowd turbulence, and thus requires an immediate control if it happens in a crowded situation. All three nodes in Figure A1 (Appendix C) signify a medium crowd density, and the two nodes indicated a very low crowd density, which is basically known as a normal situation. Therefore, it is found to be within the threshold suggested using Equation (11). Based on this, the pattern of 12 nodes using an undirected graph in real life may result in crowd abnormality occurrence. In such cases of the 12 nodes with early recognition and sensitization using the proposed context-aware framework, such crowd density can easily be controlled before it reaches a critical state. Most importantly, for example, in Appendix D, with an FNR of 2.8% for every 20 and 1000 participants (nodes), which were assumed to be monitored one node and 28 nodes, respectively, will be at risk using the proposed solution, versus six and 313 nodes respectively in the basic context-aware framework (BCF) [5]. Experimental results support activity recognition studies in the literature for both cross-validation and split [11,39]. It also identifies that RF and J48 are the best classifiers suitable for the enhanced context-aware framework (EHCAF) Figure A2 Appendix D for individual and crowd condition prediction as compared to the other classifiers investigated. In view of our findings, the limitation of this work includes an inability to develop a context-aware system to effectively implement the reduced features that are newly suggested in this research. Future work could investigate and integrate the use of this methodology to the realization of safety for human lives through viable application in real life. Also, there was an inability to handle the technicality on the part of the monitoring device functionality to identify none of the functional sensors that could hinder the smooth data acquisition of individual activity recognition for prediction.

## 6. Conclusions

This study has described the sensor signals of activity recognition that are adequate for the prediction of individual and crowd conditions. The entire approach demonstrated in this article fulfills the aim, which focused on complementing other research in human activity recognition and pervasive computing toward the mitigation of crowd abnormality in the 21st century. In this article, an enhanced context-aware framework (EHCAF) was developed. The potential of reduced features with the feature selection method based on the improved feature extraction method using SBTFD was demonstrated.

The relevant parameters were derived and applied to implement the modified algorithm for grouping participants using smartphones as nodes. Based on findings, an enhanced approach for individual and crowd condition prediction is summarized as follows: the utilization of reduced features and enhanced individual behavior estimation ($IBE_{enhcaf}$) with high accuracy and low FNR performance is achieved; a clear definition of crowd density formulation for crowd condition prediction in a crowd scenario is presented. Above all, from the previous study, the FNR is 31.3%, while in this study, it is 2.8%. Hence, an improvement of 28.5% is achieved based on the experiment. However, the limitations and gaps left by previous studies have been equally addressed. The experimental results of this article have shown significant improvement from the previous studies done by [5,11,24,39]. The methods applied to achieve the proposed enhanced approach showcased in this article support the objective of the article. In the future, the approach promises a dynamic solution that intends to explore the collection of the ground truth dataset for the purpose of mitigating disasters among individuals gathering in places such as Mecca, medina during the pilgrimage in Saudi Arabia by integrating cloud-based technology.

**Author Contributions:** Funding acquisition, A.S. and O.K.; Methodology, F.I.S.; Supervision, A.S.; R.I. and O.K.; Validation, F.I.S.; Visualization, F.I.S.; Writing – original draft, F.I.S. This article was extracted from ongoing doctoral research at Universiti Teknologi Malaysia, (UTM), 81310, Johor Bahru. The First author F.I. Sadiq has recently completed his PhD in Computer Science. This article reported one of the research contributions in his doctoral thesis and research work. The remaining authors are the supervisors of the candidate. The supervisors' comments and suggestions were valuable to the success of this manuscript preparation.

**Funding:** This research was funded by Universiti Teknologi Malaysia (UTM) under Research University Grant Vot-20H04, Malaysia Research University Network (MRUN) Vot 4L876 and the Fundamental Research Grant Scheme (FRGS) Vot 5F073 supported under Ministry of Education Malaysia for the completion of the research. The Smart Solutions in Ubiquitous Computing Environments", Grant Agency of Excellence 2019, projects No. 2204, University of Hradec Kralove, Faculty of Informatics and Management is acknowledged. The work is partially supported by the SPEV project, University of Hradec Kralove, FIM, Czech Republic (ID: 2102-2019).

**Acknowledgments:** The authors wish to thank Universiti Teknologi Malaysia (UTM) under Research University Grant Vot-20H04, Malaysia Research University Network (MRUN) Vot 4L876 and the Fundamental Research Grant Scheme (FRGS) Vot 5F073 supported under Ministry of Education Malaysia for the completion of the research. The work is partially supported by the SPEV project, University of Hradec Kralove, FIM, Czech Republic (ID: 2102-2019). We are also grateful for the support of Ph.D. student Sebastien Mambou in consultations regarding application aspects. The Smart Solutions in Ubiquitous Computing Environments", Grant Agency of Excellence 2019, projects No. 2204, University of Hradec Kralove, Faculty of Informatics and Management is acknowledged. Likewise, the Authority of Ambrose Ali University, Ekpoma, under Tertiary Education Trust Fund (TETFUND), Nigeria, is also acknowledged for the opportunity giving to the Scholar to conduct his Research leading to Doctor of Philosophy (PhD) in Computer Science in UTM.

**Conflicts of Interest:** The authors declare no conflict of interest.

## Appendix A

**Algorithm A1.** Modified algorithm for region identification and grouping of participants based on clusters using K-means with node S

1. **Set** S: node for participant's smartphone
2. **Set** Lat: Latitude
3. **Set** Long: Longitude
4. **Set** T: Time
5. **Set** SA: Sub-arealist = $[SA_1, SA_2, SA_3, \ldots, SA_n]$
6. **Set** Dist: Distance
7. **K**: Clusters of nodes into sub-areas
8. **TWindow**: Time T, set for the location of nodes a threshold
9. **Start**
10. **Input** S: Output (Lat, Long, Time)
11. **Input** Sub-area list $[SA_1, SA_2, SA_3, \ldots, SA_n$, Lat, long, T]
12. **Output** S clusters in Sub-areas, $SA_n$
13. **While** S is ready **do**
14. **For each S for participant in Sub-Arealist do**
15. Set locationUpdateWindow
16. Set minT i.e., for location manager minimum power consumption with minT
Milliseconds between location update to reserve power
17. Set minDist: as location transmission in case device moves using minDistance
meters
18. TDifference = location.getT( )- currentbestlocation.getT( )
If TDifference > TWindow then participant (node) have moved and transmit
the new location into a Crowd Controller Station (CCS) based on timestamp
change
19. **If** (Lat, Long) in location context with Sub-arealist $SA_n$ are the same,
clusters set K using Dist between the nodes S
20. Group S into $SA_1, SA_2, SA_3, \ldots, SA_n$ clusters
21. Crowdcount = S + 1
22. **End If**
23. **End If**
24. **End For**
25. **End While**
26. **End**

## Appendix B

---

**Algorithm A2:** Enhanced approach for individual and crowd condition prediction proposed to extend BCF

---

1. **IAR$_{ehcaf}$ Module**
2. *Set* S: as node for a participant using a smartphone
3. *Set* CCS: crowd controller station: stakeholder as STHD
4. *Set* IAR: Individual activity recognition
5. *Set* SBTFD: Improved feature extraction method
6. *Set* Vsi and Dsi: Flow velocity and flow direction
7. *Set* PBE: Pairwise behavior estimation
8. *Set* CCP: crowd condition prediction = 0 for all nodes using S
9. *Set* CCP as threshold using equation (11)
10. *Input* IAR sensor signals dataset D1 from CCS
11. *Execute* IAR for S using improved SBTFD
12. *Execute* dimensionality reduction using reduced features based on FSM
13. **IBE$_{ehcaf}$ Module**
14. *Cluster* node S using set K based on Algorithm 1
15. *Compute* Vsi and Dsi for each S based on Section 3.2.3
16. *Execute* PBEA using lines 12 and 15 for each class based on Figure 4
17. **CCP Module**
18. *Compute* CDD using equations 9 and 10
19. **If** the threshold satisfies condition 1, **then**
20. *Terminate* the PBE testing
21. *Else*
22. *If* the threshold satisfies condition 2, **then**
23. *Terminate* the PBE testing
24. *Else*
25. *If* the threshold satisfies condition 3, **then**
26. *Evaluate* CDD inflow, outflow and crowd turbulence
27. *Else*
28. *If* the threshold satisfies condition 4, **then**
29. *Evaluate* line 26 and set CCP = 1
30. (Send context-aware alert to S and *STHD* for safety measure)
31. *Output* context-aware alert for CCP based on line 29 using EHCAF
32. *End if*
33. *Else*
34. *Execute line 14 to 31*
35. *End if*
36. *End if*
37. *End if*
38. *End*

---

## Appendix C

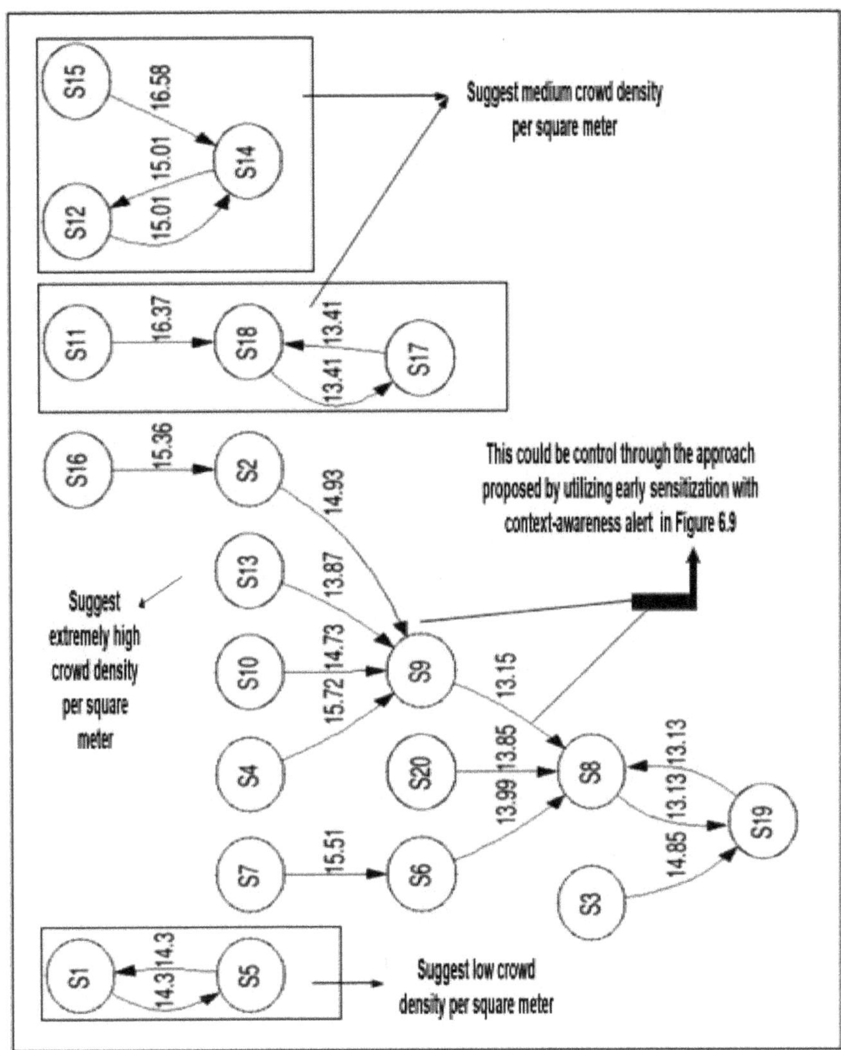


**Figure A1.** Patterns of participant behavior estimation using a disparity matrix for 20 nodes, S1–S20, for the recognition of abnormality of individual behavior per m$^2$.

## Appendix D

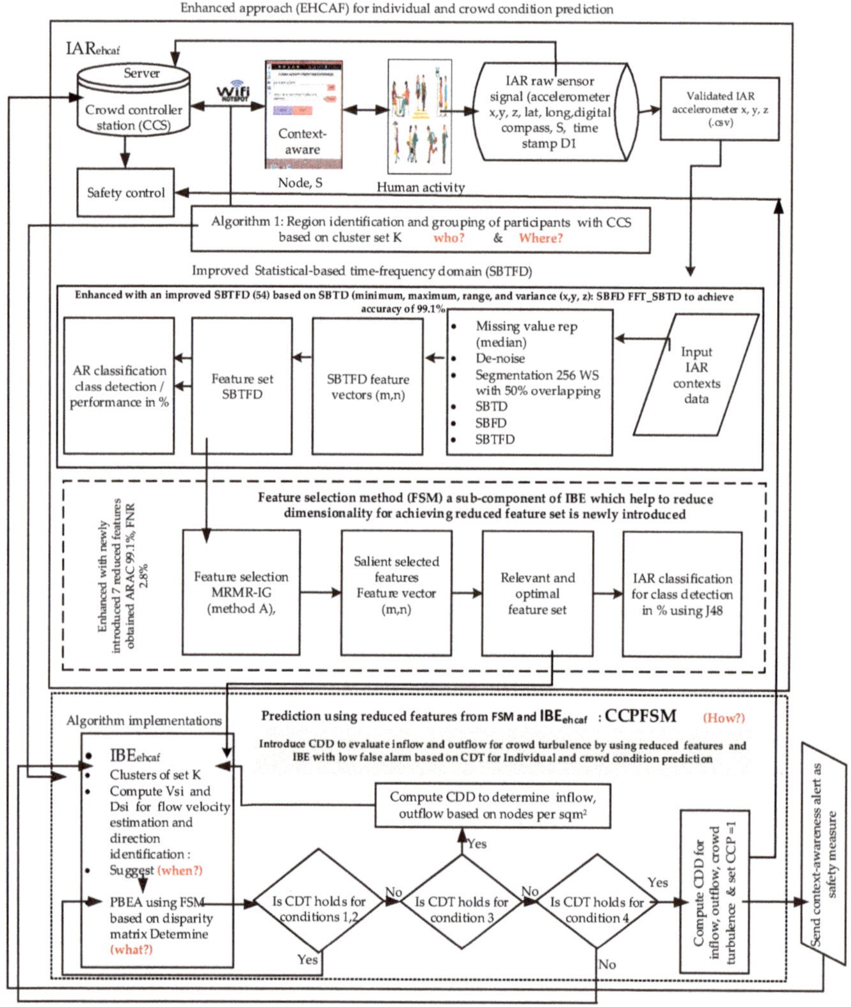



**Figure A2.** Patterns of participant behavior estimation using a disparity matrix for 20 nodes S1 to S20 for the recognition of abnormality of individual behavior per m$^2$.

## References


1. Duives, D.C.; Wang, G.; Kim, J. Forecasting pedestrian movements using recurrent neural networks: An application of crowd monitoring data. *Sensors* **2019**, *19*, 382. [CrossRef] [PubMed]
2. Li, F.; Al-Qaness, M.; Zhang, Y.; Zhao, B.; Luan, X. A robust and device-free system for the recognition and classification of elderly activities. *Sensors* **2016**, *16*, 2043. [CrossRef]
3. Zhang, D.; Peng, H.; Haibin, Y.; Lu, Y. Crowd abnormal behavior detection based on machine learning. *Inf. Technol. J.* **2013**, *12*, 1199–1205. [CrossRef]
4. Sadiq, F.I.; Selamat, A.; Ibrahim, R.; Ondrej, K. Improved feature extraction method with statistical based time frequency domain for classification oindividual activity recognition in a crowd scenario. Available online: www.news.unimas.my/conference/eventdetail/4309/-/- (accessed on 1 September 2018).

5.   Mshali, H.; Lemlouma, T.; Magoni, D. Adaptive monitoring system for e-health smart homes. *Pervasive Mob. Comput.* **2018**, *43*, 1–19. [CrossRef]

6.   Ramesh, M.V.; Shanmughan, A.; Prabha, R. Context aware ad hoc network for mitigation of crowd disasters. *Ad Hoc Netw.* **2014**, *18*, 55–70. [CrossRef]

7.   Franke, T.; Lukowicz, P.; Blanke, U. Smart crowds in smart cities: Real life, city scale deployments of a smartphone based participatory crowd management platform. *JISA* **2015**, *6*, 1–19. [CrossRef]

8.   Sadiq, F.I.; Selamat, A.; Ondrej, K.; Ibrahim, R. Impacts of feature selection on classification of individual activity recognitions for prediction of crowd disasters. *Int. J. Intell. Inf. Database Syst.*. in press.

9.   Yaseen, S.; Al-Habaibeh, A.; Su, D.; Otham, F. Real-time crowd density mapping using a novel sensory fusion model of infrared and visual systems. *Saf. Sci.* **2013**, *57*, 313–325. [CrossRef]

10.   Ngo, M.Q.; Haghighi, P.D.; Burstein, F. A crowd monitoring framework using emotion analysis of social media for emergency management in mass gatherings. Available online: https://arxiv.org/abs/1606.00751 (accessed on 8 May 2019).

11.   Lara, O.D.; Labrador, M. A survey on human activity recognition using wearable sensors. *IEEE Commun. Surv. Tutor.* **2013**, *15*, 1192–1209. [CrossRef]

12.   Pressac. Benefits of smart Sensor Technology. Available online: https://www.pressac.com/insights/benefits-of-smart-sensor-technology/ (accessed on 12 April 2019).

13.   Otebolaku, A.M.; Andrade, M.T. User context recognition using smartphone sensors and classification models. *J. Netw. Comput. Appl.* **2016**, *66*, 33–51. [CrossRef]

14.   Holgersson, P.; Åkerberg, F. Analysis of Activity Recognition and the Influence of Feature Extraction and Selection in an Android Based Device. Master's Theses, Lund University, Lund, Sweden, 2015.

15.   Khan, A.M.; Tufail, A.; Khattak, A.M.; Laine, T.H. Activity recognition on smartphones via sensor-fusion and KDA-based SVMs. *Int. J. Distrib. Sens. Netw.* **2014**, *10*, 503291. [CrossRef]

16.   Kwon, Y.; Kang, K.; Bae, C. Unsupervised learning for human activity recognition using smartphone sensors. *Expert Syst. Appl.* **2014**, *41*, 6067–6074. [CrossRef]

17.   Attal, F.; Mohammed, S.; Dedabrishvili, M.; Chamroukhi, F.; Oukhellou, L.; Amirat, Y. Physical human activity recognition using wearable sensors. *Sensors* **2015**, *15*, 31314–31338. [CrossRef] [PubMed]

18.   Figo, D.; Diniz, P.C.; Ferreira, D.R.; Cardoso, J.M.P. Preprocessing techniques for context recognition from accelerometer data. *Pers. Ubiquit. Comput.* **2010**, *14*, 645–662. [CrossRef]

19.   Reiss, A.; Hendeby, G.; Stricker, D. A Competitive approach for human activity recognition on smartphones. In Proceedings of the European Symposium on Artificial Neural Networks, Computational Intelligence and Machine Learning (ESANN 2013), Bruges, Belgium, 24–26 April 2013; pp. 1–7.

20.   Anguita, D.; Ghio, A.; Oneto, L.; Parra, X.; Reyes-Ortiz, J.L. A public domain dataset for human activity recognition using smartphones. In Proceedings of the European Symposium on Artificial Neural Networks, Computational Intelligence and Machine Learning, ESANN. 201, Bruges, Belgium, 24–26 April 2013.

21.   Kumari, S.; Mitra, S.K. Human action recognition using DFT. In Proceedings of the 2011 Third National Conference on Computer Vision, Pattern Recognition, Image Processing and Graphics, Hubli, India, 15–17 December 2011.

22.   Rahman, M. *Applications of Fourier Transforms to Generalized Functions*; WIT Press: Southampton, UK; Boston, MA, USA, 2011; pp. 1–168.

23.   Phan, T. Improving activity recognition via automatic decision tree pruning. In Proceedings of the International Joint Conference on Pervasive and Ubiquitous Computing: Adjunct Publication, Seattle, WA, USA, 13–17 September 2014; pp. 827–832.

24.   Cao, L.; Wang, Y.; Zhang, B.; Jin, Q.; Vasilakos, A.V. GCHAR: An efficient group-based context–aware human activity recognition on smartphone. *J. Parallel Distr. Comput.* **2017**, in press. [CrossRef]

25.   Zhenyu, H.; Lianwen, J. Activity Recognition from acceleration data Based on Discrete Consine Transform and SVM. In Proceedings of the 2009 IEEE International Conference on Systems, Man, and Cybernetics, San Antonio, TX, USA, 11–14 October 2009.

26.   Roggen, D.; Wirz, M.; Tröster, G.; Helbing, D. Recognition of crowd behavior from mobile sensors with pattern analysis and graph clustering methods. Available online: https://arxiv.org/abs/1109.1664 (accessed on 8 May 2019).

27.   Stisen, A.; Blunck, H.; Bhattacharya, S.; Prentow, T.S.; Kjærgaard, M.B.; Dey, A.; Sonne, T.; Jensen, M.M. Smart Devices are Different: Assessing and MitigatingMobile Sensing Heterogeneities for Activity

Recognition. In Proceedings of the 13th ACM Conference on Embedded Networked Sensor Systems, Seoul, Korea, 1–4 November 2015.

28. Chen, Y.; Shen, C. Performance analysis of smartphone-sensor behavior for human activity recognition. *IEEE Access* **2017**, *5*, 3095–3110. [CrossRef]

29. Sukor, A.A.; Zakaria, A.; Rahim, N.A. Activity recognition using accelerometer sensor and machine learning classifiers. In Proceedings of the 2018 IEEE 14th International Colloquium on Signal Processing & Its Applications (CSPA), Batu Feringghi, Malaysia, 9–10 March 2018.

30. Haritha, V. Physical Human Activity Recognition Using Machine Learning Algorithms. Available online: https://arrow.dit.ie/cgi/viewcontent.cgi?article=1117&context=scschcomdis (accessed on 8 May 2019).

31. Zainuddin, Z.; Lai, K.H.; Ong, P. An enhanced harmony search based algorithm for feature selection: Applications in epileptic seizure detection and prediction. *Comput. Electr. Eng.* **2016**, *2016*, 1–20. [CrossRef]

32. Chernbumroong, S.; Cang, S.; Yu, H. Maximum relevancy maximum complementary feature selection for multi-sensor activity recognition. *Expert Syst. Appl.* **2015**, *42*, 573–583. [CrossRef]

33. Ji, Y.; Yang, Y.; Xu, X.; Tao, H. One-shot learning based pattern transition map for action early recognition. *Signal Process.* **2018**, *143*, 364–370. [CrossRef]

34. Yasser, E.-M.; Hsieh, T.-Y.; Shivakumar, M.; Kim, D.; Honavar, V. Min-redundancy and max-relevance multi-view feature selection for predicting ovarian cancer survival using multi-omics data. *BMC Med. Genom.* **2018**, *11*, 71.

35. Chang, H.; Kang, Y.; Ahn, H.; Jang, C.; Choi, E. Context-aware Mobile platform for intellectual disaster alerts system. *Energy Procedia* **2012**, *16*, 1318–1323. [CrossRef]

36. Ravindran, R.; Suchdev, R.; Tanna, Y.; Swamy, S. Context aware and pattern oriented machine learning framework (CAPOMF) for Android. In Proceedings of the Advances in Engineering and Technology Research (ICAETR), Unnao, India, 1–2 August 2014.

37. Baldauf, M.; Dustdar, S.; Rosenberg, F. A survey on context-aware systems. *Int. J. Ad Hoc Ubiquit. Comput.* **2007**, *2*, 263–277. [CrossRef]

38. Otebolaku, A.; Lee, G.M. A framework for exploiting internet of things for context-aware trust-based personalized services. *Mob. Inf. Syst.* **2018**, *2018*. [CrossRef]

39. Bouguessa, A.; Mebarki, L.A.; Boudaa, B. Context-aware adaptation for sustaining disaster management. In Proceedings of the 12th International Symposium on Programming and Systems (ISPS), Algiers, Algeria, 28–30 April 2015.

40. Markets, A.M. *Context-Aware Computing Market Worth $120 Billion by 2018*. Available online: https://www.marketsandmarkets.com/PressReleases/context-aware-computing.asp (accessed on 8 May 2019).

41. Kayes, A.; Han, J.; Rahayu, W.; Islam, M.; Colman, A. A policy model and framework for context-aware access control to information resources. *Comput. J.* **2019**, *62*, 670–705. [CrossRef]

42. Suto, J.; Oniga, S.; Sitar, P.P. Feature analysis to human activity recognition. *Int. J. Comput. Commun. Contr.* **2017**, *12*, 116–130. [CrossRef]

43. Akhavian, R.; Behzadan, A.H. Smartphone-based construction workers' activity recognition and classification. *Automat. Constr.* **2016**, *71*, 198–209. [CrossRef]

44. Sadiq, F.I.; Selamat, A.; Ibrahim, R.; Selamat, M.H.; Krejcar, O. Stampede prediction based on individual activity recognition for context-aware framework using sensor-fusion in a crowd scenarios. *SoMeT* **2017**, *297*, 385–396.

45. Vermeesch, P. Multi-sample comparison of detrital age distributions. *Chem. Geol.* **2013**, *341*, 140–146. [CrossRef]

46. Celebi, M.E.; Kingravi, H.A.; Vela, P.A. A comparative study of efficient initialization methods for the k-means clustering algorithm. *Expert Syst. Appl.* **2013**, *40*, 200–210. [CrossRef]

47. Kruskal, J.B. Multidimensional scaling by optimizing goodness of fit to a nonmetric hypothesis. *Psychometrika* **1964**, *29*, 1–27. [CrossRef]

48. Fruin, J.J. The causes and prevention of crowd disasters. *Eng. Crowd Saf.* **1993**, 99–108.

49. Helbing, D.; Johansson, A.; Al-Abideen, H.Z. Crowd turbulence: The physics of crowd disasters. Available online: https://arxiv.org/abs/0708.3339 (accessed on 8 May 2019).

50. Rodrigues Leal Moitinho de Almeida, M. Human stampedes: A scoping review. Available online: http://digibuo.uniovi.es/dspace/bitstream/10651/39115/6/TFM_MariaRodriguesLMdeAlmeida.pdf (accessed on 8 May 2019).

51. Zheng, Y. Human activity recognition based on the hierarchical feature selection and classification framework. *J. Electr. Comput. Eng.* **2015**, *2015*, 34. [CrossRef]
52. Erdaş, Ç.B.; Atasoy, I.; Açıcı, K.; Oğul, H. Integrating features for accelerometer-based activity recognition. *Procedia Comput. Sci.* **2016**, *98*, 522–527. [CrossRef]

*Article*

# 3D CNN-Based Speech Emotion Recognition Using K-Means Clustering and Spectrograms

**Noushin Hajarolasvadi * and Hasan Demirel**

Department of Electrical and Electronics Engineering, Eastern Mediterranean University, 99628 Gazimagusa, North Cyprus, via Mersin 10, Turkey; hasan.demirel@emu.edu.tr
* Correspondence: noushin.hajarolasvadi@cc.emu.edu.tr

Received: 12 March 2019; Accepted: 4 May 2019; Published: 8 May 2019

**Abstract:** Detecting human intentions and emotions helps improve human–robot interactions. Emotion recognition has been a challenging research direction in the past decade. This paper proposes an emotion recognition system based on analysis of speech signals. Firstly, we split each speech signal into overlapping frames of the same length. Next, we extract an 88-dimensional vector of audio features including Mel Frequency Cepstral Coefficients (MFCC), pitch, and intensity for each of the respective frames. In parallel, the spectrogram of each frame is generated. In the final preprocessing step, by applying $k$-means clustering on the extracted features of all frames of each audio signal, we select $k$ most discriminant frames, namely keyframes, to summarize the speech signal. Then, the sequence of the corresponding spectrograms of keyframes is encapsulated in a 3D tensor. These tensors are used to train and test a 3D Convolutional Neural network using a 10-fold cross-validation approach. The proposed 3D CNN has two convolutional layers and one fully connected layer. Experiments are conducted on the Surrey Audio-Visual Expressed Emotion (SAVEE), Ryerson Multimedia Laboratory (RML), and eNTERFACE'05 databases. The results are superior to the state-of-the-art methods reported in the literature.

**Keywords:** speech emotion recognition; 3D convolutional neural networks; deep learning; k-means clustering; spectrograms

---

## 1. Introduction

Designing an accurate automatic emotion recognition (ER) system is crucial and beneficial to the development of many applications such as human–computer interactive (HCI) applications [1], computer-aided diagnosis systems, or deceit-analyzing systems. Three main models are in use for this purpose, namely acoustic, visual, and gestural. While a considerable amount of research and progress is dedicated to the visual model [2–5], speech as one of the most natural ways of communication among human beings is neglected unintentionally. Speech emotion recognition (SER) is useful for addressing HCI problems provided that it can overcome challenges such as understanding the true emotional state behind spoken words. In this context, SER can be used to improve human–machine interaction by interpreting human speech.

SER refers to the field of extracting semantics from speech signals. Applications such as pain and lie detection, computer-based tutorial systems, and movie or music recommendation systems that rely on the emotional state of the user can benefit from such an automatic system. In fact, the main goal of SER is to detect discriminative features of a speaker's voice in different emotional situations.

Generally, a SER system extracts features of voice signal to predict the associated emotion using a classifier. A SER system needs to be robust to speaking rate and speaking style of the speaker. It means particular features such as age, gender, and culture differences should not affect the performance of the SER system. As a result, appropriate feature selection is the most important step of designing the SER system. Acoustic, linguistic, and context information are three main categories of features used in the SER research [6]. In addition to those features, hand-engineered features including pitch, Zero-Crossing Rate (ZCR), and MFCC are widely used in many research works [6–9]. More recently, convolutional neural network (CNN) has been in use at a dramatically increasing rate to address the SER problem [2,10–13].

Since the results from deep learning methods are more promising [8,14,15], we used a 3D CNN model to predict the emotion embedded in a speech signal. One challenge in SER using multi-dimensional CNNs is the dimension of speech signal. Since the purpose of this study is to learn spectra-temporal features using a 3D CNN, one must transform the one-dimensional audio signal to an appropriate representation to be able to use it with 3D CNN. A spectrogram is a 2D visual representation of short-time Fourier transform (STFT) where the horizontal axis is the time, and the vertical axis is the frequency of signal [16]. In the proposed framework, audio data is converted into consecutive 2D spectrograms in time. The 3D CNN is especially selected because it captures not only the spectral information but also the temporal information.

To train our 3D CNN using spectrograms, firstly, we divide each audio signal to shorter overlapping frames of equal length. Next, we extract an 88-dimensional vector of commonly known audio features for each of the corresponding frames. This means, at the end of this step, each speech signal is represented by a matrix of size $n \times 88$ where $n$ is the total number of frames for one audio signal and 88 is the number of features extracted for each frame. In parallel, the spectrogram of each frame is generated by applying STFT. In the next step, we apply k-means clustering on the extracted features of all frames of each audio signal to select $k$ most discriminant frames, namely keyframes. This way, we summarize a speech signal with $k$ keyframes. Then, the corresponding spectrograms of the keyframes are encapsulated in a tensor of size $k \times P \times Q$ where $P$ and $Q$ are horizontal and vertical dimensions of the spectrograms. These tensors are used as the input samples to train and test a 3D CNN using 10-fold cross-validation approach. Each of 3D tensors is associated with the corresponding label of the original speech signal. The proposed 3D CNN model consists of two convolutional layers and a fully connected layer which extracts the discriminative spectra-temporal features of so-called tensors of spectrograms and outputs a class label for each speech signal. The experiments are performed on three different datasets, namely Ryerson Multimedia Laboratory (RML) [17] database, Surrey Audio-Visual Expressed Emotion (SAVEE) database [18] and eNTERFACE'05 Audio-Visual Emotion Database [19]. We achieved recognition rate of 81.05%, 77.00% and 72.33% for SAVEE, RML and eNTERFACE'05 databases, respectively. These results improved the state-of-the-art results in the literature up to 4%, 10% and 6% for these datasets, respectively. In addition, the 3D CNN is trained using all spectrograms of each audio file. As a second series of experiments, we used a pre-trained 2D CNN model, say VGG-16 [20] and performed transfer learning on the top layers. The results obtained from our proposed method is superior than the ones achieved from training VGG-16. This is mainly due to fewer parameters used in the freshly trained 3D CNN architecture. Also, VGG-16 is a 2D model and it cannot detect the temporal information of given spectrograms.

The main contributions of the current work are: (a) division of an audio signal to $n$ frames of equal length and selecting the $k$ most discriminant frames (keyframes) using k-means clustering algorithm where $k << n$; (b) representing each audio signal by a 3D tensor of size $k \times P \times Q$ where $k$ is the number of consecutive spectrograms corresponding to keyframes and $P$ and $Q$ are horizontal and vertical dimensions of each spectrogram; (c) Improving the ER rate for three benchmark datasets by learning spectra-temporal features of audio signal using a 3D CNN and 3D tensor inputs.

The main motivation of the proposed work is to employ 3D CNNs, which is capable of learning spectra-temporal information of audio signals. We proposed to use a subset of spectrogram frames which minimizes redundancy and maximizes the discrimination capability of the represented audio signal. The selection of such a subset provides a computationally cheaper tensor processing for comparable or improved performance.

The rest of the paper is organized as follows: In Section 2, we review the related works and describe steps of our proposed method. In Section 3, our experimental results are illustrated and compared with the state of the art in the literature. Finally, in Section 4 conclusion and future work is discussed.

## 2. Materials and Methods

Generally speaking, a SER system is composed of two parts: a preprocessing part that extracts suitable features and a classifier that employs those features to perform ER. This section overviews existing strategies in the SER research area [21,22].

### 2.1. Related Works

In a very recent work, ref. [23] proposed a robust technique of SER by embedding phoneme sequences and spectrograms. The authors represented each phoneme as an embedding numeric vector. They use two CNN models, a phoneme-based CNN model and a 2D CNN model for spectrograms. Both models have four parallel convolutional kernels. They used the Interactive Emotional Dyadic Motion Capture (IEMOCAP) database [24] and they achieved an overall accuracy of 73.9% on this corpus. Considering the high computational cost of training CNNs, the drawback of this method is employing two separate CNN models. Also, comparison with other benchmark databases is ignored.

In another recent work, Zhang et al. [25] achieved 70.4% accuracy on the same corpus, IEMOCAP. They proposed an attention-based fully convolutional neural network (FCN). FCNs can handle spectrograms with variable sizes. In fact, they turn AlexNet [26] into an FCN by removing its fully connected layers and then using it as an encoder. Later, they attach an attention layer which is followed by a SoftMax layer. They compared their results with a fine-tuned version of AlexNet and VGG-16 [20]. They reported 67.9% and 66.8% accuracy on IEMOCAP database. Also, they reported recognition rate of 66.5% and 65.3% by direct training (without fine tuning) of these two deep networks. The advantage of this work is that the preprocessing step is limited to the generation of so-called spectrograms.

Avots et al. [9] conducted a cross-corpus evaluation. They analyzed a model on the audio-visual information of SAVEE, RML and eNTERFACE'05 databases and tested the same model on AFEW database to merely show how challenging the task of recognizing emotional states in real world environment might be. They represented the emotional speech in SAVEE, RML and eNTERFACE'05 databases by a $1 \times 650$, $1 \times 1725$ and $1 \times 1570$ feature vector, respectively. Mainly, they used spectral features such as energy entropy, ZCR, and harmonic product spectrum to represent each audio signal. Then, they applied SVM classifier and achieved 77.4%, 69.3% and 50.2% for SAVEE, RML and eNTERFACE'05 databases and only 27.1% for AFEW database. One disadvantage of this work is the different feature vector size that is used for each dataset which ignores the generalization aspect of machine learning methods and makes it highly susceptible to overfitting on a specific dataset.

Torfi et al. [8] proposed a 3D CNN for cross audio-visual matching recognition. Their audio-visual recognition system couples two non-identical 3D CNN architecture. This can map a pair of speech and video input into a new representation space for evaluation of correspondence between them. The input that they used were spectrograms, as well as the first and second order derivatives of the MFEC features. They applied feature-level fusion of audio and video features and reported the area under the curve 95.4% for Lip Reading in the Wild dataset.

Badshah et al. [22] used spectrograms of a speech signal as the input for a 2D CNN. They extracted spectrograms of each speech signal and then split the spectrogram into several smaller spectrograms. These smaller spectrograms are later resized and used as the input to a 2D CNN architecture. They reported using rectangular shaped kernels for convolution layers help to capture local features effectively. They trained and evaluated their model on Berlin Emotional Database (EmoDB) [27] and obtained a weighted (overall) accuracy of 72.21%. Also, in [14], they reported that a freshly trained CNN performs better than transfer learning on AlexNet [26] for SER purpose.

Ref. [28] evaluated two types of neural networks: CNNs and long short-term memory networks. They used IEMOCAP corpus for training and evaluation. In the preprocessing step, they split each sentence longer than 3 s to shorter sub-sentences. The emotional label of the original sentence is assigned to sub-sentences. Then they calculate a spectrogram for each sub-sentence. They studied the effect of 10 Hz and 20 Hz grid resolution and they report using lower resolution yields lower accuracy. They obtained weighted accuracy of 68.8%. They also, used harmonic modeling to remove noise from spectrograms. We believe k-means clustering will select the frames which are less redundant and therefore the corresponding spectrogram of the selected frames is more informative.

Noroozi et al. [29] proposed an audio-visual ER system for video clips. They extracted 88 features including MFCC, pitch, intensity, mean, variance, etc. from the whole speech signal. No framing is performed. Then, they applied SVM and Random Forest on this feature space. They reported the weighted accuracy of 56.07% and 65.28% and 47.11% for SAVEE, RML and eNTERFACE'05 datasets using Random Forest. Results obtained by SVM were lower than the Random Forest. In another work from same author [6], they used random forests and decision trees to classify speech signals using a vector of size 14 para-linguistic features. They obtained an overall accuracy of 66.28% on SAVEE dataset.

Schluter and Grill [13] applied pitch-shifting and time-stretching as two significant methods for data augmentation of spectrograms. They used the augmented data as input to 2D CNN. One disadvantage of this work is that due to a huge number of spectrograms, they used a fixed number of weight updates which means the convergence of CNN optimizer is not guaranteed. Other researchers such as Palaz et al. [12] split a raw input signal to a sequence of frames, and report a class-base score for each frame by passing through several convolution filter stages and a multi-layer perceptron classifier.

CNN is used to learn affect-salient features for SER in the precious work of [7]. In the first step of training, the unlabeled samples are used to learn Local Invariant Features (LIF) using a sparse auto-encoder. In the second step, LIF is used as the input to a feature extractor. The weighted accuracy on SAVEE, EmoDB was 71.8% and 57.2%.

Abdel-Hamid et al. [15] proposed a limited-weight-sharing scheme that models the speech features for speech recognition systems while [11] proposed a new method for modeling speech signals using Restricted Boltzmann Machine.

*2.2. Proposed Method*

2.2.1. Preprocessing

In this study, RML, SAVEE and eNTERFACE'05 datasets are used. The preprocessing pipeline is shown in Figure 1. First, the speech signals are extracted from video clips using the FFmpeg framework. Then, each speech signal is divided to shorter overlapping frames of equal length. Each frame has 50% overlap with the previous one. This step results to division of each speech signal to $n$ frames. Depending on the length of speech signal, the length of frames differs from one audio signal to another, but all frames of one audio signal has the same length. Then, for each frame 88 commonly known audio features such as MFCC, pitch, variance, intensity, and filter-bank energies are extracted. We adopted the set of extracted features from [29]. The complete list of extracted features is shown in Table 1.

In parallel, the spectrogram of each frame is generated. A Spectrogram is simply a signal strength versus time at different frequencies and is generated by applying STFT. A sequence of overlapping Hamming windows is applied to each frame with window size of 20 ms [30], a window shift of 10 ms and hope size of 256. At the end of this step, each speech signal is represented by a matrix of size $n \times 88$ and $n$ spectrograms as shown in Figure 1. $n$ is the number of frames and matching to each frame there exist a spectrogram, i.e., each audio frame has one feature vector and a corresponding spectrogram.

In the next step, k-means clustering algorithm is applied on all extracted feature vectors of one speech signal to select $k$ most discriminant frames known as keyframes. As we mentioned before, corresponding to each of these keyframes, there exist a spectrogram. The sequence of $k$ successive spectrograms of the keyframes for one speech signal forms a 3D tensor representing that speech signal. Such tensors are used as the input samples for training our 3D CNN architecture. Label of the original speech signal is assigned to the generated 3D tensor. To find the best representative $k$, we started with $k$ is equal to 9 and we increased it in a heuristic fashion to 18 and 27. The best $k$ which maximized the accuracy over the validation set and during training is equal to 9.

**Table 1.** List of extracted features for each audio frame.

| Feature | # | Feature | # | Feature | # |
|---|---|---|---|---|---|
| Intensity | 1 | Pitch | 1 | Median | 1 |
| Standard Deviation | 1 | Mean | 1 | Harmonic mean | 1 |
| Minimum amplitude | 1 | maximum amplitude | 1 | Percentile | 1 |
| Zero-Crossing Rate (ZCR) | 1 | ΔMFCCs | 13 | MFCCs | 13 |
| ZCR Density | 1 | Formants | 20 | Autocorrelation | 1 |
| Filter-Bank Energies | 26 | Formant bandwidth | 4 | | |
| | | | | Total | 88 |

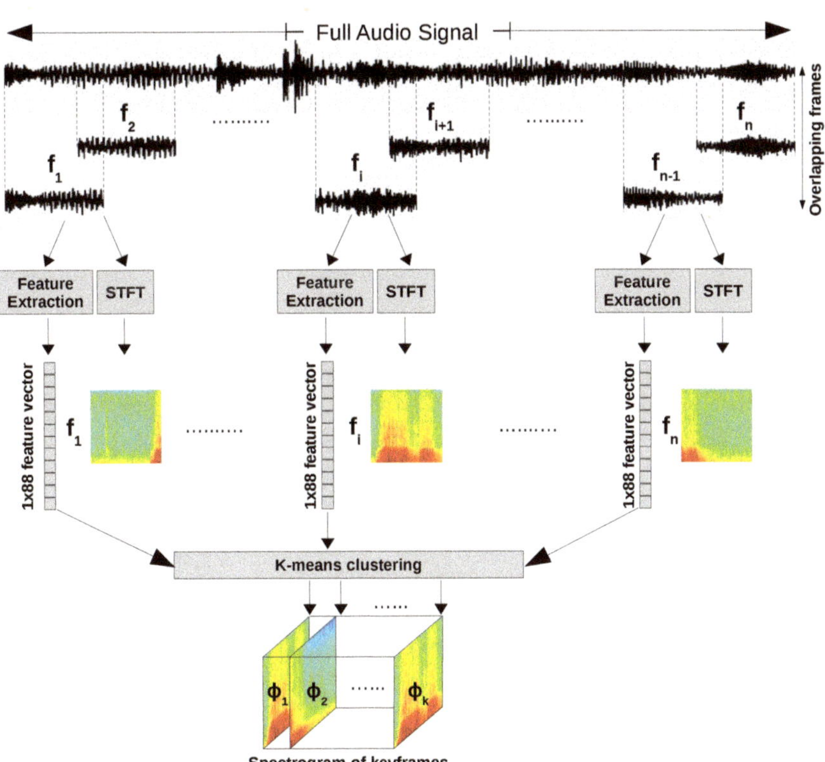


**Figure 1.** The proposed framework for preprocessing the data.

Training CNNs and especially 3D CNNs is an exhaustive and time-consuming process. As a result, summarizing the input samples (a speech signal represented by a selected sequence of spectrograms) without degrading the performance becomes highly important. For example, in [13], huge number of spectrograms is produced using hop size equal to 1. Due to high redundancy of overlapping audio frames and memory limitation, training of the CNN is performed for a fixed number of 40,000 weight updates instead of training over a full dataset. This means that not only the optimizer might not converge but also, not all the spectrograms of one audio signal is observed during training. In addition, a 3D CNN can be trained as deep as possible subject to the machine memory limit and computation affordability [31]. Thus, it is desired to handle memory limitation and reducing the computational time by summarizing input samples while preserving the performance.

In our methodology, k-means clustering algorithm addresses these problems. Because it detects the redundancy by clustering the feature vectors representing the frames of one audio signal and maximizing the distinctions between those frames. Figure 2 shows the generated clusters and their corresponding centroids. To visualize the discrimination of clusters, we applied *t*-test score on the $1 \times 88$ feature vectors of selected frames and non-selected frames of a single audio file to find the two best representative features. The *t*-test examines the differences of two populations using the mean and standard deviation of each population. The first formant and the MFCC provided the maximum difference. The k-means clustering is visualized using the selected features by *t*-test. In the following context, first we explain feature extraction and spectrogram generation in more details. Then, the proposed 3D CNN for SER is described.

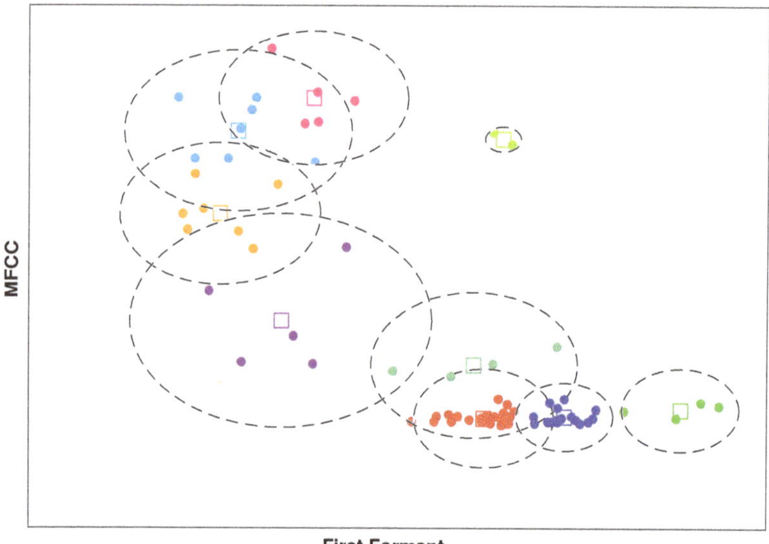


**Figure 2.** k-Means Clustering visualization for one audio sample in Angry category.

1.  Extracted Features:

Emotions can be represented using different features of speech. For example, a speaker who is angry has a faster speech rate as well as higher energy and pitch frequency. Some of the most effective features of speech for ER are duration, intonation, pitch and intensity, filter-bank energies, MFCCs, $\Delta$MFCCs, and ZCR. In this paper, we extracted 88 features proposed by [29]. The complete list of features is shown in Table 1 and for a speech signal $s$ with length $N$, they are explained in detail in Appendix A.

2.  Spectrograms:

As we mentioned before, one challenge in SER using CNNs is the dimension of speech signal. Since the purpose of this study is to learn spectra-temporal features using a 3D CNN, one must transform the one-dimensional audio signal to an appropriate representation for CNNs. One such representation is spectrogram which is the visual representation of signal strength over time at different frequencies [22]. Spectrogram is generated by applying STFT. STFT is a Fourier-based transform which determines the sinusoidal frequency and phase of local portions of a signal as it changes over time. In practice, to compute STFT, first a long time signal must be divided to shorter frames or segments of equal length. Then, by applying Fourier transform on each shorter frame, Fourier spectrum of that frame reveals. Visualizing the changing spectra as a function of time results in spectrogram [16].

In other word, the spectrogram is a visual representation of STFT where the horizontal axis represents the time and the vertical axis represents the frequency of signal in that short frame. In a spectrogram, at a particular time point and a particular frequency, dark colors illustrate the frequency in a low magnitude, whereas light colors show the frequency in higher magnitudes. Spectrograms are perfectly suitable for variety of speech analysis including SER [16]. In this work, we aim to represent each speech signal as a selected sequence of spectrograms generated by applying STFT on overlapping frames.

3.    k-means clustering:

It is an iterative, data-partitioning algorithm that assigns each sample point to exactly one of the k clusters. First, k observations are selected randomly to be the centroids of clusters. Then the distance between each sample point and the cluster-centroids are calculated. The sample point is assigned to the cluster with the closest centroid. When all sample points are assigned to exactly one of the clusters, the average of the sample points in each cluster is computed to obtain k new centroid locations. The distance calculation step and modifying the centroid location is then repeated until clusters stabilize or a maximum number of iterations is reached [32,33].

### 2.2.2. 3D CNN Architecture

The proposed architecture is a 3D CNN trained using 3D tensors. Each of these tensors contain a sequence of spectrograms for one audio signal. The proposed 3D model consists of two convolutional layers, one fully connected layer, a dropout, and a SoftMax layer. In Table 2 the spatial size of the 3D kernels is reported as $T \times H \times W$ where $T$ is the kernel size in temporal dimension, and $H$ and $W$ are the kernel sizes in height and width dimensions, respectively. By applying a 3D kernel, spectra-temporal features are extracted using a 3D convolutional operation. The complete block diagram of our proposed architecture is shown in Figure 3. We did not use any zero padding because it adds extra zero-energy coefficients which is not meaningful in local feature extraction.

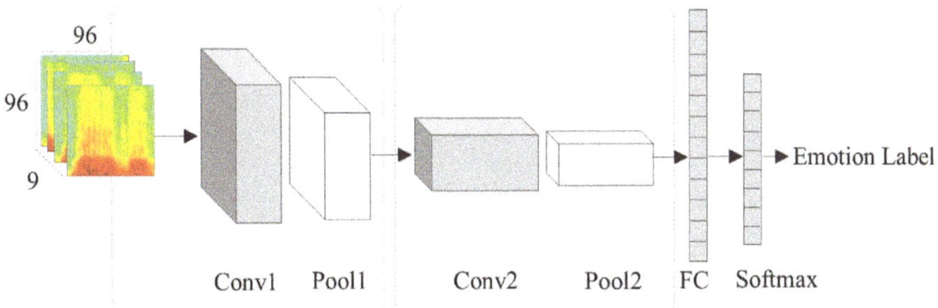


**Figure 3.** Block diagram of the proposed architecture for SER.

As we mentioned before, the best $k$ obtained equal to 9. As a result, each input sample of our proposed network is 9 consecutive spectrograms representing one emotional speech signal. All the spectrograms obtained from the pipeline explained in Section 2.2.1 are resized to $96 \times 96$ images. The first convolution layer, Conv1 has 128 kernels of size $3 \times 5 \times 1$ which are applied at strides of 1 pixel. The 3D convolutional layers extract the correlation between high-level temporal features and the spatial features of spectrograms. Conv1 uses a Parametric Rectified Linear Unit (PReLU). Following, a 3D max pooling layer with a kernel size $2 \times 2 \times 2$ (Pool1) and stride $1 \times 2 \times 1$ is used. PReLU is an activation function that is used instead of regular sigmoid ones with the aim of improving efficiency of the training process. Layer Conv2 has 256 kernels of size $3 \times 7 \times 1$ again with a moving stride of 1. Conv2 also uses PReLU as activation function. Pool2 is a 3D max pooling layer with the same kernel size and stride as Pool1. Pool2 is followed by a dropout layer with a dropout rate of 75% to avoid overfitting. Then, one fully connected (FC) layer with 64 units and a classification layer with 6 output class is used. Also, batch normalization [34] has been used to improve the training convergence.

In the proposed 3D model, we followed best experimental observations reported in [22,31,35]. In [14], it is reported that using rectangular kernels with large heights captures the local features effectively. As a result, we used a rectangular kernel of size $3 \times 5 \times 1$ and $3 \times 7 \times 1$ in the convolution layers. Also, [35] reported that using shallow temporal and moderately deep spectral kernels are optimal for the SER purpose. Thus, we employed 128 and 256 filters for convolutional layers which resulted in the best performance on the validation set. Using more than 256 filters did not help to improve the performance on the validation set. For initialization of weights and bias parameters, two methods including variance scaling [8] and random uniform distributions are tested. Initialization of both parameters with random uniform distribution resulted in a better performance on the validation set. For regularization, we used $l2$ weight regularization with setting the regularization factor to $5 \times 10^{-4}$.

**Table 2.** The resolution of the proposed 3D CNN.

| Layer | Input-Size | Output-Size | Kernel | Stride |
| --- | --- | --- | --- | --- |
| Conv1 | $9 \times 96 \times 96 \times 3$ | $7 \times 92 \times 96 \times 128$ | $3 \times 5 \times 1$ | 1 |
| Pool1 | $7 \times 92 \times 96 \times 128$ | $6 \times 46 \times 95 \times 128$ | $2 \times 2 \times 2$ | $1 \times 2 \times 1$ |
| Conv2 | $6 \times 46 \times 95 \times 128$ | $4 \times 40 \times 95 \times 256$ | $3 \times 7 \times 1$ | 1 |
| Pool2 | $4 \times 40 \times 95 \times 256$ | $3 \times 20 \times 94 \times 256$ | $2 \times 2 \times 2$ | $1 \times 2 \times 1$ |
| dropout | $3 \times 20 \times 94 \times 256$ | $3 \times 20 \times 94 \times 256$ | - | - |
| FC | 1,443,840 | 64 | - | |
| Dense | 64 | 6 | - | |
| Dense | 6 | - | - | |

## 3. Results and Discussion

Taking into account the acquisition source of the data, three general groups of emotional databases exist: spontaneous emotions, acted emotions based on invocation and simulated emotions. Sample databases recorded in natural situations such as TV shows or movies are categorized under the first group. Usually, such databases suffer from low quality due to different sources of interference. For databases under second group, an emotional state is induced using various methods such as watching emotional video clips or reading emotional context. Although psychologists prefer this type of databases, the resulted reaction to the same stimulant may differ. Also, ethically provoking strong emotions might be harmful for the subject. eNTERFACE'05 and RML are examples of this group. The last group of databases are simulated emotions with high quality recordings and still emotional state. SAVEE database is a good example of this group.

### 3.1. Dataset

Three benchmark datasets were used to conduct the experiments, namely RML, SAVEE and eNTERFACE'05. All three datasets support audio-visual modals. Several reasons have been considered while choosing the datasets. We selected databases in a way covering a variation of size to show the flexibility of our model. Firstly, all three datasets are represented for same emotional states which makes them highly comparable. It is known that distinction between two emotion categories (for example disgust and happy) with large inter-class differences is easier than two emotions with small inter-class discrepancy. In addition, having the same number of emotional states prevents misinterpretation of the experimental results. Because as the number of emotional states increase the classification task becomes more challenging.

Second, since all three datasets recorded for both the audio and the visual modals, the quality of the recorded audios is almost the same (16-bit single channel format). For example, comparing databases recorded with high acoustic quality and for the specific purpose of SER (EmoDB) with databases recorded in real environments is not preferable. Extraction of speech signals from videos for all three datasets is performed using the FFmpeg framework. Third, SAVEE, RML and eNTERFACE'05 can be categorized as small-size, mid-size, and large-size databases. Thus, the proposed model is evaluated to have a stable performance in terms of number of input samples.

The data processing pipeline explained in Section 2.2.1 is applied on each audio sample. To avoid overfitting, in all experiments, we divided the data such that 90% is used for training and 10% for test. We performed 10-fold cross-validation on the train part which means 90% of the train data is used for training and 10% for validation. Finally, the cross validated model is evaluated on the test part. The experiments are all performed for speaker-independent scenarios.

### 3.1.1. SAVEE

The SAVEE database has 4 male subjects who acted emotional videos for six basic emotions namely anger, disgust, fear, happiness, sadness, and surprise. A neutral category is recorded as well but since the other two datasets does not include neutral, we discard it. This dataset consists of 60 videos per category. 360 emotional audio samples extracted from the videos of this dataset.

### 3.1.2. RML

The RML database represented by Ryerson Multimedia Laboratory [17] includes 120 videos in each of six basic categories mentioned above from 8 subjects spoke various languages such as English, Mandarin, and Persian. A dataset of 720 emotional audio samples is obtained from this database.

### 3.1.3. eNTERFACE'05

The third dataset is eNTERFACE'05 [19] recorded from 42 subjects. All the participants spoke English and 81% of them are female. Each subject was asked to perform all six basic emotional states. Emotional states are exactly the same as SAVEE and RML. 210 audio samples per category is extracted from this dataset.

### 3.2. Experiments

To assess the proposed method, four experiments are conducted on each dataset. In the first experiment, we trained the proposed 3D CNN model using the spectrograms of selected keyframes by applying 10-fold cross-validation method. In the second experiment 3D CNN model is trained using spectrograms of all frames. In the third experiment, by means of transfer learning, we trained VGG-16 [20] using the spectrograms of keyframes. Finally, in the last experiment we trained VGG-16 using all spectrograms generated for each audio signal. Comparing the results obtained from the second and third experiment shows that k-means clustering discarded the audio frames which convey insignificant or redundant information. This can be interpreted from the results given in Tables 3–5 which does not differ notably. It is important to note that the overall accuracy results obtained from these four experiments are shown by Proposed 3D CNN$^{(1)}$, Proposed 3D CNN$^{(2)}$, VGG-16$^{(1)}$ and VGG-16$^{(2)}$ in those tables.

**Table 3.** Comparison of recognition rates among different methods for SAVEE dataset.

| Method | Audio Representation | Accuracy |
|---|---|---|
| SVM [9] | feature vector [†] | 77.4% |
| SVM [29] | feature vector [†] | 48.81% |
| Random Forest [29] | feature vector [†] | 56.07% |
| 2D CNN [7] | Spectrograms | 73.6% |
| VGG-16$^{(1)}$ | all spectrograms [*] | 49.20% |
| VGG-16$^{(2)}$ | $k$ selected spectrograms [◦] | 45.11% |
| Proposed 3D CNN$^{(1)}$ | all spectrograms [*] | 80.41% |
| Proposed 3D CNN$^{(2)}$ | $k$ selected spectrograms [◦] | 81.05% |

[†]: A feature vector of commonly known audio features like Table 1. [*]: All generated frames/spectrograms of one audio is used. [◦]: Only $k$ (9) frames/spectrograms of one audio is used.

**Table 4.** Comparison of recognition rates among different methods for RML dataset.

| Method | Audio Representation | Accuracy |
|---|---|---|
| SVM [9] | feature vector [†] | 69.30% |
| SVM [29] | feature vector [†] | 43.47% |
| Random Forest [29] | feature vector [†] | 65.28% |
| VGG-16$^{(1)}$ | all spectrograms [*] | 43.58% |
| VGG-16$^{(2)}$ | $k$ selected spectrograms [◦] | 41.17% |
| Proposed 3D CNN$^{(1)}$ | all spectrograms [*] | 71.44% |
| Proposed 3D CNN$^{(2)}$ | $k$ selected spectrograms [◦] | 77.00% |

[†]: A feature vector of commonly known audio features like Table 1. [*]: All generated frames/spectrograms of one audio is used. [◦]: Only $k$ (9) frames/spectrograms of one audio is used.

**Table 5.** Comparison of recognition rates among different methods for eNTERFACE05 dataset.

| Method | Audio Representation | Accuracy |
|---|---|---|
| SVM [9] | feature vector [†] | 50.2% |
| SVM [29] | feature vector [†] | 41.32% |
| Random Forest [29] | feature vector [†] | 47.11% |
| HMM [36] | feature vector [†] | 52.19% |
| VGG-16$^{(1)}$ | all spectrograms [*] | 33.33% |
| VGG-16$^{(2)}$ | $k$ selected spectrograms [◦] | 39.23% |
| Proposed 3D CNN$^{(1)}$ | all spectrograms [*] | 69.50% |
| Proposed 3D CNN$^{(2)}$ | $k$ selected spectrograms [◦] | 72.33% |

[†]: A feature vector of commonly known audio features like Table 1. [*]: All generated frames/spectrograms of one audio is used. [◦]: Only $k$ (9) frames/spectrograms of one audio is used.

### 3.2.1. Training the Proposed 3D CNN

The CNN architecture illustrated in Figure 3 was trained on a sequence of 9 consecutive spectrograms paired with the emotional label of the original speech sample. We train the network for 400 epochs with assuring that each input sample consists of a sequence of 9 successive spectrograms. Also, as a second experiment, the proposed 3D CNN was trained using all spectrograms of each audio signal.

Updates are performed using Adam optimizer [37], categorical cross-entropy error, mini-batches of size 32 [13] and a triangular cyclical learning rate policy by setting the initial learning rate to $1 \times 10^{-4}$, maximum learning rate to $6 \times 10^{-4}$, cycle length to 100 and step size to 50. Cycle length is the number of iterations until the learning rate returns to the initial value [38]. Step size is set to half of the cycle length. Figure 4b shows the learning rate for 400 iterations on RML dataset. As we mentioned before, to fight overfitting, we used $l2$ weight regularization with factor $5 \times 10^{-4}$. In all experiments, 90% of the data is used for training and the rest for test. This means, the model learned spectra-temporal features by applying 10-fold cross-validation on the training part of the data. Then, the trained model is evaluated using the test data.

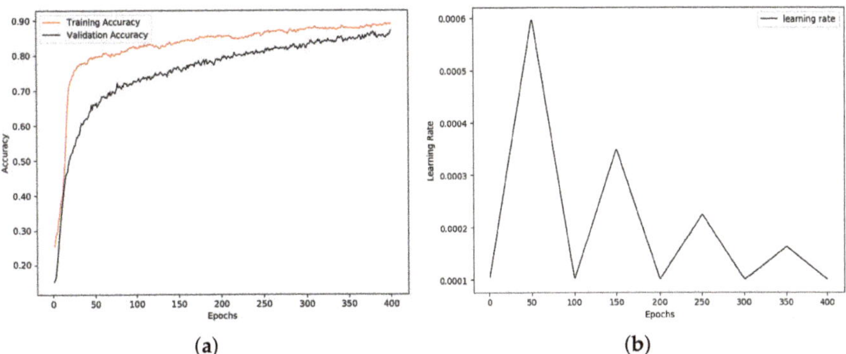

**Figure 4.** Results on RML database. (**a**) Training versus validation, accuracy improvement; (**b**) Cyclical learning rate decay.

The average accuracy on test set of SAVEE, RML and eNTERFACE'05 databases is illustrated as a confusion matrix in Tables 6–8, respectively. Clearly, the proposed method achieved superior results than the state-of-the-arts in the literature. Since the complexity of CNNs are extremely large, using discriminant input samples is of high importance especially when it comes to real-time applications. To the best of our knowledge, this is the first paper representing a whole audio signal by means of $k$ most discriminant spectrograms. This means, speech signal can be represented with fewer frames, yet preserving the accuracy. Figure 4a shows the training and validation accuracy improvement for RML dataset over 400 iterations. Also, Figure 4b shows the cyclical learning rate decay over same number of iterations and same dataset.

**Table 6.** Confusion matrix for SAVEE.

|       | Ang  | Dis  | Fea  | Hap  | Sad  | Sur  |
|-------|------|------|------|------|------|------|
| Ang   | 0.89 | 0.02 | 0    | 0    | 0    | 0.08 |
| Dis   | 0    | 0.9  | 0.1  | 0    | 0    | 0    |
| Fea   | 0    | 0    | 0.75 | 0.08 | 0.17 | 0    |
| Hap   | 0    | 0.08 | 0    | 0.75 | 0.08 | 0.08 |
| Sad   | 0.08 | 0.1  | 0    | 0    | 0.81 | 0    |
| Sur   | 0.08 | 0.08 | 0.08 | 0    | 0    | 0.76 |
|       |      |      |      |      | **Average RR%** | 81.05 |

**Table 7.** Confusion matrix for RML.

|       | Ang  | Dis  | Fea  | Hap  | Sad  | Sur  |
|-------|------|------|------|------|------|------|
| Ang   | 0.92 | 0.08 | 0    | 0    | 0    | 0    |
| Dis   | 0    | 0.67 | 0.25 | 0.08 | 0    | 0    |
| Fea   | 0    | 0.17 | 0.7  | 0.04 | 0    | 0.08 |
| Hap   | 0.08 | 0    | 0.08 | 0.75 | 0.08 | 0    |
| Sad   | 0    | 0.08 | 0.08 | 0    | 0.83 | 0    |
| Sur   | 0.08 | 0.17 | 0    | 0    | 0    | 0.75 |
|       |      |      |      |      | **Average RR%** | 77.00 |

**Table 8.** Confusion matrix for eNTERFACE′05.

|     | Ang | Dis | Fea | Hap | Sad | Sur |
| --- | --- | --- | --- | --- | --- | --- |
| Ang | 0.92 | 0.08 | 0 | 0 | 0 | 0 |
| Dis | 0 | 0.67 | 0.25 | 0.08 | 0 | 0 |
| Fea | 0 | 0.17 | 0.42 | 0.08 | 0.25 | 0.08 |
| Hap | 0.08 | 0 | 0.08 | 0.75 | 0.08 | 0 |
| Sad | 0 | 0.08 | 0.08 | 0 | 0.83 | 0 |
| Sur | 0.08 | 0.17 | 0 | 0 | 0 | 0.75 |
| | | | | | **Average RR%** | 72.33 |

### 3.2.2. Transfer Learning of VGG-16

In the next two experiments, we selected one of the well-known 2D CNNs, VGG-16 [20]. We applied transfer learning on the top layers to make it more suitable for the SER purpose. We trained the network for 400 weight updates. The initial learning rate is set to $1 \times 10^{-4}$.

In the first scenario, only the selected spectrograms of audio signals are given to VGG-16. In the second scenario, without applying k-means clustering algorithm, all generated spectrograms for each audio signal are used. In both cases, majority voting is used to make a final decision for each audio signal and assign a label to it. This means majority of labels predicted for the spectrograms of one audio is considered to be the final label for that audio signal. Both experiments under-performed the proposed 3D CNN.

This is mainly because VGG-16 is pre-trained on ImageNet dataset [39] for object detection and image classification purposes. Also, it has more complexity to adjust its weight. As a result, transfer learning was not helpful. Same conclusion has been reported by [14] and [25] for applying transfer learning on AlexNet using spectrograms. Fewer parameters in the freshly trained 3D CNN is the main reason for achieving the higher performance. The overall accuracy obtained by these experiments is compared with the state of the art in the literature in Tables 3–5 for SAVEE, RML and eNTERFACE′05 datasets, respectively.

## 4. Conclusions

In this paper, we studied the performance of 3D Convolutional Neural Networks using spectrograms. Instead of using the whole set of spectrograms corresponding to the audio frames, we selected $k$ best frames for representing the whole speech signal. We compared the results of the proposed 3D CNN with the results obtained from 2D CNNs. It shows that the proposed method performs better than the pre-trained 2D networks. Future works may include comparing with pre-trained 3D-architecture such as Res-3D and C3D or applying different types of data augmentation to improve the results by fighting the overfitting. Fusion with visual data is another direction to study the multimodal performance of 3D architectures as well as cross-correlation between different modalities.

**Author Contributions:** writing–original draft preparation, N.H.; writing–review and editing, H.D.

**Funding:** This research was funded by BAP-C project of Eastern Mediterranean University under grant number BAP-C-02-18-0001.

**Conflicts of Interest:** The authors declare no conflict of interest.

## Abbreviations

The following abbreviations are used in this manuscript:

| | |
|---|---|
| HCI | Human–Computer Interaction |
| SER | Speech Emotion Recognition |
| ZCR | Zero-Crossing Rate |
| MFCC | Mel Frequency Cepstral Coefficient |
| CNN | Convolutional Neural Network |
| STFT | Short-Term Fourier Transform |
| IEMOCAP | Interactive Emotional Dyadic Motion Capture |
| FCN | Fully Convolutional Neural Network |
| SVM | Support Vector Machine |
| RML | Ryerson Multimedia Laboratory database |
| LIF | Local Invariant Features |
| SAVEE | Surrey Audio-Visual Expressed Emotion database |
| HMM | Hidden Markov Model |

## Appendix A

In this section, we explain the extracted features given in Table 1 with more detail.

*Appendix A.1*

1.  The loudness of speech signal or the syllable peak is perceived as intensity. In another word, intensity is the power conveyed by speech signal per unit area in a direction perpendicular to that area. It can be expressed as follows:

$$I(dB) = 10 \log_{10}[\frac{I}{I_0}] \tag{A1}$$

where $I$ is the intensity and $I_0$ is the standard threshold of hearing intensity at 1000 Hz for the human ear which represented in terms of sound intensity by a value equal to $10^{(-12)}$ watts/m$^2$ [40].

2.  Pitch is known as the fundamental frequency of the speech signal. It can be measured either using statistical methods or in the time-frequency domains. It can be calculated as follows:

$$\rho_0(s) = \mathcal{F}\{log|\mathcal{F}(s.w_n^H\|s\|)|\} \tag{A2}$$

where $w_n^H$ is the Hamming window and it is defined as follows:

$$w_n^H = 0.54 - 0.46\cos(\frac{2\pi n}{L}), \qquad 1 \leq n \leq N-1 \tag{A3}$$

$L$ is the order of the filter and it is equal to filter length $-1$ [29].

3.  Mean of each frame is calculated as:

$$\mu = \frac{1}{N}\sum_{i=1}^{N} s_i \tag{A4}$$

4.  Standard deviation is extracted by calculating the following formula:

$$std = \sqrt{\frac{1}{N-1}\sum_{i=1}^{N}(s_i - \mu)^2} \tag{A5}$$

where $\mu$ is the mean of audio frame and $s_i$ shows the value of audio frame at $i$.

5. Zero-Crossing Rate (ZCR) of an audio frame is the number of times the signal passes zero or changes sign during the frame. The ZCR is expressed as below by [41]:

$$Z(n) = \frac{1}{2} \sum_{m=1} L|sgn[(m+1)] - sgn[x(m)]| \tag{A6}$$

where

$$sgn[x(m)] = \begin{cases} +1 & \text{if } x(m) \geq 0 \\ -1 & \text{if } x(m) < 0 \end{cases} \tag{A7}$$

A high ZCR is indicative of a stationary series.

6. With an input signal starting at time zero and stopping at time $T$, the probability distribution satisfies [42]:

$$P(g < u) = \frac{2}{\pi} \arcsin \sqrt{\frac{u}{T}} \tag{A8}$$

where $g$ is the last time that the signal passed zero. The density function is then:

$$P(u) = \frac{1}{\pi} \frac{1}{\sqrt{u(T-u)}} \tag{A9}$$

7. Harmonic mean is computed using the following formula:

$$\bar{m} = \frac{N}{\sum\limits_{i=1}^{N} \frac{1}{s_i}} \tag{A10}$$

8. Maximizing the inner product of the speech signal by its shifted version is another important feature that can be computed using the autocorrelation function $r(\tau)$ where $\tau$ is the time shift.

$$r(\tau) = \frac{1}{N} \sum_{0}^{N-1} s(n)s(n+\tau) \tag{A11}$$

9. In calculation of MFCC, the formula proposed by Davis et al. [43] is used.

$$MFCC_i = \sum_{\theta=1}^{N} \cos\left[i(\theta-1)\frac{\pi}{N}\right], \quad i = 1, ..., M \quad and \quad \theta = 1, ..., N \tag{A12}$$

$M$ and $N$ are the number of extracted cepstrum coefficients and number of band-pass filters, respectively. $\theta$ denotes the log energy of $\theta th$ filter.

10. Calculation of the filter-bank energies and their derivatives are performed using a first order Finite Impulse Response (FIR). An array of band-pass filters that breaks up the input signal into multiple components is called a filter bank. Each separated component carries a single frequency sub-band of the original input signal. Let the unit-sample response impulse response $h_n$ be the response of a discrete-time signal to a unit-sample impulse $\delta_n$ where $\delta_n = 1$ for $n = 0$ and $\delta_n = 0$ for $n \neq 0$. Then, for an arbitrary input signal $s_n$, the output $y_n$ is given by:

$$y_n = \sum_{i=0}^{M} \alpha_i s(n-1) + \sum_{j=1}^{N} \beta_j y(n-j) \tag{A13}$$

$\alpha_i$ and $\beta_j$ are coefficients of FIR filter and $M$ is the order of the filter function. The calculation of FBEs are as follows:

$$y(m) = \sum_{\theta=0}^{L-1} h(\theta)s[(m - \theta) \quad mod(N)], \qquad m = 0, 1, ..., N \tag{A14}$$

where $L$ is the length of the filter [29].

11. Also, $\Delta MFCC$ is obtained using the proposed formula by [44]

$$C(n) = DCT * \log(y(m)) \tag{A15}$$

where $DCT$ is the discrete cosine transform.

## References


1. Bolotnikova, A.; Demirel, H.; Anbarjafari, G. Real-time ensemble based face recognition system for NAO humanoids using local binary pattern. *Analog Integr. Circuits Signal Process.* **2017**, *92*, 467–475. [CrossRef]
2. Guo, J.; Lei, Z.; Wan, J.; Avots, E.; Hajarolasvadi, N.; Knyazev, B.; Kuharenko, A.; Junior, J.C.S.J.; Baró, X.; Demirel, H.; et al. Dominant and Complementary Emotion Recognition From Still Images of Faces. *IEEE Access* **2018**, *6*, 26391–26403. [CrossRef]
3. Schroff, F.; Kalenichenko, D.; Philbin, J. Facenet: A unified embedding for face recognition and clustering. In Proceedings of the IEEE Conference on Computer Vision and Pattern Recognition, Boston, MA, USA, 7–12 June 2015; pp. 815–823.
4. Soleymani, M.; Pantic, M.; Pun, T. Multimodal emotion recognition in response to videos. *IEEE Trans. Affect. Comput.* **2012**, *3*, 211–223. [CrossRef]
5. Kessous, L.; Castellano, G.; Caridakis, G. Multimodal emotion recognition in speech-based interaction using facial expression, body gesture and acoustic analysis. *J. Multimodal User Interfaces* **2010**, *3*, 33–48. [CrossRef]
6. Noroozi, F.; Sapiński, T.; Kamińska, D.; Anbarjafari, G. Vocal-based emotion recognition using random forests and decision tree. *Int. J. Speech Technol.* **2017**, *20*, 239–246. [CrossRef]
7. Mao, Q.; Dong, M.; Huang, Z.; Zhan, Y. Learning salient features for speech emotion recognition using convolutional neural networks. *IEEE Trans. Multimedia* **2014**, *16*, 2203–2213. [CrossRef]
8. Torfi, A.; Iranmanesh, S.M.; Nasrabadi, N.M.; Dawson, J.M. 3D Convolutional Neural Networks for Cross Audio-Visual Matching Recognition. *IEEE Access* **2017**, *5*, 22081–22091. [CrossRef]
9. Avots, E.; Sapiński, T.; Bachmann, M.; Kamińska, D. Audiovisual emotion recognition in wild. *Mach. Vis. Appl.* **2018**, 1–11. [CrossRef]
10. Kim, Y.; Lee, H.; Provost, E.M. Deep learning for robust feature generation in audiovisual emotion recognition. In Proceedings of the 2013 IEEE International Conference on Acoustics, Speech and Signal Processing (ICASSP), Vancouver, BC, Canada, 26–31 May 2013; pp. 3687–3691.
11. Jaitly, N.; Hinton, G. Learning a better representation of speech soundwaves using restricted boltzmann machines. In Proceedings of the 2011 IEEE International Conference on Acoustics, Speech and Signal Processing (ICASSP), Prague, Czech Republic, 22–27 May 2011; pp. 5884–5887.
12. Palaz, D.; Collobert, R. Analysis of cnn-based speech recognition system using raw speech as input. In Proceedings of the INTERSPEECH 2015, Dresden, Germany, 11–15 September 2015; pp. 11–15.
13. Schlüter, J.; Grill, T. Exploring Data Augmentation for Improved Singing Voice Detection with Neural Networks. In Proceedings of the 16th International Society for Music Information Retrieval Conference (ISMIR 2015), Malaga, Spain, 26–30 October 2015; pp. 121–126.
14. Badshah, A.M.; Ahmad, J.; Rahim, N.; Baik, S.W. Speech emotion recognition from spectrograms with deep convolutional neural network. In Proceedings of the 2017 International Conference on Platform Technology and Service (PlatCon), Busan, South Korea, 13–15 February 2017; pp. 1–5.
15. Abdel-Hamid, O.; Mohamed, A.R.; Jiang, H.; Deng, L.; Penn, G.; Yu, D. Convolutional neural networks for speech recognition. *IEEE/ACM Trans. Audio Speech Lang. Process.* **2014**, *22*, 1533–1545. [CrossRef]
16. Dennis, J.; Tran, H.D.; Li, H. Spectrogram image feature for sound event classification in mismatched conditions. *IEEE Signal Process. Lett.* **2011**, *18*, 130–133. [CrossRef]

17. Wang, Y.; Guan, L. Recognizing human emotional state from audiovisual signals. *IEEE Trans. Multimedia* **2008**, *10*, 936–946. [CrossRef]

18. Jackson, P.; Haq, S. *Surrey Audio-Visual Expressed Emotion (SAVEE) Database*; University of Surrey: Guildford, UK, 2014.

19. Martin, O.; Kotsia, I.; Macq, B.; Pitas, I. The eNTERFACE'05 audio-visual emotion database. In Proceedings of the 22nd International Conference on Data Engineering Workshops (ICDEW'06), Atlanta, GA, USA, 3–7 April 2006; p. 8.

20. Simonyan, K.; Zisserman, A. Very deep convolutional networks for large-scale image recognition. *arXiv* **2014**, arXiv:1409.1556.

21. Ahmad, J.; Sajjad, M.; Rho, S.; Kwon, S.I.; Lee, M.Y.; Baik, S.W. Determining speaker attributes from stress-affected speech in emergency situations with hybrid SVM-DNN architecture. *Multimedia Tools Appl.* **2018**, *77*, 4883–4907. [CrossRef]

22. Badshah, A.M.; Rahim, N.; Ullah, N.; Ahmad, J.; Muhammad, K.; Lee, M.Y.; Kwon, S.; Baik, S.W. Deep features-based speech emotion recognition for smart affective services. *Multimedia Tools Appl.* **2019**, *79*, 5571–5589. [CrossRef]

23. Yenigalla, P.; Kumar, A.; Tripathi, S.; Singh, C.; Kar, S.; Vepa, J. Speech Emotion Recognition Using Spectrogram & Phoneme Embedding. In Proceedings of the Interspeech 2018, Hyderabad, India, 2–6 September 2018.

24. Busso, C.; Bulut, M.; Lee, C.C.; Kazemzadeh, A.; Mower, E.; Kim, S.; Chang, J.N.; Lee, S.; Narayanan, S.S. IEMOCAP: Interactive emotional dyadic motion capture database. *Lang. Resour. Eval.* **2008**, *42*, 335. [CrossRef]

25. Zhang, Y.; Du, J.; Wang, Z.R.; Zhang, J. Attention Based Fully Convolutional Network for Speech Emotion Recognition. *arXiv* **2018**, arXiv:1806.01506.

26. Krizhevsky, A.; Sutskever, I.; Hinton, G.E. Imagenet classification with deep convolutional neural networks. In *Advances in Neural Information Processing Systems 25 (NIPS 2012)*; Curran Associates, Inc.: Red Hook, NY, USA, 2012; pp. 1097–1105.

27. Burkhardt, F.; Paeschke, A.; Rolfes, M.; Sendlmeier, W.F.; Weiss, B. A database of German emotional speech. In Proceedings of the Ninth European Conference on Speech Communication and Technology, Lisbon, Portugal, 4–8 September 2005.

28. Satt, A.; Rozenberg, S.; Hoory, R. Efficient Emotion Recognition from Speech Using Deep Learning on Spectrograms. In Proceedings of the INTERSPEECH 2017, Stockholm, Sweden, 20–24 August 2017; pp. 1089–1093.

29. Noroozi, F.; Marjanovic, M.; Njegus, A.; Escalera, S.; Anbarjafari, G. Audio-visual emotion recognition in video clips. *IEEE Trans. Affect. Comput.* **2017**, *10*, 60–75. [CrossRef]

30. Paliwal, K.K.; Lyons, J.G.; Wójcicki, K.K. Preference for 20–40 ms window duration in speech analysis. In Proceedings of the 2010 4th International Conference on Signal Processing and Communication Systems (ICSPCS), Gold Coast, QLD, Australia, 13–15 December 2010; pp. 1–4.

31. Tran, D.; Bourdev, L.; Fergus, R.; Torresani, L.; Paluri, M. 1175 "Learning spatiotemporal features with 3d convolutional networks". *arXiv* **2014**, *arXiv:1412.0767*.

32. Lloyd, S. Least squares quantization in PCM. *IEEE Trans. Inf. Theory* **1982**, *28*, 129–137. [CrossRef]

33. Boutsidis, C.; Zouzias, A.; Mahoney, M.W.; Drineas, P. Randomized dimensionality reduction for $k$-means clustering. *IEEE Trans. Inf. Theory* **2015**, *61*, 1045–1062. [CrossRef]

34. Ioffe, S.; Szegedy, C. Batch normalization: Accelerating deep network training by reducing internal covariate shift. *arXiv* **2015**, arXiv:1502.03167.

35. Kim, J.; Truong, K.P.; Englebienne, G.; Evers, V. Learning spectro-temporal features with 3D CNNs for speech emotion recognition. In Proceedings of the 2017 Seventh International Conference on Affective Computing and Intelligent Interaction (ACII), San Antonio, TX, USA, 23–26 October 2017; pp. 383–388.

36. Jiang, D.; Cui, Y.; Zhang, X.; Fan, P.; Ganzalez, I.; Sahli, H. Audio visual emotion recognition based on triple-stream dynamic bayesian network models. In *Proceedings of the International Conference on Affective Computing and Intelligent Interaction*; Springer: Berlin/Heidelberg, Germany, 2011; pp. 609–618.

37. Kingma, D.P.; Ba, J.L. Adam: Amethod for stochastic optimization. In Proceedings of the 3rd International Conference for Learning Representations, San Diego, CA, USA, 7–9 May 2015.

38. Smith, L.N. Cyclical learning rates for training neural networks. In Proceedings of the 2017 IEEE Winter Conference on Applications of Computer Vision (WACV), Santa Rosa, CA, USA, 24–31 March 2017; pp. 464–472.

39. Russakovsky, O.; Deng, J.; Su, H.; Krause, J.; Satheesh, S.; Ma, S.; Huang, Z.; Karpathy, A.; Khosla, A.; Bernstein, M.; et al. ImageNet Large Scale Visual Recognition Challenge. *Int. J. Comput. Vis.* **2015**, *115*, 211–252. [CrossRef]

40. Stevens, S.S. The relation of pitch to intensity. *J. Acoust. Soc. Am.* **1935**, *6*, 150–154. [CrossRef]

41. Giannakopoulos, T.; Pikrakis, A. *Introduction to Audio Analysis: A MATLAB® Approach*; Academic Press: Cambridge, MA, USA, 2014.

42. Vidyamurthy, G. *Pairs Trading: Quantitative Methods and Analysis*; John Wiley & Sons: Hoboken, NJ, USA, 2004; Volume 217.

43. Davis, S.; Mermelstein, P. Comparison of parametric representations for monosyllabic word recognition in continuously spoken sentences. *IEEE Trans. Acoust. Speech Signal Process.* **1980**, *28*, 357–366. [CrossRef]

44. Kopparapu, S.K.; Laxminarayana, M. Choice of Mel filter bank in computing MFCC of a resampled speech. In Proceedings of the 10th International Conference on Information Science, Signal Processing and their Applications (ISSPA 2010), Kuala Lumpur, Malaysia, 10–13 May 2010; pp. 121–124.

*Article*

# Learning Using Concave and Convex Kernels: Applications in Predicting Quality of Sleep and Level of Fatigue in Fibromyalgia

Elyas Sabeti [1,2,*], Jonathan Gryak [1], Harm Derksen [3], Craig Biwer [1], Sardar Ansari [1,2], Howard Isenstein [4], Anna Kratz [5] and Kayvan Najarian [1,2,6,7]

[1] Department of Computational Medicine and Bioinformatics, University of Michigan, 2800 Plymouth Rd, NCRC, Ann Arbor, MI 48109-2800, USA; gryakj@umich.edu (J.G.); cbiwer@umich.edu (C.B.); sardara@umich.edu (S.A.); kayvan@umich.edu (K.N.)

[2] Michigan Center for Integrative Research in Critical Care (MCIRCC), University of Michigan, 2800 Plymouth Rd, NCRC, Ann Arbor, MI 48109-2800, USA

[3] Department of Mathematics, University of Michigan, 2800 Plymouth Rd, Bldg. 18-163, Ann Arbor, MI 48109-2800, USA; hderksen@umich.edu

[4] Digidence, LLC 7315 Wisconsin Ave., Bethesda, MD 20814-3202, USA; hisenstein@digidence.co

[5] Department of Physical Medicine & Rehabilitation, University of Michigan, 2800 Plymouth Rd, NCRC B14 #D034, Ann Arbor, MI 48109-2800, USA; alkratz@umich.edu

[6] Department of Emergency Medicine, University of Michigan, 1500 E Medical Center Dr, Ann Arbor, MI 48109, USA

[7] Department of Electrical Engineering and Computer Science, University of Michigan, 1301 Beal Ave, Ann Arbor, MI 48109, USA

[*] Correspondence: sabeti@umich.edu

Received: 8 February 2019; Accepted: 24 April 2019; Published: 28 April 2019

**Abstract:** Fibromyalgia is a medical condition characterized by widespread muscle pain and tenderness and is often accompanied by fatigue and alteration in sleep, mood, and memory. Poor sleep quality and fatigue, as prominent characteristics of fibromyalgia, have a direct impact on patient behavior and quality of life. As such, the detection of extreme cases of sleep quality and fatigue level is a prerequisite for any intervention that can improve sleep quality and reduce fatigue level for people with fibromyalgia and enhance their daytime functionality. In this study, we propose a new supervised machine learning method called Learning Using Concave and Convex Kernels (LUCCK). This method employs similarity functions whose convexity or concavity can be configured so as to determine a model for each feature separately, and then uses this information to reweight the importance of each feature proportionally during classification. The data used for this study was collected from patients with fibromyalgia and consisted of blood volume pulse (BVP), 3-axis accelerometer, temperature, and electrodermal activity (EDA), recorded by an Empatica E4 wristband over the courses of several days, as well as a self-reported survey. Experiments on this dataset demonstrate that the proposed machine learning method outperforms conventional machine learning approaches in detecting extreme cases of poor sleep and fatigue in people with fibromyalgia.

**Keywords:** fibromyalgia; Learning Using Concave and Convex Kernels; Empatica E4; self-reported survey

---

## 1. Introduction

Fibromyalgia is medical condition characterized by widespread muscle pain and tenderness that is typically accompanied by a constellation of other symptoms, including fatigue and poor sleep [1–9]. Poor sleep, which is a cardinal characteristic of fibromyalgia, is strongly related to greater pain and fatigue, and lower quality of life [10–16]. As a result, any intervention that can improve sleep quality may enhance daytime functionality and reduce fatigue in people with fibromyalgia.

Studies of sleep in fibromyalgia often rely on self-reported measures of sleep or polysomnography. While easy to administer, self-reported measures of sleep demonstrate limited reliability and validity in terms of their correspondence with objective measures of sleep. In contrast, polysomnography is considered the gold standard of objective sleep measurement; however, it is expensive, difficult to administer, especially on a large scale, and may lack ecological validity. Autonomic nervous system (ANS) imbalance during sleep has been implicated as a mechanism underlying unrefreshed sleep in fibromyalgia. ANS activity can be assessed unobtrusively through ambulatory measures of heart rate variability (HRV) and electrodermal activity (EDA) [17,18]. Wearable devices such as the Empatica E4 are able to directly, continuously, and unobtrusively measure autonomic functioning such as EDA and HRV [19–22].

In the literature, there are few studies in which machine learning methods are used for classification or prediction of conditions related to fibromyalgia, none of which use physiological signals. A recent survey paper [23] summarizes various types of machine learning methods that have been used in pain research, including fibromyalgia. Previously, using data from 26 individuals (14 individuals with fibromyalgia and 12 healthy controls), the relative performance of machine learning methods for classification of individuals with and without pain using neuroimaging and self-reported data have been compared [24]. In another study using MRI images of 59 subjects, support vector machine (SVM) and decision tree models were used to first distinguish healthy control patients from those with fibromyalgia or chronic fatigue syndrome, and then differentiate fibromyalgia from chronic fatigue syndrome [25]. In [26], an SVM trained on fMRI images was used to distinguish fibromyalgia patients from healthy controls. The combination of fMRI with multivariate pattern analysis has also been investigated in classifying fibromyalgia patients, rheumatoid arthritis patients and healthy controls [27]. Psychopathologic features within an ADABoost classifier have also been employed for classification of patients with fibromyalgia and arthritis [28]. In another recent work [29], secondary analysis of gene expression data from 28 patients with fibromyalgia and 19 healthy controls was used to distinguish between these two groups.

In this study our immediate interest is to predict extreme cases of fatigue and poor sleep in people with fibromyalgia. For such an analysis, we use self-reported quality of sleep and fatigue severity, continuously collected data from the Empatica E4, to measure autonomic nervous system activity during sleep (Section 2). These signals are preprocessed to remove noise and other artifacts as described in Section 3.1. After preprocessing, a number of mathematical features are extracted, including various statistics, signal characteristics, and HRV features (Section 3.2). Section 4 provides a detailed description of our novel Learning Using Concave and Convex Kernels (LUCCK) machine learning method. This model, along with other conventional machine learning methods, were trained on the extracted features and used to predict extreme cases of poor sleep and fatigue, with our method yielding the best results (Section 5).

We believe this analytical framework can be readily extended to outpatient monitoring of daytime activity, with applications to assessing extreme levels of fatigue and pain, such as those experienced by patients undergoing chemotherapy.

## 2. Dataset

The data used for this study was collected from a group of 20 adults with fibromyalgia and consists primarily of a set of signals recorded by an Empatica E4 wristband over the course of seven days (removing 1 h/day for charging/download). Most (80%) participants were female with mean age = 38.79 (min-max=18–70 years). Of a possible 140 nights of sleep data, the sample had data for 119 (85%) nights. In this dataset, 19.9% of heartbeats were missing due to noisy signals or failure of the Empatica E4 in detecting beats. Data were divided into 5-min windows for HRV analysis; windows with more than 15% missing peaks were eliminated. This led to the exclusion of 30.9% of the windows.

The signals used in this analysis are each patient's blood volume pulse (BVP), 3-axis accelerometer, temperature, and EDA. In addition to these recordings, each subject self-reported his or her wake and sleep times, as well as self-assessed his or her level of fatigue and quality of sleep every morning. These data are labeled by self-reported quality of sleep (1 to 10, 1 being the worst) and level of fatigue (from 1 to 10, 10 indicating the highest level of fatigue).

## 3. Signal Processing: Preprocessing, Filtering, and Feature Extraction

The schematic diagram of Figure 1 represents our approach to analyzing the BVP and accelerometer signals in the fibromyalgia dataset. During preprocessing, we remove noise from the input signals and format them for future processing (via the Epsilon Tube filter). Once the BVP and accelerometer signals are fully processed, they along with the EDA and temperature signals can then be analyzed and features can be extracted, which in turn leads to the application of machine learning. The final output is a prediction model to which new data can be fed.

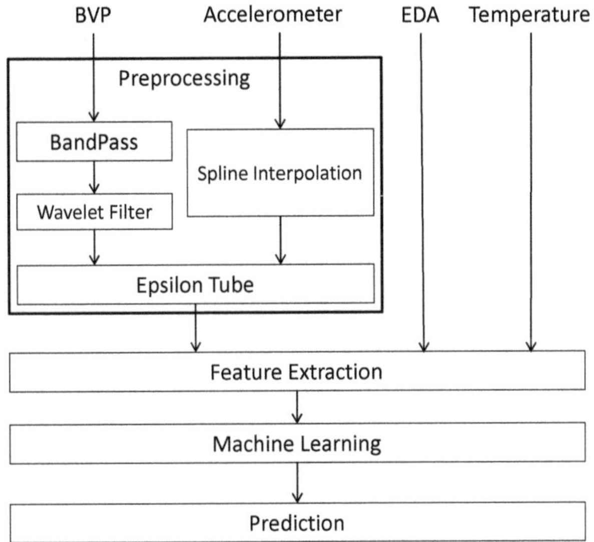


**Figure 1.** Schematic Diagram of the Proposed Processing System for BVP, accelerometer, EDA and temperature signals.

### 3.1. Preprocessing

To begin, the raw signals are extracted per patient according to his or her reported wake and sleep times. These are then split into two groups: awake and asleep. For each patient and day, the awake data is paired with the following night's data and ensuing morning's self-assessed level of fatigue and quality of sleep.

Our approach to preprocessing BVP signals consists of a bandpass filter (to remove both the low-frequency components and the high-frequency noise), a wavelet filter (to help reduce motion artifacts while maintaining the underlying rhythm), and Epsilon Tube filtering. In order to least perturb the true BVP signal, we chose the Daubechies mother wavelet of order 2 ('db2') as it closely resembles the periodic shape of the BVP signal. Other wavelets were also considered but ultimately discarded. Once we selected a mother wavelet, we performed an eight-level deconstruction of the input BVP signal. By setting threshold values for each level of detail coefficients (Table 1) and using the results to reconstruct the original signal, we were able to significantly reduce the amount of noise present without compromising the measurement integrity of the underlying physiological values. Utilizing this filter on a number of test cases showed that the threshold values produced consistently useful results regardless of the input, meaning tailored interactions are not required for each signal.

**Table 1.** Chosen coefficient thresholds for the 8-level wavelet decomposition.

| Detail Coefficients Level | Threshold |
|:---:|:---:|
| 8 | 94.38 |
| 7 | 147.8 |
| 6 | 303.1 |
| 5 | 329.9 |
| 4 | 90.16 |
| 3 | 30.67 |
| 2 | 0 |
| 1 | 0 |

The accelerometer data was upsampled from 32 Hz to 64 Hz via spline interpolation to match the sampling frequency of the BVP signal. The other signals (temperature and EDA) were left unfiltered. We then use these preprocessed signals as input into our main filtering approach (Epsilon Tube), the output of which is then used for feature extraction (Section 3.2).

After filtering of the BVP signal and interpolation of the accelerometer signal, the Epsilon Tube filter [30] is the final component of the preprocessing stage. As discussed in [30], since the BVP signal (and generally any impedance-plethysmography-based measurements) is very susceptible to motion artifact, reduction of this noise is a crucial part of the filtering process. This method uses the synchronized accelerometer data to estimate the motion artifact of BVP signal while leaving the periodic component intact. Let $b_t$ represent BVP values at time $t$, $A$ a matrix whose rows are the accelerometer signals, and $\mathbf{w}$ the vector of Epsilon Tube filter coefficients. Given the tube radius $\epsilon$, the error of $b_t$ estimation, i.e., $y_t(A, \mathbf{w})$, is zero if the point $b_t$ falls inside the tube

$$|b_t - y_t(A, \mathbf{w})|_\epsilon = \max\{0, |b_t - y_t(A, \mathbf{w})| - \epsilon\}.$$

The Epsilon Tube filter is formulated as a constrained optimization problem that can be expressed as

$$\min \sum_{t=0}^{N-1} \zeta_t + \sum_{t=0}^{N-1} \zeta_t' - cR(s, A, \mathbf{w}); \tag{1}$$

subject to

$$b_t - y_t(A, \mathbf{w}) \leq \epsilon + \zeta_t \qquad t = 0, ..., N-1;$$
$$y_t(A, \mathbf{w}) - b_t \leq \epsilon + \zeta_t' \qquad t = 0, ..., N-1;$$
$$\zeta_t \geq 0, \qquad \zeta_t' \geq 0 \qquad t = 0, ..., N-1;$$

where $N$ is the length of BVP signal, $\zeta_t$ and $\zeta_t'$ are slack variables, $R(s, A, \mathbf{w})$ is the regularization term and $c$ is a designated parameter that adjusts the trade-off between the two objectives. More information about the Epsilon Tube filter can be found in [30]. Taking both the BVP and accelerometer signals as input, the method assumes periodicity in the BVP signal and looks for a period of inactivity at the beginning of the data to use as a template for the rest of the signal. To achieve this, the calmest section of the accelerometer signal (as determined by the longest stretch during which the values never exceed one standard deviation from the mean of the signal) is found. The signal is then shifted so this period of inactivity is at the beginning, and the BVP signal is also shifted to ensure the timestamps remain aligned. The shifted signals are then fed into the Epsilon Tube algorithm, and the resulting output is used for feature extraction.

### 3.2. Feature Extraction

Once the BVP and accelerometer signals are processed, the full signal set is used for feature extraction. There are 91 features extracted from each of the following signals:

- Denoised (filtered) BVP signal, i.e., the output of the Epsilon Tube algorithm, with sampling frequency of 64 Hz.
- Low-band, mid-band, and high-band pass filters applied to the denoised BVP signal.
- Interpolated accelerometer signal, from 32 HZ to 64 Hz.
- Tube sizes from the Epsilon Tube filtering method, another output of the Epsilon Tube algorithm that has the time-varying tube size signal.
- Temperature signal, with sampling frequency of 4 Hz.
- EDA signal, with sampling frequency of 4 Hz.
- The calculated breaths per minute (BPM) signal based on the denoised BVP signal.
- The calculated HRV signal based on the denoised BVP signal.

The extracted features are listed in Table 2. These are extracted from both the awake and the sleep signals, resulting in a full feature set consisting of 182 features. When feature selection is performed using Weka's information gain algorithm [31] on the first four subjects, the only feature ranked consistently near the top is the average of the BVP signal after being run through a mid-band bandpass filter.

**Table 2.** The list of features extracted from all signals.

| Signals | Features |
| --- | --- |
| Denoised BVP | Mean, Standard deviation, Variance, Power, Median, Frequency with the highest peak, Amplitude of the frequency with highest peak, FFT power, Mean of FFT amplitudes, Mean of the FFT frequencies, Median of FFT amplitudes (11 features) |
| Low-band denoised BVP | Mean, Standard deviation, Variance, Power, Median, Frequency with the highest peak, Amplitude of the frequency with highest peak, FFT power, Mean of FFT amplitudes, Mean of the FFT frequencies, Median of FFT amplitudes (11 features) |
| Mid-band denoised BVP | Mean, Standard deviation, Variance, Power, Median, Frequency with the highest peak, Amplitude of the frequency with highest peak, FFT power, Mean of FFT amplitudes, Mean of the FFT frequencies, Median of FFT amplitudes (11 features) |
| High-band denoised BVP | Mean, Standard deviation, Variance, Power, Median, Frequency with the highest peak, Amplitude of the frequency with highest peak, FFT power, Mean of FFT amplitudes, Mean of the FFT frequencies, Median of FFT amplitudes (11 features) |
| Tube size | Mean, Standard Deviation, Variance, Power (4 features) |
| Interpolated accelerometer | Mean, Standard Deviation, Variance, Power (4 features) |
| Temperature signal | Mean, Standard Deviation, Variance, Power (4 features) |
| EDA signal | Mean, Standard Deviation, Variance, Power (4 features) |
| BPM signal | Maximum, Minimum, Range, Mean, Standard deviation, Power (6 features) |
| HRV | The Kubios Standard HRV feature set [32] (25 features) |

## 4. Machine Learning: Learning Using Concave and Convex Kernels

The final step in the analysis pipeline is the creation of a model that can be used to predict the extreme cases of quality of sleep or level of fatigue for people with fibromyalgia. As detailed in Section 5, in addition to testing a number of conventional machine learning methods, we tested a novel supervised machine learning called Learning Using Concave and Convex Kernels (LUCCK). A key factor in the classification of complex data is the ability of the machine learning algorithm to use vital, feature-specific information to detect settled and complex patterns of changes in the data. The LUCCK method does this by employing similarity functions (defined below) to capture and quantify a model for each of the features separately. The similarity functions are parametrized so that the concavity or convexity of the function within the feature space can be modified as desired. Once the similarity functions and attendant parameters are chosen, the model uses this information to reweight the importance of each feature proportionally during classification.

*4.1. Notation*

In this section, $\mathbf{x} \in \mathbb{R}^n$ is a real-valued vector of features such that $\mathbf{x} = (x_1, \ldots, x_n)$, and $x_i$ is a real-valued (scalar) feature. Throughout this section, we consider $d$ classes, $n$ features and $m$ (data) samples; also the indexes $k = 1, \ldots, d$; $i = 1, \ldots, n$; and $j = 1, \ldots, m$ are used for classes, features and samples respectively. Additionally, $j = 1, \ldots, m_k$ refers to $m_k < m$ samples in class $C_k$.

*4.2. Classification Using a Similarity Function*

An instructive model for comparison to the Learning Using Concave and Convex Kernels method is the $k$-nearest neighbors algorithm [33–35] and weighted $k$-nearest neighbors algorithm [36]. In $k$-nearest neighbors, a test sample $\mathbf{x}$ is classified by comparing it to the $k$ nearest training samples in each class. This can make the classification sensitive to a small subset of samples. Instead, LUCCK classifies test data by comparing it to *all* training data, properly weighted according to their distance to $\mathbf{x}$, which is determined by a similarity function. One major difference between LUCCK and weighted $k$-nearest neighbors is that our approach is based on a similarity function that can be highly non-convex. A fat-tailed (relative to a Gaussian) distribution is more realistic for our data, given that there is a small but non-negligible chance that large errors may occur during measurement, resulting in a large deviation in the values of one or more of the features. The LUCCK method allows for large deviations in a few of the features with only a moderate penalty. Methods based on convex notions of similarity or distance (such as the Mahalanobis distance) are unable to deal adequately with such errors.

Suppose that the feature space is comprised of real-valued vectors $\mathbf{x} \in \mathbb{R}^n$. A *similarity function* is a function $Q : \mathbb{R}^n \to \mathbb{R}$ that measures the closeness of $\mathbf{x}$ to the origin, and satisfies the following properties:

1. $Q(\mathbf{x}) > 0$ for all $\mathbf{x} \in \mathbb{R}^n$;
2. $Q(\mathbf{x}) = Q(-\mathbf{x})$ for all $\mathbf{x} \in \mathbb{R}^n$;
3. $Q(\lambda \mathbf{x}) > Q(\mathbf{x})$ if $\mathbf{x} \in \mathbb{R}^n$ is non-zero and $|\lambda| < 1$.

The value $Q(\mathbf{x} - \mathbf{y})$ measures the closeness between the vectors $\mathbf{x}$ and $\mathbf{y}$. Using the similarity function $Q(\mathbf{x})$, a classification algorithm can be created as follows:

The set of training data $C$ is a subset of $\mathbb{R}^n$ and is a disjoint union of $d$ classes: $C = C_1 \cup C_2 \cup \cdots \cup C_d$. Let $m = |C|$ be the cardinality of $C$ and define $m_k = |C_k|$ for all $k$ so that $m = m_1 + \cdots + m_d$. To measure the proximity of a feature vector $\mathbf{x}$ to a set $Y$ of training samples, we simply add the contributions of each of the elements in $Y$:

$$R(\mathbf{x}, Y) = \sum_{\mathbf{y} \in Y} Q(\mathbf{x} - \mathbf{y}). \tag{2}$$

A vector $\mathbf{x}$ is classified in class $C_k$, where $k$ is chosen such that $R(\mathbf{x}, C_k)$ is maximal. This classification approach can also be used as the maximum a posteriori estimation (details can be found in Appendix A).

*4.3. Choosing the Similarity Function*

The function $Q(\mathbf{x})$ has to be chosen carefully. Let $Q(\mathbf{x})$ be defined as the product

$$Q(\mathbf{x}) = \prod_{i=1}^{n} Q_i(x_i), \tag{3}$$

where $\mathbf{x} = (x_1, \ldots, x_n) \in \mathbb{R}^n$ and $Q_i(x_i)$ only depends on the $i$-th feature. The function $Q_i(x_i)$ is again a similarity function satisfying the properties $Q_i(-x_i) = Q_i(x_i) > 0$ for all $x \in \mathbb{R}$, and $Q(x) > Q(y)$

whenever $|x| < |y|$. After normalization, the $Q, Q_1, Q_2, \ldots, Q_n$ can be considered as probability density functions. As such, the product formula can be interpreted as instance-wise independence for the comparison of training and test data. In the naive Bayes method, features are assumed to be independent globally [37]. Summing over all instances in the training data allows for features to be independent in our model.

Next we need to choose the functions $Q_1, \ldots, Q_n$. One could choose $Q_i(x_i) = e^{-\gamma_i x^2}$, so that

$$Q(\mathbf{x}) = e^{-(\gamma_1 x_1^2 + \cdots + \gamma_n x_n^2)}$$

is a Gaussian kernel function (up to a scalar). However, this does not work well in practice:

- One or more of the features is prone to large errors —The value of $Q(\mathbf{x} - \mathbf{y})$ is close to 0 even if $\mathbf{x}$ and $\mathbf{y}$ only differ significantly in a few of the features. This choice of $Q(\mathbf{x})$ is therefore very sensitive to small subsets of bad features.
- The curse of dimensionality—For the training data to properly represent the probability distribution function underlying the data, the number of training vectors should be exponential in $n$, the number of features. In practice, it usually is much smaller. Thus, if $\mathbf{x}$ is a test vector in class $C_k$, there may not be a training vector $\mathbf{y}$ in $C_k$ for which $Q(\mathbf{x} - \mathbf{y})$ is not small.

Consequently, let

$$Q_i(x_i) = (1 + \lambda_i x^2)^{-\theta_i}, \tag{4}$$

for some parameters $\lambda_i, \theta_i > 0$. The function $Q_i(x_i)$ can behave similarly to the Cauchy distribution. This function has a "fat tail": as $x \to \infty$ the rate that $Q_i(x_i)$ goes to 0 is much slower than the rate at which $e^{-\gamma_i x^2}$ goes to 0. We have

$$Q(\mathbf{x}) = \prod_{i=1}^{n} (1 + \lambda_i x_i^2)^{-\theta_i}. \tag{5}$$

The function $Q$ has a finite integral if $\theta_i > \frac{1}{2}$ for all $i$, though this is not required. Three examples of this function can be found in Appendix B.

### 4.4. Choosing the Parameters

Values for the parameters $\lambda_1, \lambda_2, \ldots, \lambda_n$ and $\theta_1, \theta_2, \ldots, \theta_n$ must be chosen to optimize classification performance. The value of $\log(Q_i(x_i)) = -\theta_i \log(1 + \lambda_i x^2)$ is the most sensitive to changes in $x$ when

$$\frac{\partial}{\partial x} \log(1 + \lambda_i x^2) = \frac{2\lambda_i x}{1 + \lambda_i x^2}$$

is maximal. An easy calculation shows that this occurs when $x = \lambda_i^{-\frac{1}{2}}$. Since the value $\lambda_i$ directly controls the wideness of $Q_i(x_i)$'s tail, it is reasonable to choose a value for $\lambda_i^{-\frac{1}{2}}$ that is close to the standard deviation of the $i$-th feature. Suppose that the set of training vectors is

$$C = \{\mathbf{x}^{(1)}, \mathbf{x}^{(2)}, \ldots, \mathbf{x}^{(m)}\} \subseteq \mathbb{R}^n,$$

where $\mathbf{x}^{(j)} = (x_1^{(j)}, \ldots, x_n^{(j)})$ for all $j$.

Let $s = (s_1, \ldots, s_n)$, where

$$s_i = \mathrm{std}(x_i^{(1)}, x_i^{(2)}, \ldots, x_i^{(m)})$$

be the standard deviation of the $i$-th feature. Let

$$\lambda_i = \frac{\Lambda}{s_i^2}$$

where $\Lambda$ is some fixed parameter.

Next we choose the parameters $\theta_1, \ldots, \theta_n$. We fix a parameter $\Theta$ that will be the average value of $\theta_1, \ldots, \theta_n$. If we use only the $i$-th feature, then we define

$$R_i(\mathbf{x}, Y) = \sum_{\mathbf{y} \in Y} (1 + \lambda_i(x_i - y_i)^2)^{-\Theta}$$

for any set $Y$ of feature vectors. For $\mathbf{x}$ in the class $C_k$, $\frac{1}{m_i - 1} R_i(\mathbf{x}, C_k \setminus \{\mathbf{x}\})$ gives the average value of $(1 + \lambda_i(x_i - y_i)^2)^{-\Theta}$ over $\mathbf{y} \in C_k \setminus \{\mathbf{x}\}$. The quantity $\frac{1}{m_k - 1} R_i(\mathbf{x}, C_k \setminus \{\mathbf{x}\}) - \frac{1}{m-1} R_i(\mathbf{x}, C \setminus \{\mathbf{x}\})$ measures how much closer $x_i$ is to samples in the class $C_k$ than to vectors in the set $C$ of all feature vectors except $\mathbf{x}$ itself. This value measures how well the $i$-th feature can classify $\mathbf{x}$ as lying in $C_k$ as opposed to some other class. If we sum over all $\mathbf{x} \in C$ and ensure that the result is non-negative we obtain

$$\alpha_i = \max\left\{0, \sum_{k=1}^{d} \sum_{\mathbf{x} \in C_k} \left( \frac{R_i(\mathbf{x}, C_k \setminus \{\mathbf{x}\})}{m_k - 1} - \frac{R_i(\mathbf{x}, C \setminus \{\mathbf{x}\})}{m - 1} \right) \right\}. \tag{6}$$

The $\theta_1, \ldots, \theta_n$ can be chosen so that they have the same ratios as $\alpha_1, \ldots, \alpha_n$ and sum up to $n\Theta$:

$$\theta_i = \frac{n\alpha_i \Theta}{\sum_{i=1}^{n} \alpha_i}. \tag{7}$$

In terms of complexity, if $n$ is the number of features and $m$ is the number of training samples then the complexity of the proposed method would be $O(n \times m^2)$.

*4.5. Reweighting the Classes*

Sometimes a disproportionate number of test vectors are classified as belonging to a particular class. In such cases one might get better results after reweighting the classes. The weights $\omega_1, \omega_2, \ldots, \omega_d$ can be chosen so that all are greater than or equal to 1. If $p$ is a probability vector, then we can reweight it to a vector

$$W_\omega(p) = (p'_1, \ldots, p'_d)$$

where

$$p'_l = \frac{\omega_l p_l}{\sum_{k=1}^{d} \omega_k p_k}.$$

If the output of the algorithm consists of the probability vectors $p(\mathbf{x}^{(1)}), \ldots, p(\mathbf{x}^{(m)})$ the algorithm can be modified so that it yields the output $W_\omega(p(\mathbf{x}^{(1)})), \ldots, W_\omega(p(\mathbf{x}^{(m)}))$. A good choice for the weights $\omega_1, \ldots, \omega_d$ can be learned by using a portion of the training data. To determine how well a *training* vector $\mathbf{x} \in C$ can be classified using the remaining training vectors in $C \setminus \{\mathbf{x}\}$, we define

$$\tilde{p}_k(\mathbf{x}) = \frac{R(\mathbf{x}, C_k \setminus \{\mathbf{x}\})}{R(\mathbf{x}, C \setminus \{\mathbf{x}\})}.$$

The value $\tilde{p}_k(\mathbf{x})$ is an estimate for the probability that $\mathbf{x}$ lies in the class $C_k$, based on all feature vectors in $C$ except $\mathbf{x}$ itself. We consider the effect of reweighting the probabilities $\tilde{p}_k(\mathbf{x})$, by

$$\tilde{p}'_k(\mathbf{x}) = \frac{\omega_i \tilde{p}(\mathbf{x})}{\sum_{i=1}^{d} \omega_i \tilde{p}_i(\mathbf{x})}.$$

If $\mathbf{x}$ lies in the class $C_k$, then the quantity

$$\max\{\widetilde{p}_1'(\mathbf{x}),\ldots,\widetilde{p}_d'(\mathbf{x})\} - \widetilde{p}_k'(\mathbf{x})$$

measures how badly $\mathbf{x}$ is misclassified if the reweighting is used. The total amount of misclassification is

$$\sum_{k=1}^{d}\sum_{\mathbf{x}\in C_k}\left(\max\{\widetilde{p}_1'(\mathbf{x}),\ldots,\widetilde{p}_d'(\mathbf{x})\} - \widetilde{p}_k'(\mathbf{x})\right) =$$

$$\sum_{k=1}^{d}\sum_{\mathbf{x}\in C_k}\left(\frac{\max\{\omega_1\widetilde{p}_1(\mathbf{x}),\ldots,\omega_d\widetilde{p}_d(\mathbf{x})\} - \omega_k\widetilde{p}_k(\mathbf{x})}{\sum_{l=1}^{d}\omega_l\widetilde{p}_l(\mathbf{x})}\right).$$

We would like to minimize this over all choices of $\omega_1,\ldots\omega_d \geq 1$. As this is a highly nonlinear problem, making optimization difficult, we instead minimize

$$\sum_{k=1}^{d}\sum_{\mathbf{x}\in C_k}\left(\max\{\omega_1\widetilde{p}_1(\mathbf{x}),\ldots,\omega_d\widetilde{p}_d(\mathbf{x})\} - \omega_k\widetilde{p}_k(\mathbf{x})\right) =$$

$$\sum_{\mathbf{x}\in C}\max\{\omega_1\widetilde{p}_1(\mathbf{x}),\ldots,\omega_d\widetilde{p}_d(\mathbf{x})\} - \sum_{k=1}^{d}\omega_k\sum_{\mathbf{x}\in C_k}\widetilde{p}_k(\mathbf{x}).$$

instead. This minimization problem can be solved using linear programming, i.e., by minimizing the quantity

$$\sum_{j=1}^{m}z^{(j)} - \sum_{k=1}^{d}\omega_k\sum_{\mathbf{x}\in C_k}\widetilde{p}_k(\mathbf{x}).$$

for the variables $\omega_1,\ldots,\omega_d$ and new variables $z^{(1)},\ldots,z^{(m)}$ under the constraints that

$$z^{(j)} \geq \omega_k\widetilde{p}(\mathbf{x}^{(j)})$$

and

$$\omega_k \geq 1$$

for all $k$ and $j$ with $1 \leq k \leq d$ and $1 \leq j \leq m$.

## 5. Experiments

In this section, the performance of LUCCK is first compared with other common machine learning methods using four conventional datasets, after which its performance on the fibromyalgia dataset is evaluated.

### 5.1. UCI Machine Learning Repository

In this set of experiments, LUCCK in compared to some well-known classification methods on a number of datasets downloaded from the University of California, Irvine (UCI) Machine Learning Repository [38]. Each method was tested on each dataset using 10-fold cross-validation, with the average performance and execution time across all folds provided in Table 3. Table 4 contains the average values for accuracy and time across all four datasets.

**Table 3.** Comparison of our proposed method (LUCCK) with other machine learning methods in terms of accuracy and running time, averaged over 10 folds.

| Dataset | Method | Accuracy (%) | Time (s) |
|---|---|---|---|
| Sonar (208 samples) | LUCCK | 87.42 | 1.5082 |
|  | 3-NN | 81.66 | 0.0178 |
|  | 5-NN | 81.05 | 0.0178 |
|  | Adaboost | 82.19 | 1.0239 |
|  | SVM | 81.00 | 0.0398 |
|  | Random Forest (10) | 78.14 | 0.1252 |
|  | Random Forest (100) | 83.39 | 1.1286 |
|  | LDA | 74.90 | 0.0343 |
| Glass (214 samples) | LUCCK | 82.56 | 0.3500 |
|  | 3-NN | 68.72 | 0.0161 |
|  | 5-NN | 67.04 | 0.0162 |
|  | Adaboost | 50.82 | 0.5572 |
|  | SVM | 35.57 | 0.0342 |
|  | Random Forest (10) | 75.31 | 0.1062 |
|  | Random Forest (100) | 79.24 | 0.9319 |
|  | LDA | 63.28 | 0.0155 |
| Iris (150 samples) | LUCCK | 95.93 | 0.1508 |
|  | 3-NN | 96.09 | 0.0135 |
|  | 5-NN | 96.54 | 0.0135 |
|  | Adaboost | 93.82 | 0.4912 |
|  | SVM | 96.52 | 0.0143 |
|  | Random Forest (10) | 94.81 | 0.0889 |
|  | Random Forest (100) | 95.29 | 0.7686 |
|  | LDA | 98.00 | 0.0122 |
| E. coli (336 samples) | LUCCK | 87.61 | 0.5937 |
|  | 3-NN | 85.08 | 0.0190 |
|  | 5-NN | 86.43 | 0.0193 |
|  | Adaboost | 74.13 | 0.6058 |
|  | SVM | 87.53 | 0.0448 |
|  | Random Forest (10) | 84.56 | 0.1075 |
|  | Random Forest (100) | 87.34 | 0.9265 |
|  | LDA | 81.46 | 0.0182 |

**Table 4.** Model accuracy with standard deviation and execution time for each model, averaged across the four UCI datasets.

| Method | Accuracy (%) | Time (s) |
|---|---|---|
| LUCCK | 88.38 $\pm$ 5.55 | 0.6507 |
| 3-NN | 82.89 $\pm$ 11.27 | 0.0166 |
| 5-NN | 82.77 $\pm$ 12.29 | 0.0167 |
| Adaboost | 75.24 $\pm$ 18.18 | 0.6695 |
| SVM | 75.16 $\pm$ 27.15 | 0.0333 |
| Random Forest (10) | 83.21 $\pm$ 8.65 | 0.1070 |
| Random Forest (100) | 86.32 $\pm$ 6.84 | 0.9389 |
| LDA | 79.41 $\pm$ 14.49 | 0.0201 |

*5.2. Fibromyalgia Dataset*

In this study, we have created a model that can be used to predict the quality of sleep or level of fatigue for people with fibromyalgia. The labels are self-assessed scores ranging from 1 to 10. Attempts to develop a regression model showed less promise than the results from a binary split. The most likely reason for this failure of the linear regression model is the nature of self-reported scores, especially those related to patient assessment of their level of pain. This fact is primarily due to the differences in individual levels of pain-tolerance. In previous studies [39,40], proponents of neural "biomarkers" argued that self-reported scores are unreliable, making objective markers of pain imperative. In another study [24], self-reported scores were found to be reliable only for extreme cases of pain and fatigue. Consequently, in this study, binary classification of extreme cases of fatigue and poor sleep is investigated. In this situation, a cutoff value is selected: patients that chose a value less than the threshold are placed in one group, while those that chose a value above the threshold are placed in another. As such, the values >8 are chosen for extreme cases of fatigue, and the values <4 are chosen for extreme cases of poor sleep quality. In this way, binary classifications are possible (>8 vs. <8 for fatigue and >4 vs. <4 for sleep). Using the extracted feature set, machine learning algorithms are applied and tested using 10-fold cross-validation. This is done in a way so as to prevent the data from any one patient being in multiple folds: all of a given patient's data are including entirely in a single fold. In addition, in order to address possibly imbalanced data during fold creation, random undersampling is performed to ensure the ratio between the two classes is not less than 0.3 (this rate is chosen since the extreme cases are at most 30 percent of the [1,10] interval of self-reported scores). This prevents the methods from developing a bias towards the larger class.

### 5.2.1. Results with Conventional Machine Learning Methods

A number of conventional machine learning models listed in Table 5 were applied to the extracted data in this study. As can be seen, many major machine learning methods were tested. For each of these methods, various configurations were tested, and the best sets of parameters were chosen using cross-validation (hyperparameter optimization). For instance, we used the combination of AdaBoost with different types of standard methods such as Decision Stump and Random Forest in order to explore the possibility of improving the performance of these methods via boosting. The $k$-nearest neighbor method with $k = 7$ was used in this experiment. For the weighted $k$-nearest neighbor method [36], the inversion kernel (inverse distance weights) with $k = 7$ resulted in the best performance. For the Neural Network algorithm, the Weka (Waikato Environment for Knowledge Analysis) [41] multilayer perceptron with two hidden layers was used. The results of using these machine learning approaches for prediction of extreme sleep quality (cutoff of 4) and fatigue level (cutoff of 8) are presented in Table 5. As shown in this table, the AdaBoost method based on random forest yielded the best results for quality of sleep (based on area under the receiver operating characteristic curve, or AUROC). For level of fatigue, the neural network was the best performing model.

### 5.2.2. Results with Our Machine Learning Method: Machine Learning Using Concave and Convex Kernels

In addition to the aforementioned conventional methods, we also used our machine learning approach that resulted in superior performance compared to the standard machine learning methods discussed above. Recall that in the Learning Using Concave and Convex Kernels algorithm, test data is classified by comparing it to all training data, properly weighted according to information extracted from each of the features (see Section 4 for further details). The results of applying our method to fibromyalgia are presented in Table 5, with cutoff values of 4 and 8 for quality of sleep and level of fatigue, respectively. As can be seen, LUCCK was able to vastly outperform other models on the fatigue outcome; however, the improvement on sleep outcome was not significant. This disparity is likely due to the different feature spaces for the sleep and fatigue outcomes. In general, the feature space for fatigue is significantly more dispersed, due to there being more samples (during daytime) and also that daytime activity negatively affects the signal quality, increasing dispersion. In contrast, signals (and their associated features) recorded during sleep are of better quality. This leads to the better prediction result for sleep in all methods used. Our proposed LUCCK algorithm can ameliorate the nature of the fatigue feature space, as it is specifically designed to reduce the effect of training data for which there is a large deviation from test data. As such, LUCCK was able to vastly outperform other models on the fatigue outcome. We should note that while the cohort size in this study seems to be limited, the continuous recording of physiological signals for seven days and nights created a comprehensive dataset. Additionally, similar to $k$-NN and its weighted version (and unlike SVM and neural network models), LUCKK can be trained even with few samples, which is one advantage of the proposed algorithm.

**Table 5.** Results of conventional machine learning methods.

| Method | Sleep | | Fatigue | |
|---|---|---|---|---|
| | Accuracy (%) | AUROC | Accuracy (%) | AUROC |
| AdaBoost - Decision Stump | 62.07 | 0.63 | 46.64 | 0.55 |
| AdaBoost - Random Forest | 59.97 | 0.65 | 51.24 | 0.55 |
| K-Nearest Neighbor | 60.55 | 0.55 | 51.88 | 0.53 |
| Weighted K-Nearest Neighbor | 65.27 | 0.62 | 68.05 | 0.51 |
| Neural Network | 63.47 | 0.64 | 54.80 | 0.59 |
| Random Forest | 63.32 | 0.63 | 52.46 | 0.57 |
| Support Vector Machine | 64.47 | 0.50 | 55.94 | 0.50 |
| LUCCK | 66.95 | 0.66 | 87.59 | 0.68 |

## 6. Conclusions and Discussion

In this study we primarily focused on prediction of the extreme cases of fatigue and poor sleep. As such, we have created preprocessing/conditioning methods that have the ability to improve the quality of parts of the signals with low quality due to motion artifact and noise. In addition, we identified a set of mathematical features that are important in extracting patterns from physiological signals that can distinguish poor and good clinical outcomes for applications such as fibromyalgia. Additionally, we showed that our proposed machine learning method outperformed the standard methods in predicting the outcomes such as fatigue and sleep quality. Generally, our proposed framework (preprocessing, mathematical features, and proposed machine learning method) can be employed in any study that involves prediction using BVP, HRV and EDA signals.

**Author Contributions:** Conceptualization, H.D.; Data curation, C.B. and S.A.; Formal analysis, E.S.; Funding acquisition, H.I. and K.N.; Supervision, A.K. and K.N.; Writing—original draft, E.S.; Writing—review & editing, J.G. and K.N.

**Funding:** This research was funded by Care Progress LLC through National Science Foundation grant number 1562254.

**Conflicts of Interest:** The authors declare no conflict of interest.

**Patents:** The epsilon tube filter is covered by US Patent 10,034,638, for which Kayvan Najarian is a named inventor.

## Abbreviations

The following abbreviations are used in this manuscript:

| | |
|---|---|
| LUCCK | Learning Using Concave and Convex Kernels |
| BVP | Blood Volume Pulse |
| EDA | Electrodermal Activity |
| ANS | Autonomic Nervous System |
| HRV | Heart Rate Variability |
| FFT | Fast Fourier transform |
| BPM | Breaths Per Minute |
| AUROC | Area Under Receiver Operating Characteristic Curve |

## Appendix A. Classification as Maximum a Posteriori Estimation

The classification approach suggested in Section 4.2 can also be viewed in terms of probability density functions. Suppose that $\int_{\mathbb{R}^n} Q(\mathbf{x}) = e$ with $0 < e < \infty$. The function

$$f_C(\mathbf{x}) = \frac{R(\mathbf{x}, C)}{me} = (me)^{-1} \sum_{\mathbf{y} \in C} Q(\mathbf{x} - \mathbf{y}) \tag{A1}$$

is therefore a probability density function. This probability density function is an estimation for the probability distribution from which the training data were taken.

We have

$$f_C = p(C_1)f_{C_1} + \cdots + p(C_d)f_{C_d}$$

where

$$f_{C_k}(\mathbf{x}) = \frac{R(\mathbf{x}, C_k)}{m_k e} = (m_k e)^{-1} \sum_{\mathbf{y} \in C_k} Q(\mathbf{x} - \mathbf{y}) \tag{A2}$$

is a probability density function for the training data in class $C_k$ for $k = 1, 2, \ldots, d$ and $p(C_k) := \frac{m_k}{m}$ is the probability that a randomly chosen training vector lies in $C_k$. $f_C$ can be considered as a mixture of the probability density functions $f_{C_1}, \ldots, f_{C_d}$. Suppose that $\mathbf{x} \in \mathbb{R}^n$ is taken from the distribution $f_{C_k}$ with probability $p(C_k)$, then the distribution for $\mathbf{x}$ is $f_C$. Given the outcome $\mathbf{x}$, the probability that it was taken from the distribution $f_{C_k}$ is

$$p_k(\mathbf{x}) := \frac{p(C_k)f_{C_k}(\mathbf{x})}{f_C(\mathbf{x})} = \frac{R(\mathbf{x}, C_k)}{R(\mathbf{x}, C)}.$$

This shows that the classifying scheme is the maximum a posteriori estimation. Instead of classifying a feature vector, the probability vector

$$p(\mathbf{x}) = (p_1(\mathbf{x}), p_2(\mathbf{x}), \ldots, p_d(\mathbf{x}))$$

can be given as output. The formula

$$p_k(\mathbf{x}) = \frac{R(\mathbf{x}, C_k)}{R(\mathbf{x}, C)}$$

is well-formed, even if $Q(\mathbf{x})$ does not have a finite integral, which may be the case in some examples.

## Appendix B. Examples

**Example A1.** *Suppose that there is only one feature, i.e., $n = 1$, then $Q(x)$ can be defined as*

$$Q(x) = (1 + \lambda_1 x^2)^{-\theta_1},$$

*whose graph at various values of $\theta$ and $\lambda$ is depicted in Figure A1:*

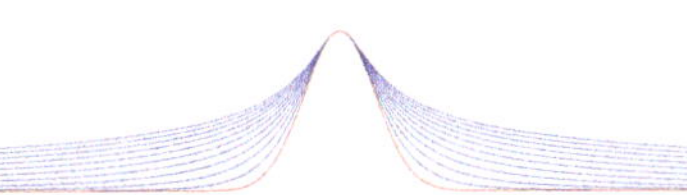

**Figure A1.** $Q(x) = (1 + \lambda_1 x^2)^{-1/\lambda_1}$ with for $\lambda_1 = 0.4, 0.8, \ldots, 4$ (blue curves) and $\lambda_1 = 0$ (red curve).

*As $\lambda_1$ goes to zero, the function converges to the normal distribution $e^{-x^2}$ (the red curve in Figure A1).*

**Example A2.** *Suppose that $n = 2$, then $Q(\mathbf{x})$ is defined as*

$$Q(x_1, x_2) = (1 + x_1^2)^{-1}(1 + x_2^2)^{-1},$$

*with $\theta_1 = \theta_2 = \lambda_1 = \lambda_2 = 1$. $Q(\mathbf{x})$ is depicted in Figure A2 at various level curves for $Q(\mathbf{x}) = \alpha$, with $0 < \alpha < 1$.*

*The Equation $Q(x_1, x_2) = \alpha$ is a closed curve. Such a curve can be thought of as the set of all points that have a given distance to the origin. We observe that for $\alpha \geq \frac{1}{4}$, the neighborhood*

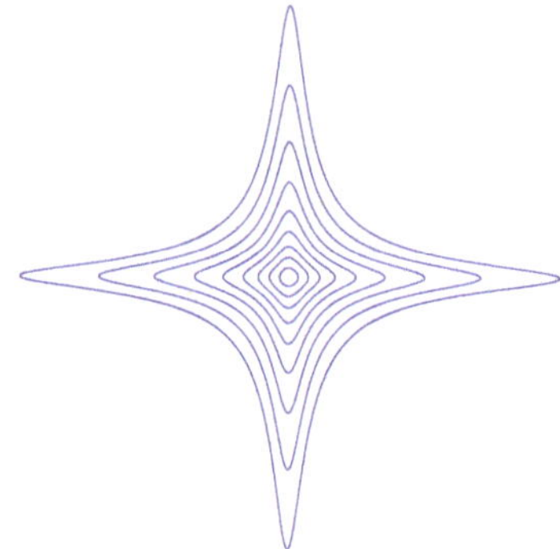

**Figure A2.** $Q(\mathbf{x}) = (1+x_1^2)^{-1}(1+x_2^2)^{-1} = \alpha$ with $0 < \alpha < 1$.

$$\{x \in \mathbb{R}^2 \mid Q(x) > \alpha\}$$

*of the origin is convex, but for $\alpha < \frac{1}{4}$ it is not.*

**Example A3.** *Consider the case when $n = 2$ and $\theta_1 = 1$, $\theta_2 = 2$, $\lambda_1 = 1$ and $\lambda_2 = \frac{1}{2}$, then $Q(\mathbf{x})$ is defined as*

$$Q(x_1, x_2) = (1 + 2x_1^2)^{-\frac{1}{2}}(1 + x_2^2)^{-1}.$$

$Q(\mathbf{x})$ *is depicted in Figure A3 at various level curves for $Q(\mathbf{x}) = \alpha$, with $0 < \alpha < 1$.*

**Figure A3.** $Q(\mathbf{x}) = (1 + 2x_1^2)^{-\frac{1}{2}}(1 + x_2^2)^{-1} = \alpha$ with $0 < \alpha < 1$.

*For small values of $\mathbf{x}$, the function $Q$ is equally sensitive to $x_1$ and $x_2$. However, if $\mathbf{x}$ is large, then $Q$ is more sensitive to $x_2$.*

## References


1. Moldofsky, H. The significance of dysfunctions of the sleeping/waking brain to the pathogenesis and treatment of fibromyalgia syndrome. *Rheumatic Dis. Clin.* **2009**, *35*, 275–283. [CrossRef]
2. Moldofsky, H. The significance of the sleeping–waking brain for the understanding of widespread musculoskeletal pain and fatigue in fibromyalgia syndrome and allied syndromes. *Joint Bone Spine* **2008**, *75*, 397–402. [CrossRef] [PubMed]
3. Horne, J.; Shackell, B. Alpha-like EEG activity in non-REM sleep and the fibromyalgia (fibrositis) syndrome. *Electroencephalogr. Clin. Neurophysiol.* **1991**, *79*, 271–276. [CrossRef]
4. Burns, J.W.; Crofford, L.J.; Chervin, R.D. Sleep stage dynamics in fibromyalgia patients and controls. *Sleep Med.* **2008**, *9*, 689–696. [CrossRef]
5. Belt, N.; Kronholm, E.; Kauppi, M. Sleep problems in fibromyalgia and rheumatoid arthritis compared with the general population. *Clin. Expe. Rheumatol.* **2009**, *27*, 35.
6. Landis, C.A.; Lentz, M.J.; Rothermel, J.; Buchwald, D.; Shaver, J.L. Decreased sleep spindles and spindle activity in midlife women with fibromyalgia and pain. *Sleep* **2004**, *27*, 741–750. [CrossRef]

7. Stuifbergen, A.K.; Phillips, L.; Carter, P.; Morrison, J.; Todd, A. Subjective and objective sleep difficulties in women with fibromyalgia syndrome. *J. Am. Acad. Nurse Pract.* **2010**, *22*, 548–556. [CrossRef] [PubMed]

8. Theadom, A.; Cropley, M.; Parmar, P.; Barker-Collo, S.; Starkey, N.; Jones, K.; Feigin, V.L.; BIONIC Research Group. Sleep difficulties one year following mild traumatic brain injury in a population-based study. *Sleep Med.* **2015**, *16*, 926–932. [CrossRef]

9. Buskila, D.; Neumann, L.; Odes, L.R.; Schleifer, E.; Depsames, R.; Abu-Shakra, M. The prevalence of musculoskeletal pain and fibromyalgia in patients hospitalized on internal medicine wards. *Semin. Arthritis Rheum.* **2001**, *30*, 411–417. [CrossRef] [PubMed]

10. Theadom, A.; Cropley, M.; Humphrey, K.L. Exploring the role of sleep and coping in quality of life in fibromyalgia. *J. Psychosom. Res.* **2007**, *62*, 145–151. [CrossRef]

11. Theadom, A.; Cropley, M. This constant being woken up is the worst thing–experiences of sleep in fibromyalgia syndrome. *Disabil. Rehabil* **2010**, *32*, 1939–1947. [CrossRef] [PubMed]

12. Stone, K.C.; Taylor, D.J.; McCrae, C.S.; Kalsekar, A.; Lichstein, K.L. Nonrestorative sleep. *Sleep Med. Rev.* **2008**, *12*, 275–288. [CrossRef]

13. Harding, S.M. Sleep in fibromyalgia patients: Subjective and objective findings. *Am. J. Med. Sci.* **1998**, *315*, 367–376. [PubMed]

14. Landis, C.A.; Frey, C.A.; Lentz, M.J.; Rothermel, J.; Buchwald, D.; Shaver, J.L. Self-reported sleep quality and fatigue correlates with actigraphy in midlife women with fibromyalgia. *Nurs. Res.* **2003**, *52*, 140–147. [CrossRef]

15. Fogelberg, D.J.; Hoffman, J.M.; Dikmen, S.; Temkin, N.R.; Bell, K.R. Association of sleep and co-occurring psychological conditions at 1 year after traumatic brain injury. *Arch. Phys. Med Rehabil.* **2012**, *93*, 1313–1318. [CrossRef]

16. Towns, S.J.; Silva, M.A.; Belanger, H.G. Subjective sleep quality and postconcussion symptoms following mild traumatic brain injury. *Brain Injury* **2015**, *29*, 1337–1341. [CrossRef] [PubMed]

17. Trinder, J.; Kleiman, J.; Carrington, M.; Smith, S.; Breen, S.; Tan, N.; Kim, Y. Autonomic activity during human sleep as a function of time and sleep stage. *J. Sleep Res.* **2001**, *10*, 253–264. [CrossRef]

18. Baharav, A.; Kotagal, S.; Gibbons, V.; Rubin, B.; Pratt, G.; Karin, J.; Akselrod, S. Fluctuations in autonomic nervous activity during sleep displayed by power spectrum analysis of heart rate variability. *Neurology* **1995**, *45*, 1183–1187. [CrossRef]

19. Sano, A.; Picard, R.W.; Stickgold, R. Quantitative analysis of wrist electrodermal activity during sleep. *Int. J. Psychophysiol.* **2014**, *94*, 382–389. [CrossRef]

20. Sano, A.; Picard, R.W. Comparison of sleep-wake classification using electroencephalogram and wrist-worn multi-modal sensor data. In Proceedings of the 36th Annual International Conference of the IEEE Engineering in Medicine and Biology Society, Chicago, IL, USA , 26–30 August 2014; Vol. 2014, p. 930.

21. Sano, A.; Picard, R.W. Recognition of sleep dependent memory consolidation with multi-modal sensor data. In Proceedings of the 2013 IEEE International Conference on Body Sensor Networks, Cambridge, MA, USA, 6–9 May 2013; pp. 1–4.

22. Sano, A.; Picard, R.W. Toward a taxonomy of autonomic sleep patterns with electrodermal activity. In Proceedings of the 2011 Annual International Conference of the IEEE Engineering in Medicine and Biology Society, Boston, MA, USA, 30 August–3 September 2011.

23. Lötsch, J.; Ultsch, A. Machine learning in pain research. *Pain* **2018**, *159*, 623. [CrossRef] [PubMed]

24. Robinson, M.E.; O'Shea, A.M.; Craggs, J.G.; Price, D.D.; Letzen, J.E.; Staud, R. Comparison of machine classification algorithms for fibromyalgia: Neuroimages versus self-report. *J. Pain* **2015**, *16*, 472–477. [CrossRef]

25. Sevel, L.; Letzen, J.; Boissoneault, J.; O'Shea, A.; Robinson, M.; Staud, R. (337) MRI based classification of chronic fatigue, fibromyalgia patients and healthy controls using machine learning algorithms: A comparison study. *J. Pain* **2016**, *17*, S60. [CrossRef]

26. López-Solà, M.; Woo, C.W.; Pujol, J.; Deus, J.; Harrison, B.J.; Monfort, J.; Wager, T.D. Towards a neurophysiological signature for fibromyalgia. *Pain* **2017**, *158*, 34. [CrossRef] [PubMed]

27. Sundermann, B.; Burgmer, M.; Pogatzki-Zahn, E.; Gaubitz, M.; Stüber, C.; Wessolleck, E.; Heuft, G.; Pfleiderer, B. Diagnostic classification based on functional connectivity in chronic pain: model optimization in fibromyalgia and rheumatoid arthritis. *Acad. Radiol.* **2014**, *21*, 369–377. [CrossRef] [PubMed]

28. Garcia-Zapirain, B.; Garcia-Chimeno, Y.; Rogers, H. Machine Learning Techniques for Automatic Classification of Patients with Fibromyalgia and Arthritis. *Int. J. Comput. Trends Technol.* **2015**, *25*, 149–152. [CrossRef]

29. Lukkahatai, N.; Walitt, B.; Deandrés-Galiana, E.J.; Fernández-Martínez, J.L.; Saligan, L.N. A predictive algorithm to identify genes that discriminate individuals with fibromyalgia syndrome diagnosis from healthy controls. *J. Pain Res.* **2018**, *11*, 2981. [CrossRef]

30. Ansari, S.; Ward, K.; Najarian, K. Epsilon-tube filtering: Reduction of high-amplitude motion artifacts from impedance plethysmography signal. *IEEE J. Biomed. Health Inf.* **2015**, *19*, 406–417. [CrossRef] [PubMed]

31. Frank, E.; Hall, M.; Holmes, G.; Kirkby, R.; Pfahringer, B.; Witten, I.H.; Trigg, L. Weka-a machine learning workbench for data mining. In *Data Mining and Knowledge Discovery Handbook*; Springer: Boston, MA, USA, 2009; pp. 1269–1277.

32. Tarvainen, M.P.; Niskanen, J.P.; Lipponen, J.A.; Ranta-Aho, P.O.; Karjalainen, P.A. Kubios HRV–heart rate variability analysis software. *Comput. Methods Programs Biomed.* **2014**, *113*, 210–220. [CrossRef] [PubMed]

33. Altman, N.S. An introduction to kernel and nearest-neighbor nonparametric regression. *Am. Stat.* **1992**, *46*, 175–185.

34. Dudani, S.A. The distance-weighted k-nearest-neighbor rule. *IEEE Trans. Sys. Man Cybern.* **1976**, *4*, 325–327. [CrossRef]

35. Beyer, K.; Goldstein, J.; Ramakrishnan, R.; Shaft, U. When is "nearest neighbor" meaningful? In *International Conference on Database Theory*; Springer, Berlin/Heidelberg, Germany, 1999; pp. 217–235.

36. Hechenbichler, K.; Schliep, K. Weighted k-nearest-neighbor techniques and ordinal classification. 2004. Available online: https://epub.ub.uni-muenchen.de/1769/ (accessed on 24 April 2019).

37. Mitchell, T. *Machine Learning*; McGraw-Hill International, Ed.; McGraw-Hill: New York, NY, USA, 1997.

38. Dua, D.; Graff, C. UCI Machine Learning Repository. 2017. Available online: https://ergodicity.net/2013/07/ (accessed on 24 April 2019).

39. Apkarian, A.V.; Hashmi, J.A.; Baliki, M.N. Pain and the brain: specificity and plasticity of the brain in clinical chronic pain. *Pain* **2011**, *152*, S49. [CrossRef] [PubMed]

40. Wartolowska, K. How neuroimaging can help us to visualise and quantify pain? *Eur. J. Pain Suppl.* **2011**, *5*, 323–327. [CrossRef]

41. Hall, M.; Frank, E.; Holmes, G.; Pfahringer, B.; Reutemann, P.; Witten, I.H. The WEKA data mining software: An update. *ACM SIGKDD Explor. Newslett.* **2009**, *11*, 10–18. [CrossRef]

*Article*

# Action Recognition Using Single-Pixel Time-of-Flight Detection


Ikechukwu Ofodile [1,†], Ahmed Helmi [1,†], Albert Clapés [2,†], Egils Avots [1,†], Kerttu Maria Peensoo [3,†], Sandhra-Mirella Valdma [3,†], Andreas Valdmann [3,†], Heli Valtna-Lukner [3,†], Sergey Omelkov [3,†], Sergio Escalera [2,4,†], Cagri Ozcinar [5,†] and Gholamreza Anbarjafari [1,6,7,*,†]

1   iCv Lab, Institute of Technology, University of Tartu, 50411 Tartu, Estonia; ike@icv.tuit.ut.ee (I.O.); ahmed@icv.tuit.ut.ee (A.H.); ea@icv.tuit.ut.ee (E.A.)
2   University of Barcelona, 08007 Barcelona, Spain; aclapes@cvc.uab.es (A.C.); sergio@maia.ub.es (S.E.)
3   Institute of Physics, University of Tartu, 50411 Tartu, Estonia; kerttumariapeensoo@gmail.com (K.M.P.); sandhra91@gmail.com (S.-M.V.); andreas.valdmann@gmail.com (A.V.); heli.lukner@ut.ee (H.V.-L.); sergey.omelkov@ut.ee (S.O.)
4   The Computer Vision Centre, 08193 Barcelona, Spain
5   Trinity College Dublin, Dublin 2, Ireland; ozcinarc@scss.tcd.ie
6   Department of Electrical and Electronic Engineering, Hasan Kalyoncu University, Gaziantep 27000, Turkey
7   Institute of Digital Technologies, Loughborough University London, London E15 2GZ, UK
*   Correspondence: shb@icv.tuit.ut.ee; Tel.: +372-737-4855
†   All authors contributed equally to this work.


Received: 14 January 2019; Accepted: 15 April 2019; Published: 18 April 2019

**Abstract:** Action recognition is a challenging task that plays an important role in many robotic systems, which highly depend on visual input feeds. However, due to privacy concerns, it is important to find a method which can recognise actions without using visual feed. In this paper, we propose a concept for detecting actions while preserving the test subject's privacy. Our proposed method relies only on recording the temporal evolution of light pulses scattered back from the scene. Such data trace to record one action contains a sequence of one-dimensional arrays of voltage values acquired by a single-pixel detector at 1 GHz repetition rate. Information about both the distance to the object and its shape are embedded in the traces. We apply machine learning in the form of recurrent neural networks for data analysis and demonstrate successful action recognition. The experimental results show that our proposed method could achieve on average 96.47% accuracy on the actions walking forward, walking backwards, sitting down, standing up and waving hand, using recurrent neural network.


**Keywords:** single pixel single photon image acquisition; time-of-flight; action recognition

---

## 1. Introduction

Action is a spatiotemporal sequence of patterns [1–6]. The ability to detect movement and recognise human actions and gestures would enable advanced human to machine interaction in wide scope of novel applications in the field of robotics from autonomous vehicles, surveillance for security or care-taking to entertainment.

In the field of machine vision, the majority of effort has been put into recognising human action from video sequences [7–9], because overwhelmingly imaging devices mimic the human-like perception of the surroundings and video format is most widely available. Videos are a sequence of two-dimensional intensity patters, captured by using an imaging lens projecting the scene to a two-dimensional detector array (a charge coupled device (CCD) device, for example). Unlike living creatures, the ever growing field of robotics has run into major difficulties while trying to recognise objects, their actions, and   their distances from two-dimensional images.  Processing the data is computationally demanding, and depth information is not unanimously retrievable.

Deep neural networks, due to their high accuracy, are widely used in many of the computer vision applications such as emotion recognition [10–16], biometric recognition [17–20], personality analysis [21,22], and activity analysis [5,23,24]. Depending on the nature of the data, different structures can be used [25,26]. In this work, we deal with time-series data, i.e., we handle temporal information. For this purpose, we are mainly focused on recurrent neural networks (RNN) and long-short term (LSTM) algorithms.

In addition to colour and intensity, incident light can be characterised by its propagation direction(s), spectral content and temporal evolution in the case of pulsed illumination.  Light also carries information about its source, and each medium, refraction, reflection and scattering event it has encountered or traversed.  This enables various uncommon ways to characterise the scene. The rapid advancements in optoelectronics and availability of sufficient computational power enable innovative imaging and light capturing concepts, which serve as the ground for action detection. For example, a detector, capable of registering evolution of backscattered light with a high temporal resolution in a wide dynamic range, would be able to detect even objects hidden from the direct line of sight [27–29]. Along the same vein, several alternative light-based methods have been developed for resolving depth information or 3D map of the surroundings (some examples can be found in [30–34]) giving 3D information in voxel format about the scene, which is also suitable for action detection [35]. Combining the fundamental understanding of light propagation and computational neural networks for the data reconstruction, it appears that objects or even persons can be detected using a single pixel detector registering temporal evolution of the back-scattered light pulse [36].

In this work, which is a feasibility study of a novel setup and methodology for conducting action recognition, we propose and demonstrate an action recognition scheme based on a single-pixel direct time-of-flight detection.  We use NAO robots in a controlled environment as a test subject. We illuminate a scene with a diverging picosecond laser pulse, (30 ps duration) and detect the temporal evolution of back scattered light with a single pixel multiphoton detector of 600 ps temporal resolution. Our data contains one-dimensional time sequences presenting the signal strength (proportional to the number of detected photons) versus arrival time. Information about both the distance to the object and its shape are embedded in the traces. We apply machine learning in the form of recurrent neural networks for data analysis and demonstrate successful action recognition.

The following list summarises the contributions of our work:

- Introduce an unexplored data modality for action recognition scenarios.  In contrast to other depth-based modalities, our single-pixel light pulses are not visually interpretable which makes it a more privacy preserving solution.
- Provide a manually annotated dataset of 550 robot action sequences, some of them containing obstacle objects.
- Apply multi-layer bi-directional recurrent neural networks for the recognition task as an initial machine learning benchmarking baseline.
- Present an extensive set of experiments on the recognition of several action classes. We demonstrate how the learning models are able to extract proper action features and generalise several action concepts from captured data.

The rest of the paper is organised as follows: in Section 2, related works to single-pixel, single-photon acquisition and action recognition are reviewed. Section 3 describes the data collection and the details of the setup. In Section 4, the details of the proposed deep neural network algorithm used for action recognition are described. The experimental results and discussions are provided in Section 5. Finally, the work is concluded in Section 6.

## 2. Related Work

The filed of motion analysis was firstly inspired by intensity images and progresses towards depth images, which are more robust in comparison to intensity images. In the case of action recognition, the most useful are sensors that provide depth map. Nevertheless, data are processed to extract human silhouettes, body parts, skeleton and pose of the person, which in turn are used as features for machine learning methods to classify actions. These sensors have drawn much interest for human activity related research and software development.

### 2.1. Depth Sensors

Depth images provide the 3D structure of the scene, and can significantly simplify tasks such as background subtraction, segmentation, and motion estimation. With the recent advances in depth sensor hardware, such as time-of-flight (ToF) cameras, research based on depth imagery has appeared. Three main depth sensing technologies are applied in computer vision research: stereo cameras, time-of-flight (ToF) cameras and structured light.

1. Stereo cameras infer the 3D structure of a scene from two images from different viewpoints. The depth map is created using information about the camera setup (stereo triangulation) [37].
2. A time-of-flight (ToF) camera estimates distance to an object surface using active light pulses from a single camera, whose time to reflect from the object give the distance. Such devices use a sinusoidally modulated infra-red light signal, and distance is estimated using the phase shift of the reflected signal on CMOS or CCD detector. The most commercially know device that uses this technology is Kinect 2 [38,39], which provides depth map of $512 \times 424$ pixels at 30 frames per second.
3. Structured light sensors [40] such as Kinect 1 [41] which was released in November 2010 by Microsoft. Kinect 1 consists of an RGB camera and a depth sensor. The depth sensor provides depth map of $320 \times 240$ pixels at 30 frames per second.

Similar to ToF depth cameras, action can be encoded in a laser pulse, which is captured by single-pixel cameras. The contents of the scene are encoded in time-series data. When using single pixel camera setups, processing steps such as pose estimation is not necessary. The acquired time series data are usable for machine learning tasks without any modification or additional processing.

### 2.2. Sing-Pixel Single-Photon Acquisition

Recent advances in photonics offer various innovative approaches for three-dimensional imaging [42]. Among those is time-of-flight imaging, which enables detection and tracking of objects. This involves illuminating the scene with diverging light pulses shorter than 100 ps. The light is scattered back from the scene and its flight time is detected with respective accuracy. Flight time $t$ of light multiplied by the speed $c$ of light directly gives the distance the light pulse has travelled from the source to the detector. Often, the laser source and detector are nearby and the value $ct$ equals twice the distance of the object. Compared to time-of-flight ranging used in LiDARs (Light Detection And Ranging device), the principles introduced here utilise the knowledge of light propagation and are potentially capable of achieving higher spatial resolution.

In early experiments, 50 fs pulse duration mode-locked Ti:Sapphire near-infrared (NIR) laser and streak camera of 15 ps temporal resolution with array matrix were used to detect movement in occluded environment or to recover the 3D shape of an object behind direct line of sight [28,43]. (Using light pulses as short as 50 fs was not necessary; this is a widely spread ultrashort pulse laser source available in photonics labs.) The reconstruction of the object shape required data traces from various viewing angles and mathematical back-projection. In the scope of current research, the non-line-of-sight illumination can be seen as a method of efficiently diverging the incident laser pulse on the scene. There has been several suggestions to use more widely accessible hardware by replacing expensive and fragile streak camera with single photon avalanche diode (SPAD) [29], or to construct a setup based on modulated laser diodes and single pixel photonic mixer device [44].

In proof-of-principle experiment [45] a single pixel SPAD detector (the actual 32 × 32 pixels were used for statistics and to speed up the measurements, such device was and early prototype at the time) was used and ca. 50 ps temporal resolution was utilised to demonstrate the ability to detect linear movement of a non-line-of-sight object. Again, ultrashort 10 fs pulse duration Ti:Sapphire laser with carrier wavelength in NIR region was used. Instead of recording the shape of the object, the shape of its reflection on a screen (a floor) was recorded and position of the object was derived from geometry. Replacing the detector array by three single-pixel SPAD detectors, real-time movement of an object was traced [46]. In this experiment, pulsed NIR diode was used instead of Ti:Sapphire laser. The integration time for single-photon detector was reduced from approximately 3 s to 1 s. In consequent papers, the table-top scenes are scaled up to detect a human [36,47]. Significance of the solution presented in [36] relies on artificial neural network machine learning algorithms for data analysis instead of deterministic tools used before. As a result, the team led by Daniele Faccio was able to distinguish between several standing position of a human and distinct between three different persons by analysing merely one-dimensional trace of SPAD detector.

### 2.3. Action Recognition

Most action recognition and monitoring systems use images with high enough quality where a person can be identified. When considering commercial applications, such systems invade human privacy. The identification factor can be removed by blurring or obscuring the images, downscaling, using encryption and IT solutions to keep the stored data safe. Nevertheless, at some point data is available in a format where people can be identified and can be mishandled due to breach of security, selling private data for commercial purposes or by request from governmental authorities.

One way of removing the privacy concerns is to use devices which by default use low resolution images, hence eliminating privacy issues at the data acquisition step. For such purpose, researchers are developing methods for action recognition using single pixel and low-resolution cameras. A privacy preserving method was proposed Jia and Radke [48] to track a person and estimate pose of a person using a network of ceiling-mounted time-of-flight sensors. Tao et al. [49] based their solution on a network of ceiling-mounted binary passive infrared sensors to recognise a set of daily activities. Kawashima et al. [50] used extremely low-resolution (16 × 16 pixels) infrared sensors to monitor a person constantly day and night without privacy concerns. Ji Dai et al. [51] studied the privacy implications using virtual space for action recognition. They studied Kinect 2 resolutions from 100 × 100 pixels down to 1 × 1 and their effect on action recognition methods. To address privacy issues, Xu et al. [52] proposed a fully-coupled two-stream spatiotemporal architecture for reliable human action recognition on extremely low resolution (e.g., 12 × 16 pixel) videos.

In this research work, we develop a new methodology for action recognition without using any data which can rise a privacy issue. Such a system can be highly used in places such as nursery and hospitals where recognition of actions might be important without violating the privacy rights of people in the environment.

### 2.4. Data Interpretability

In comparison to devices such as Kinect, the depth map provides enough information about a person's body shape and height, and facial features to visually identify the person and his/her actions in the scene. In the proposed experimental setup, we recorded a kind of a depth map, but it was recorded with a single pixel detector. Hence, the trace has no spatial resolution, which would enable identifying a person or an object directly through detecting above mentioned properties. The spatial properties of the scene are imprinted into the temporal evolution of the recorded trace. In the case an action takes place, characteristic temporal evolution pattern is imprinted to the recorded trace. The recorded 1D time series containing temporal evolution of back scattered light (timestamped detected photon amplitudes) is enough to recognise human actions when interpreted using machine learning algorithms. In the case of a static scene, there is no change in the consequent temporal traces, indicating that no actions are taking place. In addition, the data footprint of a 1D data trace is smaller than that of a depth map. This enables rapid processing times. In the case of using Kinect, the data processing pipeline contains human interpretable data that could be used for unlawful purposes, but in the proposed setup such possibility does not exist.

## 3. Collected Data

In this research, for data collection, we created a special setup. Figure 1 shows the general data collection setup, including the placement of the laser and the detector sensor. The data has been collected under the control environment where a NAO V4 humanoid robot was placed in a black box with dimensions of $800 \times 800 \times 1200$ mm$^3$ (W $\times$ H $\times$ L) and was used to conduct some pre-defined actions. The scene was illuminated by Fianium supecontinuum laser source (SC400-2-PP) working at 1 MHz rate. The scatterer ensured that the whole scene Was illuminated at once, without any scanning or other moving parts required. The reflected light from the scene was collected by a Hamamatsu R10467U-06 hybrid photodetector (HPD) with spectral sensitivity range of 220–650 nm. The neutral density filter (OD2) was used in front of the detector to prevent HPD damage due to overexposure. The signal from the HPD was boosted by a Hamamatsu C10778 preamplifier (37 dB, inverting) and then directly digitised by LeCroy WaveRunner 6100a (1 GHz, 10 Gs/s) oscilloscope. The HPD was used in a pulse current (multiphoton) mode, therefore the time resolution of the system was determined by its single-photon pulse response of 600 ps FWHM. The oscilloscope worked in a sequence acquisition mode, recording 200 traces from subsequent trigger events during one sequence, with average frame rate of five sequences per second. The traces within one sequence were averaged to improve signal-to-noise ratio due to both electronic noise and photon statistics. The usage of multiphoton detection mode allowed greatly reducing acquisition time per frame, although with a lower time resolution, unlike the single photon detection used in [46]. The oscilloscope traces in the form of reflected light intensity versus time in nanosecond scale contained all the relevant information about the scene in a non-human-readable form, thus preserving privacy. The series of such traces recorded a 5 fps therefore contain the information about motion.

Various experiments were performed using one- and two-robot setups. A short summary can be seen in Table 1, More detailed description of the tasks can be found in the following sections.

**Table 1.** Summary of the performed actions.

| | One-Robot | | | | | Two-Robot | | |
|---|---|---|---|---|---|---|---|---|
| **Task** | **Walk Forward** | **Walk Reverse** | **Sit Down** | **Stand Up** | **Hand Wave** | **Object Setup** | **Same Action** | **Different Actions** |
| Repetitions | 125 | 125 | 50 | 50 | 50 | 156 | 70 | 20 |

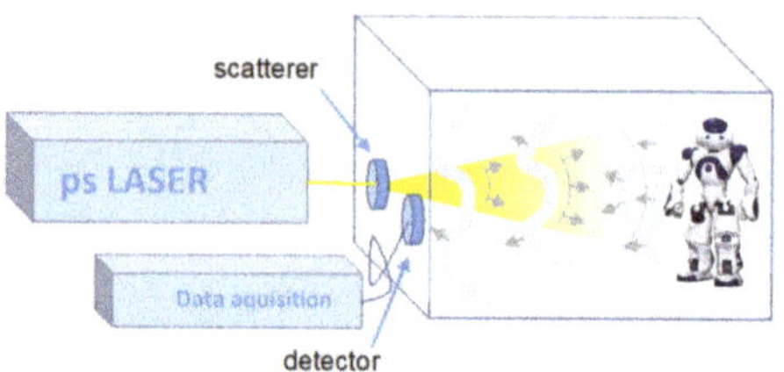


**Figure 1.** The data collection setup: Fianium laser delivers 30 ps duration light pulses. The collimated laser beam is directed to a scatterer, which creates divergent speckle pattern (giving divergence of 40 degree apex angle) inside the box, which are directed to the black box specially designed for the robot. Scattering illumination will reduce potential interference effects at the detector and, using controlled speckle pattern could be used to increase the lateral resolution. The light scattered from the moving object (NAO V4) and the walls is detected using single-pixel hybrid photodetector (HPD), which detects the temporal evolution of back scattered light.

*3.1. ONE-Robot Setup*

Initially, for acquiring training data, only one robot was used. Experiments were divided into the following categories:

1.  **Directional walk**: We specified three starting points (A, B, and C) and three end points (a, b, and c) inside the box. The robot walked from starting points at 70 cm distance to corresponding end points, and vice versa. In addition, two diagonal directions, from Point A to Point c and from Point C to Point a, were travelled both forward and reverse, as illustrated in Figure 2. All action were repeated 25 times for each point per each direction. These walking actions are shown in Tables 2 and 3.
2.  **Sitting down (sd) from standing up pose and Standing up (su) from sitting down pose**: We specified five areas where the robot was located, as illustrated in Figure 3. These actions were repeated 10 times per area and in each repetition, the position of the robot was nearly the same. Summary of performed tasks is shown in Table 4.
3.  **Waving right hand (hw) for 3 s**: This action was repeated 25 times in Areas 2 and 5 (see Table 4).
4.  **Include both object and robot**: An object was placed in the environment while the robot was doing the six tasks, which are listed in Table 5 (see Figure 4). Each task was repeated 12 times.

**Table 2.** Forward (F) movement.

| Task | A1 | A2 | B1 | C1 | C2 |
|---|---|---|---|---|---|
| Repetitions | 25 | 25 | 25 | 25 | 25 |
| Start location | A | A | B | C | C |
| Stop location | a | c | b | c | a |

**Table 3.** Reverse (R) movement.

| Task | A1 | A2 | B1 | C1 | C2 |
|---|---|---|---|---|---|
| Repetitions | 25 | 25 | 25 | 25 | 25 |
| Start location | a | c | b | c | a |
| Stop location | A | A | B | C | C |

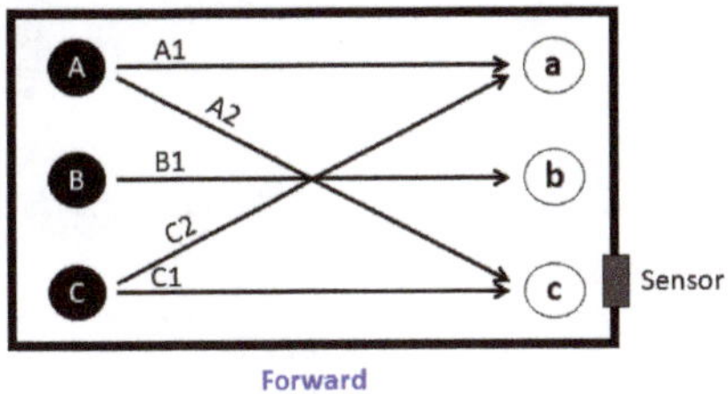



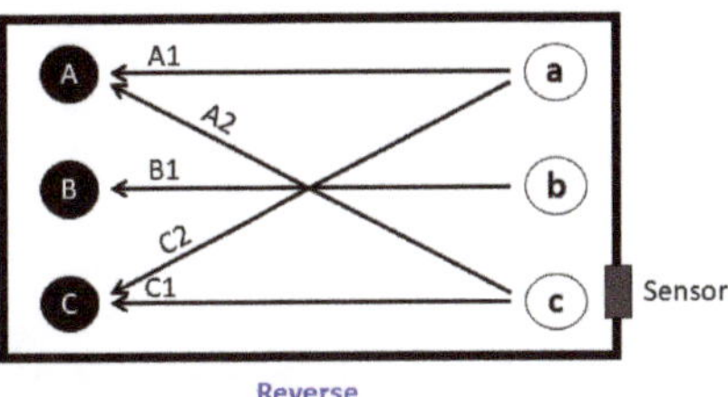



**Figure 2.** Start and endpoints, showing paths of the robot during directional walk.

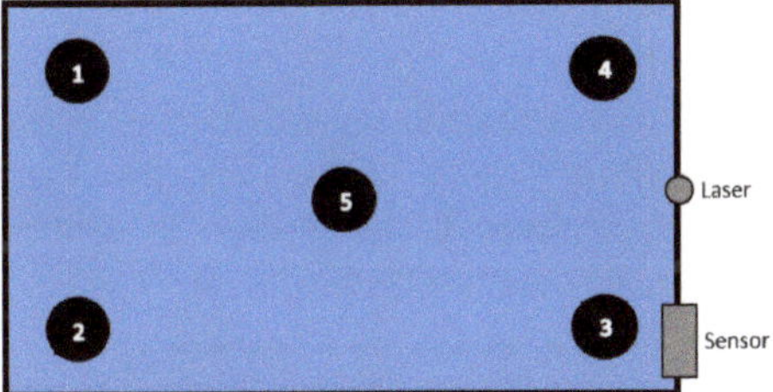



**Figure 3.** Positions of sitting down and standing up actions.

**Table 4.** Tasks performed in specific locations.

| Task | Sit Down | | | | | Stand Up | | | | | Hand-Wave | |
|---|---|---|---|---|---|---|---|---|---|---|---|---|
| Repetitions | 10 | 10 | 10 | 10 | 10 | 10 | 10 | 10 | 10 | 10 | 25 | 25 |
| Location | 1 | 2 | 3 | 4 | 5 | 1 | 2 | 3 | 4 | 5 | 2 | 5 |
| Action | sd | sd | sd | sd | sd | su | su | su | su | su | hw | hw |

**Table 5.** Tasks in presence of an object.

| Task | 1 | 2 | 3 | 4 | 5 | 6 * |
|---|---|---|---|---|---|---|
| Object location | Figure 4a | Figure 4b | Figure 4c | Figure 4d | Figure 4e | Figure 4f |
| **Walk Forward** | A to a | C to c | A to c | C to a | B to b | hw |
| Repetitions | 12 | 12 | 12 | 12 | 12 | 12 |
| **Walk Reverse** | a to A | c to C | a to C | c to A | b to B | su/sd |
| Repetitions | 12 | 12 | 12 | 12 | 12 | 12/12 |

* In Task 6, the robot did not go forward or reverse, but performed hand-wave, stand-up and sit down action, where each action was repeated 12 times.

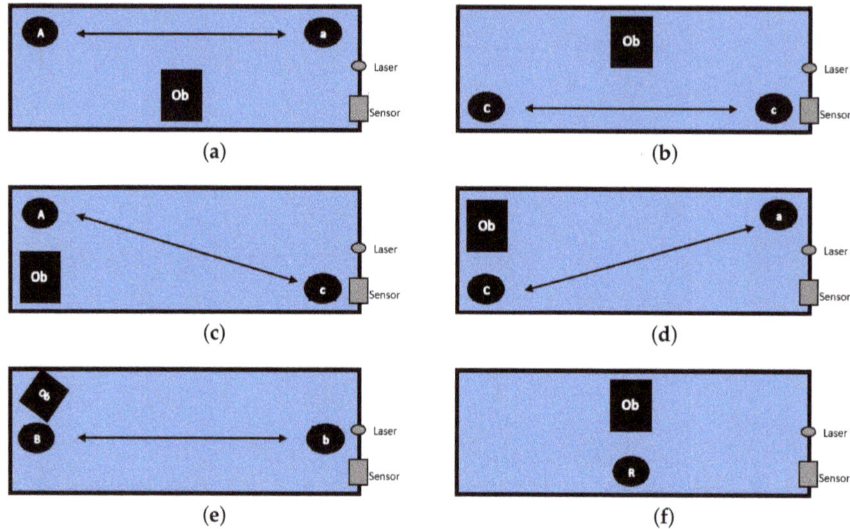

**Figure 4.** Positions of object during various robot actions.

*3.2. Two-Robot Setup*

We also devised new setup with two NAO V4 humanoid robots. Firstly, one robot was standing still at Position 1 and the other robot at Position 2 walks forward and reverse to Positions 3 and 4, as shown in Figure 5. This action was repeated 10 times. In the next experiment, both robots performed actions simultaneously. Performed actions are listed in Table 6).

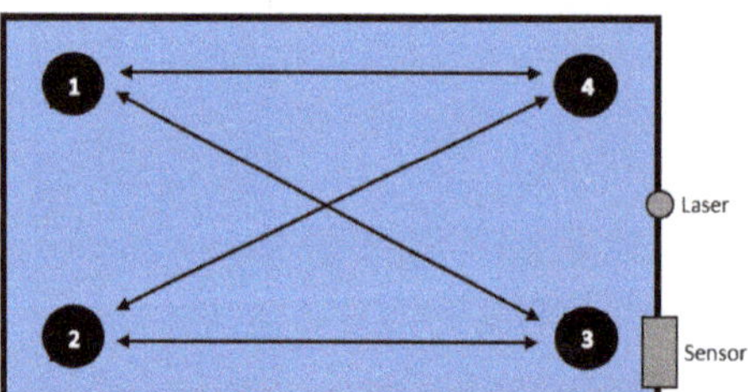

**Figure 5.** Position of two robots during actions.

**Table 6.** Actions performed by two robots.

| Repetition | Robot 1 Action | Position | Robot 2 Action | Position |
|---|---|---|---|---|
| 10 | Forward | 1 to 4 | Forward | 2 to 3 |
| 10 | Sit Down | 1 | Sit Down | 2 |
| 10 | Stand Up | 1 | Stand Up | 2 |
| 10 | Sit Down | 3 | Sit Down | 4 |
| 10 | Stand Up | 3 | Stand Up | 4 |
| 10 | Hand-Wave | 1 | Hand-Wave | 2 |
| 10 | Hand-Wave | 3 | Hand-Wave | 4 |
| 10 | Stand | 1 | Forward | 2 to 3 |
| 10 | Stand | 1 | Forward | 2 to 4 |

In Figure 6, we illustrate a few examples of preprocessed data, which was used in training. Columns correspond to different actions and the rows are different examples. That is, the sequences consisted of a time series of 500-dimensional vectors.

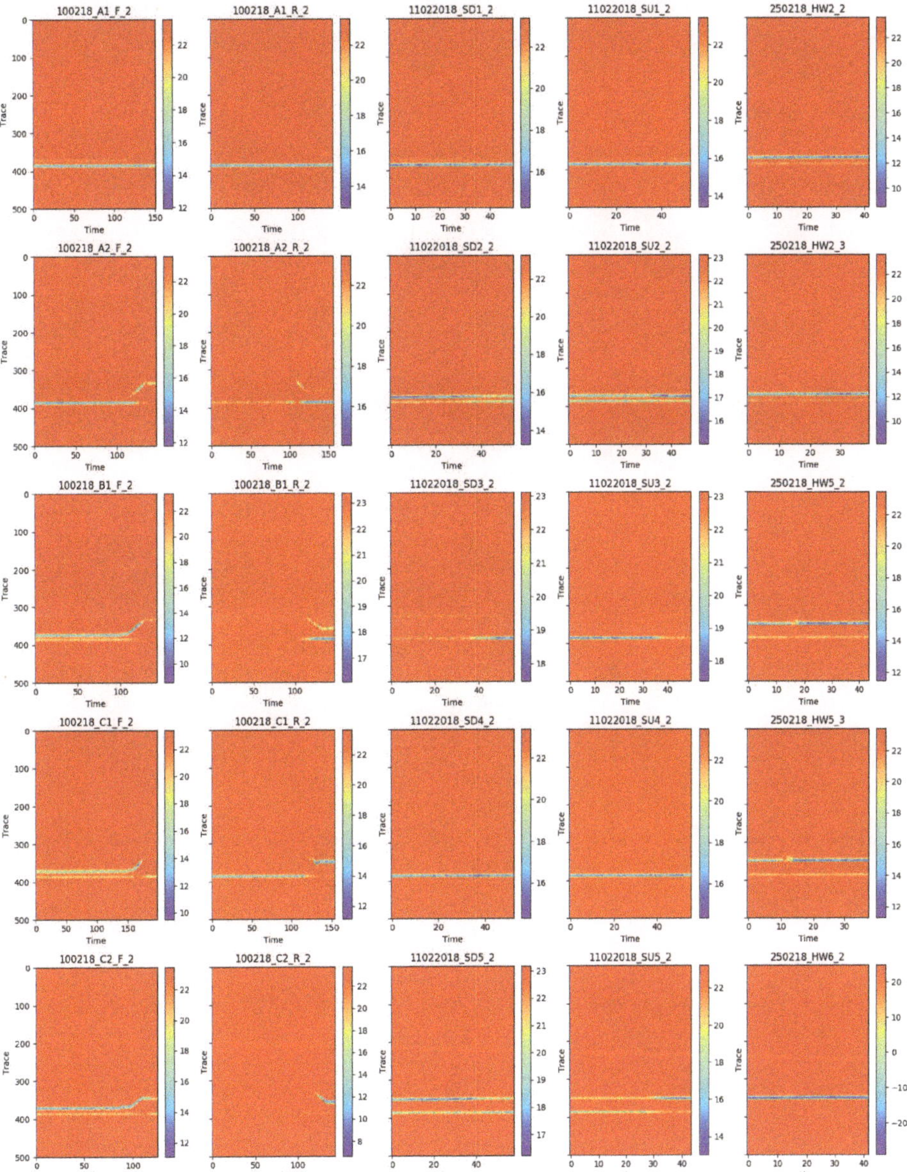

**Figure 6.** Visualisation of the traces throughout time (x-axis). Columns correspond to different actions (respectively, forward walking, reverse walking, sitting down, standing up, and waving), whereas rows correspond to different examples. The titles on the subplots correspond to the sequence files in the dataset.

## 4. Method

We chose a recurrent neural network (RNN) as our baseline. Recurrent nets are able to model multivariate time-series—in our case, time-of-flight measurements—and output a class prediction by considering the whole temporal sequence. In particular, our choice was a RNN with Gated-Recurrent Unit (GRU) cells. These cells can retain long-temporal information using internal gates and a set of optimisable parameters.

### 4.1. Gated-Recurrent Unit

We briefly introduce GRUs following the notation from [53]. Let $\mathbf{x} = (\mathbf{x}_1, \ldots, \mathbf{x}_t, \ldots, \mathbf{x}_T)$, $\mathbf{x}_t \in \mathbb{R}^n$ be a sequence of $T$ observations and $y \in C$ its ground truth class label. At each time step $t$, a GRU cell receives $\mathbf{x}_t$ and outputs an activation $h_t \in \mathbb{R}^m$ response

$$h_t^j = (1 - z_t^j)h_{t-1}^j + z_t^j \tilde{h}_t^j \tag{1}$$

by combining activation at previous time step $h_{t-1}^j$ and a candidate activation from the current time step $\tilde{h}_t^j$.

The trade-off factor $z_t^j$, namely *update gate*, is calculated as

$$z_t^j = \sigma(W_z \mathbf{x}_t + U_z \mathbf{h}_{t-1})^j, \tag{2}$$

where $W_z \in \mathbb{R}^{m \times n}$ and $U_z \in \mathbb{R}^{m \times m}$ are optimisable parameters shared across all $t$ and $\sigma$ a sigmoid function that outputs values in the interval (0,1).

In its turn, the *candidate activation* is calculated

$$\tilde{h}_t^j = \tanh(W \mathbf{x}_t + U(\mathbf{r}_t \odot \mathbf{h}_{t-1}))^j, \tag{3}$$

where $\odot$ is the element-wise product of two vectors and $\mathbf{r}_t$ also known as *reset gate*. Note that $W \in \mathbb{R}^{m \times n}$ and $U \in \mathbb{R}^{m \times m}$ are different sets of parameters from $W_z$ and $U_z$.

Similar to the update gate $\mathbf{z}_t$, the *reset gate* is

$$r_t^j = \sigma(W_r \mathbf{x}_t + U_r \mathbf{h}_{t-1})^j. \tag{4}$$

Finally, the last GRU activation at time $T$ is input to a dense layer with softmax activation function. From the dense layer, the logit value $z^i$ is computed by

$$z^i = \sum_j w_s^{ij} h_T^j, \tag{5}$$

where $W_s = (w_s^{ij})$ are the softmax layer weights. Then, the softmax activation function can be applied to output the sequence classification label

$$\hat{y}^i = \frac{e^{z^i}}{\sum_i e^{z_i}}. \tag{6}$$

### 4.2. Bidirectional GRU and Stacked Layers

Bidirectional recurrent networks consist of two independent networks processing the temporal information in the two temporal dimensions, forward and reverse, so their activation outputs are concatenated. The input of the reverse recurrent network is simply the reversed input sequence. The logit value computation becomes

$$z^i = \sum_j w_s^{ij} [h_{\text{fw},T}^j, h_{\text{rv},0}^j],$$ (7)

where $[\cdot, \cdot]$ is the concatenation of forward and reverse GRUs activations.

In addition, GRU layers can be stacked to form a deeper GRU architecture. The first GRU layer receives as input the sequence of observations **x**, whereas each subsequent layers are fed with activation outputs from the previous layer. We finally apply the softmax dense layer to the activations of the deepest stacked layer.

### 4.3. Baseline

Our architecture is a two-layer bidirectional GRU, each GRU with 512 neurons (experimentally chosen). The size of the softmax dense layer is the number of classes $|C|$. Figure 7 illustrates the architecture.

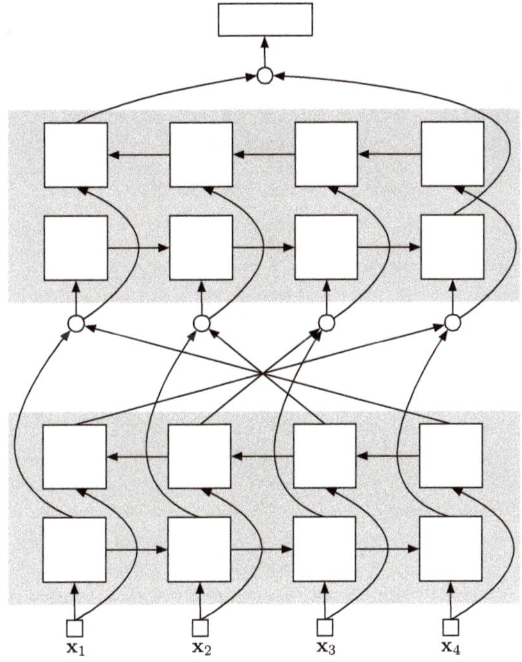


**Figure 7.** The two-layer bidirectional GRU baseline architecture. Arrays represent information flow, grey rectangles are bidirectional GRU layers, and circles represent the concatenation operation.

## 5. Experimental Results and Discussion

### 5.1. Learning Model Details and Code Implementation

Among different RNN cells, we chose Gated-Recurrent Units (GRU) for our baseline architecture. Compared to other recurrent cells, such as Long-Term Short Memory (LSTM) cells, these require a reduced number of parameters while still retaining long-term temporal information and providing highly competitive performance [54]. GRU is also often chosen over LSTM because hidden states are fully exposed and hence easier to interpret.

For the model computations, we entirely relied on GPU programming. In particular, our implementation is based on Keras [55], a GPU-capable deep-learning library written in Python. As for the GPU device itself, we utilised an NVIDIA Titan Xp with 12 GB of GDDR5X memory.

### 5.2. Ablation Experiments on GRU Architectures

To determine the best GRU architecture, we first performed a set of binary classification experiments on the following actions: forward (walking), reverse (walking), sit-down, standing up, and handwaving. We report the performance in terms of accuracy (averaging accuracies over a 10-fold cross validation). In Table 7, we illustrate the ablation experiments on different multi-layer and bidirectional GRU architectures with fixed hidden layer size to 64 neurons. For each architecture and target action, we trained a different GRU model for 25 epochs, which was enough to avoid under-fitting in the most complex model (two-layer biGRU).

In particular, the most complex model, two-layer biGRU, was the one that provided the best result. This showed how both multiple and bidirectional layers can help to model single-pixel time-of-light data sequences. In particular, adding a second stacked layer provided a +5.09% improvement over one single layer, whereas the bidirectionality increased accuracy by 4.4%. The +8.37% gain from using both showed how those two architecture variations are highly complementary when dealing with our data.

**Table 7.** Comparison on GRU models with multiple layers and/or bidirectionality. In this ablation, we defined a set of five binary problems: forward, reverse, sit-down, stand-up, and hand-wave actions. The results reported are class-weighted accuracies averaged over a 10-fold cross validation. The "Average" column is the average of performances on binary problems.

|  | Forward | Reverse | Sit-Down | Stand-Up | Handwave | Average |
|---|---|---|---|---|---|---|
| GRU (1-layer, 64-hidden) | 87.28 | 83.85 | 76.48 | 78.15 | 0.945 | 84.05 |
| GRU (two-layer, 64-hidden) | 88.48 | 90.06 | 85.75 | 86.41 | 94.99 | 89.14 |
| biGRU (1-layer, 64-hidden) | 89.28 | 87.62 | 82.49 | 86.85 | 96.02 | 88.45 |
| biGRU (two-layer, 64-hidden) | 91.42 | 91.08 | 90.07 | 92.51 | 97.01 | 92.42 |

Next, using a two-layer biGRU, we performed another set of ablation experiments on hidden layer sizes: $\{32, 64, 128, 256, 512\}$. Since the hidden layer size drastically affects the number of parameters to optimise during the training stage, each model was trained during a different number of epochs: $\{10, 25, 50, 100, 200, 400\}$, respectively. Results are shown in Table 8.

The largest model, i.e., 512 hidden layer neurons, performed the best. Its +5.62% gain with respect to the smallest two-layer biGRU model with 32 neurons demonstrated room for improvement from using more complex models despite the presumed simplicity of single-pixel time-of-flight time-series. However, we discarded further increasing the hidden size because of computational constraints: enlarging the hidden layer causes an exponential grow of the number of parameters to train. In particular, a model with 32 hidden neurons consisted of 121 K parameters, whereas 512 hidden neurons increased the size up to 7.8 M (and 37.7 M in the case of 1024 neurons); this and the saturation of accuracy discouraged us to keep enlarging the hidden layer size.

Before further experimentation with GRU recurrent nets, we compared the best performing model to its analogous LSTM variant (two-layer biLSTM with 512 hidden layer neurons). In Table 9, we show how GRU could obtain competitive performance with LSTM. The marginal improvement of 0.56% obtained by LSTM requires a substantial increment of the number of parameters, especially when considering larger models. In the case of 512 hidden size, LSTM has 2.6M additional parameters to optimise when compared to the GRU version. For further experiments, we stuck to the biGRU (two-layer, 512 hidden neurons) architecture.

**Table 8.** Hidden layer size experiments on five binary problems (see Columns 2–6). The results reported are class-weighted accuracies averaged over a 10-fold cross validation. The "Average" column is the average of performances on binary problems.

|  | Forward | Reverse | Sit-Down | Stand-Up | Handwave | Average |
|---|---|---|---|---|---|---|
| biGRU (two-layer, 32-hidden) | 89.97 | 89.78 | 87.89 | 89.47 | 95.92 | 90.61 |
| biGRU (two-layer, 64-hidden) | 91.42 | 91.08 | 90.07 | 92.51 | 97.01 | 92.42 |
| biGRU (two-layer, 128-hidden) | 93.10 | 93.63 | 92.98 | 95.32 | 97.70 | 94.55 |
| biGRU (two-layer, 256-hidden) | 93.76 | 94.17 | 94.34 | 95.52 | 98.89 | 95.34 |
| biGRU (two-layer, 512-hidden) | 94.94 | 95.20 | 95.02 | 96.70 | 99.29 | 96.23 |

**Table 9.** GRU versus LSTM on 5 binary problems (see Columns 2–6). The results reported are class-weighted accuracies averaged over a 10-fold cross validation. The "Average" column is the average of performances on binary problems.

|  | Forward | Reverse | Sit-Down | Stand-Up | Handwave | Average |
|---|---|---|---|---|---|---|
| biGRU (two-layer, 64-hidden) | 91.42 | 91.08 | 90.07 | 92.51 | 97.01 | 92.42 |
| biLSTM (two-layer, 64-hidden) | 91.56 | 96.91 | 92.02 | 89.84 | 94.58 | 92.98 |

## 5.3. Final Experiments

After having fixed the final GRU model architecture to tow bidirectional stacked layers with 512 hidden neurons, we performed evaluated its performance in multiclass classification and also other experiments to ensure the generalisation capabilities of our approach.

### 5.3.1. Multiclass Classification

To evaluate the missclassifications and potential confusion among classes from our previous binary problems, we first defined a multiclass problem with labels those same labels: {F, R, sd, su, hw}, where F is forward, R is reverse, sd is sit down, su is stand up, and hw is hand-wave. In this five-class problem, the model was able to correctly predict 92.67% of actions (see column "Actions" in Table 10). As shown in Figure 8a, the confusion is introduced by the semantically similar classes, either forward and reverse or sit-down and stand-up. The hand-wave classification was almost perfect, only confused once as a reverse instance in 50 hw examples.

The second and third experiments were intended to classify the walking path. The former was not distinguishing walking direction. We hence defined two separate sets of labels, {A1, A2, B1, C1, C2} and {FA1, FA2, FB1, FC1, FC2, RA1, RA2, RB1, RC1, RC2}, respectively, where letters (F) and (R) before action label are used to distinguish between action in forward or reverse direction. As shown in Table 10, the model performed similarly in the two cases, with slightly worse performance not considering the walking direction (86.23%) than when doing so (86.65%). Figure 8 shows the confusion matrices for these two experiments.

Finally, in the fourth and last experiment, we labelled the setup in which the action was occurring with labels {1, . . . , 6}, which correspond to tasks listed in Table 5. In this experiment, the robot had to perform various tasks, in addition to performing an action an object is also present in the same environment. Its location and performed actions can be seen in Figure 4. The accuracies obtained from those are summarised in Table 10, while Figure 8 illustrates class confusions.

**Table 10.** Classification on four multiclass problems obtained by biGRU (two-layer, 512-hidden) baseline. The results reported are class-weighted accuracies averaged over a 10-fold cross validation.

| Actions | Path | Directed-Path | Setup |
|---|---|---|---|
| 92.67 | 86.23 | 86.65 | 90.00 |

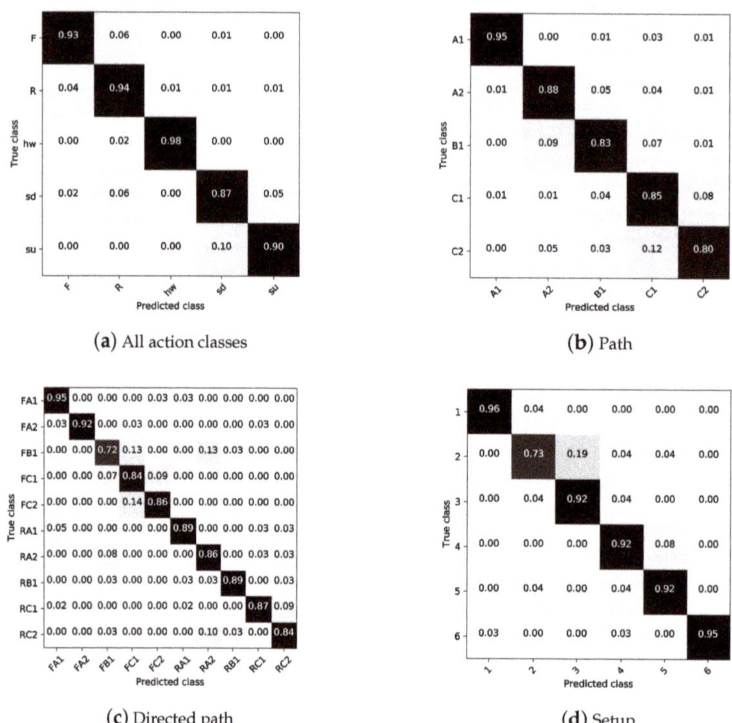

(**a**) All action classes

(**b**) Path

(**c**) Directed path

(**d**) Setup

**Figure 8.** Confusion matrices (row-wise normalised) from multiclass classification experiments from Table 10.

### 5.3.2. Model Generalisation on Actions and Two Robots

In this section, we evaluate the generalisation capabilities of the models when learning from single-pixel time-of-flight patterns.

Each action was captured a certain amount of repetitions. During this repetition, the path in walking actions (forward and reverse) or initial position (sit-down, stand-up, and hand-wave) were varied. In this experiment, we wanted to take this into account and try to learn by excluding from training the all the repetitions of one action to assert the model is not overfitting due to repetitions being very similar patterns. For that, we changed our validation procedure to leave-one-rep set-out, i.e., we predicted a repetition set all at once in the test set, and did not use repetitions from the same respect during training. Results are presented in Table 11. If we compare to those to results from the same model, i.e., biGRU (two-layer, 512-hidden), in Table 10, we can observe there was no drop in accuracy, but a slight improvement—probably due to both the generalisation capabilities and the fact that we could use more data to train across folds.

**Table 11.** Leave-one-rep set-out cross-validation (LOROCV) experiment using biGRU (two-layer, 512-hidden). These are the same as those from last row in Table 8, but using LOROCV instead of 10-fold CV.

| | F | R | sd | su | hw | Average |
|---|---|---|---|---|---|---|
| biGRU (two-layer, 512-hidden, 10fCV) | 94.94 | 95.20 | 95.02 | 96.70 | 99.29 | 96.23 |
| biGRU (two-layer, 512-hidden, LOROCV) | 96.77 | 95.14 | 93.36 | 97.11 | 100.0 | 96.47 |

All sequences were with just one robot performing actions. A separate set of sequences was used to test action classification when two robots were present, as shown in Table 12. These sequences were only used in the test phase (only one-robot sequences were used for training). In particular, we analysed three different scenarios: (1) one robot acted, while the other one stood still; (2) the two robots performed the same action; and (3) each robot performed a different action.

From results in Scenario (1), we observed the standing-up robot did not interfere in the other action category prediction. In fact, the model failed to predict stand-up action since the other actions presented a more dominant motion pattern that interfere in the stand-up pattern learned from one-robot actions.

**Table 12.** Two-robot experiments in three different scenarios: one robot standing up while other performing a particular action, the two robots performing the same action, and the two performing each a different action. Each scenario is a separate test set with a different number of examples. In brackets, the number of positive examples for each class in each scenario. Since positive/negative classes are, we report class-weighted accuracies (%).

| | #{Examples} | F | R | sd | su | hw |
|---|---|---|---|---|---|---|
| One robot standing up and sitting down | 100 | 80.00 (50) | 88.00 (50) | 95.00 (0) | 15.00 (100) | 100.00 (0) |
| Same two actions | 70 | 25.00 (10) | 72.86 (10) | 75.00 (20) | 54.00 (20) | 50.00 (20) |
| Two different actions | 20 | 50.00 (10) | 100.00 (0) | 95.00 (10) | 55.00 (10) | 55.00 (10) |

## 5.4. Discussion

In this paper, we propose a concept for detection actions while preserving the test subjects (NAO V4 robot) privacy. Our concept relies on recording only the temporal evolution of light pulses scattered back from the scene. Such data trace to record one action contains sequence of one-dimensional arrays of voltage values acquired by the single-pixel detector after amplifying and detection by the data acquisition system at 6 GHz repetition rate. The data trace is very compact and easy to process, compared to videos, containing sequences of 2D images.

The data volume reduction is achieved by controlled illumination and single pixel detector without any spatial resolution. The scene was illuminated with a diverging, speckled light pulse of 30 picosecond ($30 \times 10^{-12}$ ps) duration. The method would also work in different scenes, where most of the objects are static.

Compared to 2D images, hardly any information about the colours, object, their shapes and positions could be retrieved from the data traces by classical method. Although quite similar to the neural networks, a human can distinguish the actions and perhaps also clearly differentiate moving directions from the data traces.

The research in hand clearly articulates the core properties of movement—it imprints a temporal evolution to even most simple data trace. Owing to the interdisciplinary approach through combining the tools of photonics (modern, application oriented optics and light detection) and computer science, one is capable of reducing the data rate. The result has high potential to provide cost effective surveillance systems to aid societies to look after of public order, and take care of young, elderly and injured members.

The photonics and data acquisition schemes used in this experiment are unlikely to become widespread owing to their high cost and other features. However, detectors and laser systems capable of providing suitable illumination and detection properties in affordable price range are being developed and will enter the market in near future.

## 6. Conclusions

This research work proposed a new methodology for action recognition while preserving the test subjects privacy. The proposed method uses only the temporal evolution of light pulses scattered back from the scene. Advanced machine learning algorithms, namely RNN and LSTM, were adopted for data analysis and demonstrated successful action recognition. The experimental results show that our proposed method could achieve high recognition rate for five actions, namely walking forward, walking reverse, sitting down, standing up, and waving hand, with an average recognition rate of 96.47%. In this work, we additionally studied action recognition when multiple concurrent actors are present in the scene.

In future work, we will conduct further experiments, including more complex actions, such as running, jumping, and head movements. We are planning to record higher number of samples to conduct a better generalisation capabilities of our proposed approach.

**Author Contributions:** Conceptualization, H.V.-L., S.E., C.O. and G.A.; Data curation, I.O., A.H., K.M.P., S.-M.V. and S.O.; Funding acquisition, G.A.; Investigation, H.V.-L., S.O., S.E. and G.A.; Methodology, A.C., E.A., A.V., H.V.-L., S.E. and G.A.; Software, A.C.; Writing—original draft, A.H. and G.A.; Writing—review & editing, I.O., E.A., S.-M.V., H.V.-L., S.O., S.E., C.O. and G.A.

**Funding:** This work was partially supported by Estonian Research Council Grants (PUT638, PUT1075, PUT1081), The Scientific and Technological Research Council of Turkey (TÜBITAK) (Project 1001-116E097), the Estonian Centre of Excellence in IT (EXCITE) funded by the European Regional Development Fund, the Spanish Project TIN2016-74946-P (MINECO/FEDER, UE) and CERCA Programme/Generalitat de Catalunya. This project received funding from the European Union's Horizon 2020 research and innovation programme under the Marie Sklodowska-Curie grant agreement No. 665919. This work was partially supported by ICREA under the ICREA Academia programme.

**Acknowledgments:** We gratefully acknowledge the support of NVIDIA Corporation with the donation of the Titan Xp and V GPUs used for this research.

**Conflicts of Interest:** The authors declare no conflict of interest.

## References


1. Fernando, B.; Gavves, E.; Oramas, J.M.; Ghodrati, A.; Tuytelaars, T. Modeling video evolution for action recognition. In Proceedings of the IEEE Conference on Computer Vision and Pattern Recognition, Boston, MA, USA, 7–12 June 2015; pp. 5378–5387.

2. Nasrollahi, K.; Escalera, S.; Rasti, P.; Anbarjafari, G.; Baro, X.; Escalante, H.J.; Moeslund, T.B. Deep learning based super-resolution for improved action recognition. In Proceedings of the IEEE 2015 International Conference on Image Processing Theory, Tools and Applications (IPTA), Orleans, France, 10–13 November 2015; pp. 67–72.

3. Haque, M.A.; Bautista, R.B.; Noroozi, F.; Kulkarni, K.; Laursen, C.B.; Irani, R.; Bellantonio, M.; Escalera, S.; Anbarjafari, G.; Nasrollahi, K.; et al. Deep Multimodal Pain Recognition: A Database and Comparison of Spatio-Temporal Visual Modalities. In Proceedings of the 2018 13th IEEE International Conference on Automatic Face & Gesture Recognition (FG 2018), Xi'an, China, 15–19 May 2018; pp. 250–257.

4. Ponce-López, V.; Escalante, H.J.; Escalera, S.; Baró, X. Gesture and Action Recognition by Evolved Dynamic Subgestures. In Proceedings of the BMVC, Swansea, UK, 7–10 September 2015; pp. 129.1–129.13.

5. Wan, J.; Escalera, S.; Anbarjafari, G.; Escalante, H.J.; Baró, X.; Guyon, I.; Madadi, M.; Allik, J.; Gorbova, J.; Lin, C.; et al. Results and Analysis of ChaLearn LAP Multi-modal Isolated and Continuous Gesture Recognition, and Real Versus Fake Expressed Emotions Challenges. In Proceedings of the IEEE International Conference on Computer Vision, Venice, Italy, 22–29 October 2017; pp. 3189–3197.

6. Corneanu, C.; Noroozi, F.; Kaminska, D.; Sapinski, T.; Escalera, S.; Anbarjafari, G. Survey on Emotional Body Gesture Recognition. *IEEE Trans. Affect. Comput.* **2018**. [CrossRef]

7. Turaga, P.; Chellappa, R.; Subrahmanian, V.S.; Udrea, O. Machine recognition of human activities: A survey. *IEEE Trans. Circuits Syst. Video Technol.* **2008**, *18*, 1473. [CrossRef]

8. Jahromi, M.N.; Bonderup, M.B.; Asadi-Aghbolaghi, M.; Avots, E.; Nasrollahi, K.; Escalera, S.; Kasaei, S.; Moeslund, T.B.; Anbarjafari, G. Automatic Access Control Based on Face and Hand Biometrics in a Non-Cooperative Context. In Proceedings of the 2018 IEEE Winter Applications of Computer Vision Workshops (WACVW), Lake Tahoe, NV, USA, 15 March 2018; pp. 28–36.

9. Sapiński, T.; Kamińska, D.; Pelikant, A.; Ozcinar, C.; Avots, E.; Anbarjafari, G. Multimodal Database of Emotional Speech, Video and Gestures. In Proceedings of the International Conference on Pattern Recognition, Beijing, China, 20–24 August 2018; Springer: Berlin/Heidelberg, Germany, 2018; pp. 153–163.

10. Kim, Y.; Lee, H.; Provost, E.M. Deep learning for robust feature generation in audiovisual emotion recognition. In Proceedings of the 2013 IEEE International Conference on Acoustics, Speech and Signal Processing (ICASSP), Vancouver, BC, Canada, 26–31 May 2013; pp. 3687–3691.

11. Lusi, I.; Junior, J.C.J.; Gorbova, J.; Baró, X.; Escalera, S.; Demirel, H.; Allik, J.; Ozcinar, C.; Anbarjafari, G. Joint challenge on dominant and complementary emotion recognition using micro emotion features and head-pose estimation: Databases. In Proceedings of the 2017 12th IEEE International Conference on Automatic Face & Gesture Recognition (FG 2017), Washington, DC, USA, 30 May–3 June 2017; pp. 809–813.

12. Avots, E.; Sapiński, T.; Bachmann, M.; Kamińska, D. Audiovisual emotion recognition in wild. *Mach. Vis. Appl.* **2018**, 1–11. [CrossRef]

13. Noroozi, F.; Marjanovic, M.; Njegus, A.; Escalera, S.; Anbarjafari, G. Fusion of classifier predictions for audio-visual emotion recognition. In Proceedings of the IEEE 2016 23rd International Conference on Pattern Recognition (ICPR), Cancun, Mexico, 4–8 December 2016; pp. 61–66.

14. Guo, J.; Lei, Z.; Wan, J.; Avots, E.; Hajarolasvadi, N.; Knyazev, B.; Kuharenko, A.; Junior, J.C.S.J.; Baró, X.; Demirel, H.; et al. Dominant and Complementary Emotion Recognition From Still Images of Faces. *IEEE Access* **2018**, *6*, 26391–26403. [CrossRef]

15. Grobova, J.; Colovic, M.; Marjanovic, M.; Njegus, A.; Demire, H.; Anbarjafari, G. Automatic hidden sadness detection using micro-expressions. In Proceedings of the 2017 12th IEEE International Conference on Automatic Face & Gesture Recognition (FG 2017), Washington, DC, USA, 30 May–3 June 2017; pp. 828–832.

16. Kulkarni, K.; Corneanu, C.; Ofodile, I.; Escalera, S.; Baró, X.; Hyniewska, S.; Allik, J.; Anbarjafari, G. Automatic recognition of facial displays of unfelt emotions. *IEEE Trans. Affect. Comput.* **2018**. [CrossRef]

17. Parkhi, O.M.; Vedaldi, A.; Zisserman, A. Deep face recognition. In Proceedings of the BMVC, Swansea, UK, 7–10 September 2015; Volume 1, p. 6.

18. Schroff, F.; Kalenichenko, D.; Philbin, J. Facenet: A unified embedding for face recognition and clustering. In Proceedings of the IEEE Conference on Computer Vision and Pattern Recognition, Boston, MA, USA, 7–12 June 2015; pp. 815–823.

19. Haamer, R.E.; Kulkarni, K.; Imanpour, N.; Haque, M.A.; Avots, E.; Breisch, M.; Nasrollahi, K.; Escalera, S.; Ozcinar, C.; Baro, X.; et al. Changes in facial expression as biometric: A database and benchmarks of identification. In Proceedings of the 2018 13th IEEE International Conference on Automatic Face & Gesture Recognition (FG 2018), Xi'an, China, 15–19 May 2018; pp. 621–628.

20. Tertychnyi, P.; Ozcinar, C.; Anbarjafari, G. Low-quality fingerprint classification using deep neural network. *IET Biom.* **2018**, *7*, 550–556. [CrossRef]

21. Zhang, C.L.; Zhang, H.; Wei, X.S.; Wu, J. Deep bimodal regression for apparent personality analysis. In Proceedings of the European Conference on Computer Vision, Amsterdam, The Netherlands, 8–16 August 2016; Springer: Berlin/Heidelberg, Germany, 2016; pp. 311–324.

22. Gorbova, J.; Avots, E.; Lüsi, I.; Fishel, M.; Escalera, S.; Anbarjafari, G. Integrating Vision and Language for First-Impression Personality Analysis. *IEEE MultiMedia* **2018**, *25*, 24–33. [CrossRef]

23. Yang, J.; Nguyen, M.N.; San, P.P.; Li, X.; Krishnaswamy, S. Deep Convolutional Neural Networks on Multichannel Time Series for Human Activity Recognition. In Proceedings of the Twenty-Fourth International Joint Conference on Artificial Intelligence, Buenos Aires, Argentina, 25–31 July 2015; Volume 15, pp. 3995–4001.

24. Ma, M.; Fan, H.; Kitani, K.M. Going deeper into first-person activity recognition. In Proceedings of the IEEE Conference on Computer Vision and Pattern Recognition, Las Vegas, NV, USA, 27–30 June 2016; pp. 1894–1903.

25. Ordóñez, F.J.; Roggen, D. Deep convolutional and lstm recurrent neural networks for multimodal wearable activity recognition. *Sensors* **2016**, *16*, 115. [CrossRef]

26. Ma, X.; Dai, Z.; He, Z.; Ma, J.; Wang, Y.; Wang, Y. Learning traffic as images: A deep convolutional neural network for large-scale transportation network speed prediction. *Sensors* **2017**, *17*, 818. [CrossRef]

27. Kirmani, A.; Hutchison, T.; Davis, J.; Raskar, R. Looking around the corner using transient imaging. In Proceedings of the 2009 IEEE 12th International Conference on Computer Vision, Kyoto, Japan, 29 September–2 October 2009; pp. 159–166.

28. Velten, A.; Willwacher, T.; Gupta, O.; Veeraraghavan, A.; Bawendi, M.G.; Raskar, R. Recovering three-dimensional shape around a corner using ultrafast time-of-flight imaging. *Nat. Commun.* **2012**, *3*, 745. [CrossRef]

29. Buttafava, M.; Zeman, J.; Tosi, A.; Eliceiri, K.; Velten, A. Non-line-of-sight imaging using a time-gated single photon avalanche diode. *Opt. Express* **2015**, *23*, 20997–21011. [CrossRef]

30. Besl, P.J. Active optical range imaging sensors. In *Advances in Machine Vision*; Springer: Berlin/Heidelberg, Germany, 1989; pp. 1–63.

31. Antipa, N.; Kuo, G.; Heckel, R.; Mildenhall, B.; Bostan, E.; Ng, R.; Waller, L. DiffuserCam: Lensless single-exposure 3D imaging. *Optica* **2018**, *5*, 1–9. [CrossRef]

32. Gatti, A.; Brambilla, E.; Bache, M.; Lugiato, L.A. Ghost imaging with thermal light: Comparing entanglement and classicalcorrelation. *Phys. Rev. Lett.* **2004**, *93*, 093602. [CrossRef]

33. Shapiro, J.H. Computational ghost imaging. *Phys. Rev.* **2008**, *78*, 061802. [CrossRef]

34. Sun, M.J.; Edgar, M.P.; Gibson, G.M.; Sun, B.; Radwell, N.; Lamb, R.; Padgett, M.J. Single-pixel three-dimensional imaging with time-based depth resolution. *Nat. Commun.* **2016**, *7*, 12010. [CrossRef]

35. Li, W.; Zhang, Z.; Liu, Z. Action recognition based on a bag of 3d points. In Proceedings of the 2010 IEEE Computer Society Conference on Computer Vision and Pattern Recognition Workshops (CVPRW), San Francisco, CA, USA, 13–18 June 2010; pp. 9–14.

36. Caramazza, P.; Boccolini, A.; Buschek, D.; Hullin, M.; Higham, C.; Henderson, R.; Murray-Smith, R.; Faccio, D. Neural network identification of people hidden from view with a single-pixel, single-photon detector. *arXiv* **2017**, arXiv:1709.07244.

37. Sanchez-Riera, J.; Čech, J.; Horaud, R. Action recognition robust to background clutter by using stereo vision. In Proceedings of the European Conference on Computer Vision, Florence, Italy, 7–13 October 2012; Springer: Berlin/Heidelberg, Germany, 2012; pp. 332–341.

38. Zhang, Z. Microsoft kinect sensor and its effect. *IEEE Multimed.* **2012**, *19*, 4–10. [CrossRef]

39. Papadopoulos, G.T.; Axenopoulos, A.; Daras, P. Real-time skeleton-tracking-based human action recognition using kinect data. In Proceedings of the International Conference on Multimedia Modeling, Dublin, Ireland, 6–10 January 2014; Springer: Berlin/Heidelberg, Germany, 2014; pp. 473–483.

40. Fofi, D.; Sliwa, T.; Voisin, Y. A comparative survey on invisible structured light. In *Machine Vision Applications in Industrial Inspection XII*; International Society for Optics and Photonics: San Diego, CA, USA, 2004; Volume 5303, pp. 90–99.

41. Smisek, J.; Jancosek, M.; Pajdla, T. 3D with Kinect. In *Consumer Depth Cameras for Computer Vision*; Springer: Berlin/Heidelberg, Germany, 2013; pp. 3–25.

42. Faccio, D.; Velten, A. A trillion frames per second: The techniques and applications of light-in-flight photography. *Rep. Prog. Phys.* **2018**, *81*, 105901. [CrossRef] [PubMed]

43. Pandharkar, R.; Velten, A.; Bardagjy, A.; Lawson, E.; Bawendi, M.; Raskar, R. Estimating motion and size of moving non-line-of-sight objects in cluttered environments. In Proceedings of the 2011 IEEE Conference on Computer Vision and Pattern Recognition (CVPR), Colorado Springs, CO, USA, 20–25 June 2011; pp. 265–272.

44. Heide, F.; Hullin, M.B.; Gregson, J.; Heidrich, W. Low-budget transient imaging using photonic mixer devices. *ACM Trans. Graph. (ToG)* **2013**, *32*, 45. [CrossRef]

45. Gariepy, G.; Tonolini, F.; Henderson, R.; Leach, J.; Faccio, D. Detection and tracking of moving objects hidden from view. *Nat. Photonics* **2016**, *10*, 23–26. [CrossRef]

46. Warburton, R.E.; Chan, S.; Gariepy, G.; Altmann, Y.; McLaughlin, S.; Leach, J.; Faccio, D. Real-Time Tracking of Hidden Objects with Single-Pixel Detectors. In *Imaging Systems and Applications*; Optical Society of America: San Diego, CA, USA, 2016; p. IT4E–2.

47. Chan, S.; Warburton, R.E.; Gariepy, G.; Leach, J.; Faccio, D. Non-line-of-sight tracking of people at long range. *Opt. Express* **2017**, *25*, 10109–10117. [CrossRef]

48. Jia, L.; Radke, R.J. Using time-of-flight measurements for privacy-preserving tracking in a smart room. *IEEE Trans. Ind. Inform.* **2014**, *10*, 689–696. [CrossRef]

49. Tao, S.; Kudo, M.; Nonaka, H. Privacy-preserved behavior analysis and fall detection by an infrared ceiling sensor network. *Sensors* **2012**, *12*, 16920–16936. [CrossRef] [PubMed]
50. Kawashima, T.; Kawanishi, Y.; Ide, I.; Murase, H.; Deguchi, D.; Aizawa, T.; Kawade, M. Action recognition from extremely low-resolution thermal image sequence. In Proceedings of the 2017 14th IEEE International Conference on Advanced Video and Signal Based Surveillance (AVSS), Lecce, Italy, 29 August–1 September 2017; pp. 1–6.
51. Dai, J.; Saghafi, B.; Wu, J.; Konrad, J.; Ishwar, P. Towards privacy-preserving recognition of human activities. In Proceedings of the 2015 IEEE International Conference on Image Processing (ICIP), Quebec City, QC, Canada, 27–30 September 2015; pp. 4238–4242.
52. Xu, M.; Sharghi, A.; Chen, X.; Crandall, D.J. Fully-Coupled Two-Stream Spatiotemporal Networks for Extremely Low Resolution Action Recognition. In Proceedings of the 2018 IEEE Winter Conference on Applications of Computer Vision (WACV), Lake Tahoe, NV, USA, 12–15 March 2018; pp. 1607–1615.
53. Cho, K.; Van Merriënboer, B.; Bahdanau, D.; Bengio, Y. On the properties of neural machine translation: Encoder-decoder approaches. *arXiv* **2014**, arXiv:1409.1259.
54. Chung, J.; Gulcehre, C.; Cho, K.; Bengio, Y. Empirical evaluation of gated recurrent neural networks on sequence modeling. *arXiv* **2014**, arXiv:1412.3555.
55. Chollet, F. Keras. 2015. Available online: https://github.com/fchollet/keras (accessed on 4 February 2019).

*Article*

# Supervisors' Visual Attention Allocation Modeling Using Hybrid Entropy

**Haifeng Bao, Weining Fang *, Beiyuan Guo and Peng Wang**

State Key Lab Rail Traff Control & Safety, School of Mechanical, Electronic and Control Engineering, Beijing Jiaotong University, 100044 Beijing, China; 12116341@bjtu.edu.cn (H.B.); byguo@bjtu.edu.cn (B.G.); c2t53y48@gmail.com (P.W.)

* Correspondence: wnfang@bjtu.edu.cn; Tel.: +86-139-1108-5123

Received: 26 March 2019; Accepted: 10 April 2019; Published: 12 April 2019

**Abstract:** With the improvement in automation technology, humans have now become supervisors of the complicated control systems that monitor the informative human–machine interface. An alyzing the visual attention allocation behaviors of supervisors is essential for the design and evaluation of the interface. Supervisors tend to pay attention to visual sections with information with more fuzziness, which makes themselves have a higher mental entropy. Supervisors tend to focus on the important information in the interface. In this paper, the fuzziness tendency is described by the probability of correct evaluation of the visual sections using hybrid entropy. The importance tendency is defined by the proposed value priority function. The function is based on the definition of the amount of information using the membership degrees of the importance. By combining these two cognitive tendencies, the informative top-down visual attention allocation mechanism was revealed, and the supervisors' visual attention allocation model was built. The Building Automatic System (BAS) was used to monitor the environmental equipment in a subway, which is a typical informative human–machine interface. An experiment using the BAS simulator was conducted to verify the model. The results showed that the supervisor's attention behavior was in good agreement with the proposed model. The effectiveness and comparison with the current models were also discussed. The proposed attention allocation model is effective and reasonable, which is promising for use in behavior analysis, cognitive optimization, and industrial design.

**Keywords:** attention allocation; attention behavior; hybrid entropy; information entropy

## 1. Introduction

With the improvement in automation technology, the role of humans in complicated control systems is changing from that of operators to supervisors [1]. More and more information is being displayed on human–machine interfaces, but human attention ability is limited. Therefore, the limited attention resources of supervisors are precious and important. Most system failures and operational accidents are due to the lack of visual attention to relevant information [2]. An alyzing the visual attention behaviors and revealing the visual attention allocation mechanism are important for the design and evaluation of human–machine interfaces (HMIs). HMIs with an ergonomic design that align with the attention behaviors of the supervisors are useful for system safety, error evaluation, and accident prevention [3–5].

Attention behavior has many aspects, such as hearing, vision, and touch. Among them, vision is important to supervisors during the task of monitoring. Humans have a complex selective visual attention behavior that scans the scene both in a rapid, bottom-up, salience-driven manner as well as a slower, top-down, task-dependent manner [6]. The visual attention to bottom-up salient information is a rapid process that has limited effects on task-dependent attention allocation. Supervisory behavior is a long-term attention allocation mechanism for familiar scenes. The top-down task-driven factors occupy the majority of the attention strategy during supervisory tasks.

Many factors can affect attention behaviors. Salience-driven factors depend on visual features such as salience, blinking, shape, and colors [7–9]. The task-driven factors in the supervisory task depend on the task features such as urgency, expectation, effort, and value [10–13]. Supervisors always comprehensively consider the above task-factors during the task process, then establish the priority of the information. In the task, the importance of the displayed information is mainly considered by the supervisors. Matsuka proved that human learners do not always optimize attention; one reason they fail to do so is that, under certain conditions, the cost of information retrieval or use may affect the attention strategy adopted by the learners [14]. Therefore, in familiar procedural tasks, supervisors acquire system information based on their experience and previously acquired knowledge due to the top-down attention strategy.

The determination of information priorities is complicated and fuzzy in the cognitive process. The uncertainty of the information may produce significant anxiety in supervisors, who tend to pay attention to the information sections that can reduce that indeterminacy. The attention to information is a reduction of the entropy of the HMI. This complicated cognitive behavior was described as mental entropy processing by Wanyan [15]. Even though mental entropy theory has some limitations, it was used successfully in modeling the cognitive process for information processing in the human brain. Supervisors tend to pay attention to the visual section which has a higher information value. Therefore, the membership degrees of the importance of the information sections based on fuzzy theory could be feasibly used to quantify its value. These two selective cognitive mechanisms have been shown to synergistically affect attention behaviors [16,17].

Efficient HMIs help their users accomplish their tasks with minimal workload and fatal errors. The visual attention model is useful for the design and optimization of these interfaces [18,19]. The layout of a T-type HMI on aircraft was constructed by Fitts by analyzing the pilots' visual attention behavior [20]. The visual attention model predicted the users' selective attention behavior in supervisory tasks, which was beneficial in staff training [21]. One important aspect of on-the-job training of supervisors is to make them pay attention to the right section at the right time. Using the model, the researchers evaluated the mental workload and situation awareness of the user, which provided information about the conditions of the user's current mental status [22,23]. This model can also guide task analysis and contribute to task optimization [12]. Overall, the visual attention allocation model is useful and promising.

At present, evaluating visual attention is easily accomplished by tracking eye gaze in or after the supervisory task [24]; however, predicting the visual attention allocation behaviors before the task is challenging. We aimed to build an effective, accurate, and quantified model in visual attention allocation based on the related works.

## 2. Related Works

In previous studies, researchers proposed many valuable attention allocation models to predict the supervisory behavior of supervisors in informative HMIs. Based on saliency-based image recognition, the predictive attention model was built which considered the bottom-up attention mechanism of humans [6,9,25,26]. The observable information on the screen could be recognized using deep learning to predict the attention behavior [27–29]. These bottom-up models help us reveal the basic attention mechanism that how humans react to images. Wickens developed the SEEV model of scanning behavior considering the task-driven factors [10–12]. This model considers the salience, effort, expectancy, and value (SEEV) associated with each visual section. The model was improved to NT-SEEV, to predict the notice ability (NT for notice) of events that occurred in the context of routine task-driven scanning across large-scale visual environments [30]. Many researchers worked on the quantitation and computation of multiple factors in SEEV [31–33]. SEEV and its improved models consider both the bottom-up and top-down attention mechanisms of humans. However, due to the different chosen factors and computational methods, the results of the above models have varied significantly.

Some researchers computed attention allocation using gaze data based on fuzzy theory [34,35]. However, this involved a post analysis method that could not predict the attention allocation strategy. Senders considered the human operator as a monitor and controller in the system [36]. The model argues that humans are information processors and supervisory behavior is a data processing process. The model describes the strategy of humans when selecting their attention focus in an informative HMI. Sheridan distinguished the time interval of the supervisor when processing the information and the proposed model assumed that the operator controls the most valuable information with each sample [37]. Visual information processing is fuzzy in the human brain. Lin introduced a novel fractional-order chaotic phase synchronization model for visual selection and shifting [38]. The model uses two chaotic network layers to simulate the human cognitive system and solves the processing of the natural image in the brain, which was useful for the proposed model in this article. Junshan used multiscale entropy analysis of human operating behavior, which is a post-analysis method to determine the human dynamics [39]. Pan extended the influence model to incorporate dynamical parameters to a social system, which allowed us to uncover important shifts between actors. The model is instructive in attention shift behavior [40].

Based on the above work, Matsui researched attention allocation using fuzzy theory and quantified the selective attention mechanism of the information using hybrid entropy [41,42]. Wanyan et al. [15] and Wu et al. [16] applied detection efficiency factors and fatigue factors to Matsui's fuzzy model for pilots. Considering multiple factors in the SEEV model, Wu and Wanyan developed the attention model under multi-factor conditions [17]. This was an attempt to integrate the SEEV model and the fuzzy model. Based on subjective expected utility theory (SEU), a human is an optimal information processing processor [43]. The comprehensive consideration of the theory aimed to maximize the acquisition of the important information and minimize the fuzziness of the scene. The above attention allocation models based on fuzzy theory usually involved two main factors: information value and information fuzziness [15–17,41,42].

The above models used the membership degrees of the importance of the information (value: 0–1) expressing the information value. However, the drawback of the application of membership degrees without processing was that the attention allocation ratio did not increase when the information value increased. This means that a high information value might not lead to a high attention allocation ratio. In this aspect, the above models based on fuzzy theory need to be improved. In this study, we tried to solve this problem and demonstrate that our improvement is reasonable and effective.

The proposed attention allocation model was built based on the work of Matsui's and Wanyan et al.'s models [15,16,42]. The information value is presented by the proposed value priority function using the membership degrees of the importance and information amounts. Using the theory of hybrid entropy, the proposed model expresses the supervisors' fuzzy cognition of the information processing in the human brain. Combining these two cognitive processes, an increasing attention allocation model was built along with the increasing information value. The BAS system is a typical interface used by supervisors to monitor the environmental equipment in subway systems. We conducted an experiment using a BAS simulator, which showed that the proposed model is effective. Compared with Matsui's and Wanyan's model, the proposed model has several advantages and reasonable improvements. We think that our proposed model has potential for applications in behavior analysis, cognitive optimization, and industrial ergonomic design.

## 3. Methods

### 3.1. Value Priority Function

The supervisory task involves monitoring and controlling a large amount of system information. The information on the monitors can be partitioned into several visual displays and independent meaningful sections, creating $I_i$:

$$I_i = (I_1, I_2, \ldots, I_n) \tag{1}$$

The attention allocation model aims to predict the attention behavior of the supervisor. The attention allocation is the ratio $A_i$ of the virtual attention time required to focus on the information $I_i$ to the total virtual attention time for the whole task, as shown in Equation (2). The proposed attention allocation model aims to build the mapping relationship between $I_i$ and $A_i$ before the supervisory task:

$$A_i = (A_1, A_2, \ldots, A_n) \tag{2}$$

Based on the research of Wickens, the attention behaviors of a skillful operator are rarely affected by the bottom-up channel unless the bottom-up factors have independent meaning [11]. Subsequent research supported this view [14]. Thus, the extension of this theory tried to consider multiple factors in particular scenes.

During a familiar procedural task, the supervisor of the system would have previously evaluated the information value based on their knowledge and training. However, the priority is fuzzy to recognize. Based on fuzzy theory, the membership degree of the information importance is considered the information value $V_i$ to every information $I_i$, as shown in Equation (3). For a task, the membership degrees of the importance for the informative sections are certain values. Usually, the values are provided by experts in the field who are familiar with the task [15,16]:

$$V_i = (V_1, V_2, \ldots, V_n) \tag{3}$$

Matsui and Wanyan et al. considered these membership degrees as the information value [15,16,42]. The possible values are 0–1. In this research, we wanted to build a visual attention allocation model with a higher attention ratio to the higher information value. Therefore, the information value, $V_i$, needed to be improved to value the priority of the information, $V_i{}'$, which ranges from 0 to positive infinity.

Considering supervisors as the information processor, the information value $V_i$ of the sections should be converted with its information entropy. Usually, the information amount, $H_i$, in Equation (4), presents the information sections when event $i$ occurs, which is related to the probability that the certain information $Pr_i$ occurs:

$$H_i = -\ln Pr_i \tag{4}$$

The definition of information amounts shows:

(1) The information amount is monotonically decreases, which means that a low probabilistic event occurs with a high amount of information.

(2) The information amount tends to be 0 when the probability of the occurrence of the event tends to be 1, which means that the inevitable event carries no amount of information.

The improvement in the information value $V_i$ needs to consider the following cognitive behaviors:

(1) Supervisors pay more attention to information sections with a higher information value. This means that the ratio of the attention allocation is a monotonic increasing function of the information value.

(2) The ratio of the attention allocation to the information (from 0 to 1) tends to be 1 when the information value (from 0 to 1) tends to be 1. This means that the information valued at 1, places the highest requirements on attention resources. If the supervisor transfers their attention to other information sections, it leads to a serious failure.

Referring to similarities to the definition of information amounts and cognitive behaviors, we propose a value priority function $F(V_i)$, to manage the information value $V_i$. The improved information value is value priority $V_i'$, as shown in Equation (5), and represents the tendency where supervisors tend to pay more attention to the more important information:

$$V_i' = F(V_i) = -\ln(1 - V_i) \tag{5}$$

### 3.2. Information Fuzziness Tendency

The psychological and physiological states of the supervisor affect attention behavior. $P_i$ represents the probability that the supervisor will correctly process the information (Equation (6)). When they have a higher probability of correctly evaluating the information, the supervisor pays more attention to this information [15,16]:

$$P_i = (P_1, P_2, \ldots, P_n) \tag{6}$$

This uncertain evaluation of the information $P_i$ is caused by the fuzzy information value $V_i$. Based on fuzzy theory, the ambiguities of information can be quantified by hybrid entropy. The hybrid entropy $S$ represents the cognition fuzzy level, which involves the informative probabilistic entropy $H_{prob}$ and the informative binary entropy $H_{bin}$:

$$\begin{aligned} S &= H_{prob} + H_{bin} = \sum_{i=1}^{n} -P_i \ln P_i + \sum_{i=1}^{n} P_i h(V_i) \\ h(V_i) &= -V_i ln V_i - (1 - V_i) ln(1 - V_i) \end{aligned} \tag{7}$$

The supervisor is the optimal processer of the information when they have the highest attention cognition. That is, the best cognitive state occurs when the hybrid entropy $S$ reaches the maximum. The supervisor can process the most amount of information they can based on SEU theory [43] and in that case, the $S$ will reach the $S_{max}$. On this condition, $S = S_{max}$, we calculated the probability of the correct evaluation of $P_i$ based on Equation (7) using the Lagrange multiplier with constraints. Finally, the critical points $P_i$ was calculated using Equation (8). The calculation of the critical points can be found in the current research [15]:

$$P_i = \frac{\exp h(V_i)}{\sum_{i=1}^{n} \exp h(V_i)}, \text{ For } S \text{ reaches the maximum}|S = S_{max} \tag{8}$$

When the hybrid entropy $S$ reaches the maximum, humans become the best processer of information based on the maximum entropy principle. This means that the human optimally processes the information to decrease the uncertainty of the HMI. $S_{max}$ quantifies this ability, called mental entropy (ME).

The probability of the correct evaluation $P_i$ presents the tendency of supervisors to pay more attention to more fuzzy information [15,16].

### 3.3. Attention Allocation Model

According to the above-mentioned analysis, the cognitive process of the information in the supervisory task involves two channels. The supervisors process the information value based on their previous cognition and knowledge, while they process the information fuzziness based on the psychological and physiological state of the supervisor. Combining these two channels, we can obtain the information cognitive evaluation $C_i$ using Equation (9). Finally, the cognitive process is defined by the probability of the correct evaluation $P_i$ and the information value $V_i$:

$$C_i = P_i V_i' = P_i F(V_i) = -P_i \ln(1 - V_i) \tag{9}$$

Kleinman defined the attention allocation $A_i$ as the ability to process the information [44]. Based on information science, he considered humans the optimal multiple processors to process the information channel $I_i$. The subsequent research adopted this idea as the foundation of the attention allocation model and defined the attention allocation $A_i$, which showed that the information cognitive evaluation $C_i$ determines the final attention allocation strategy. The final attention allocation model for the supervisors can be represented as:

$$A_i = \frac{C_i}{\sum_{i=1}^{n} C_i} = \frac{-P_i \ln(1 - V_i)}{\sum_{i=1}^{n} -P_i \ln(1 - V_i)} \tag{10}$$

Figure 1 shows the framework of the proposed visual attention allocation model for the supervisors and shows how to build the model and the dependent theories.

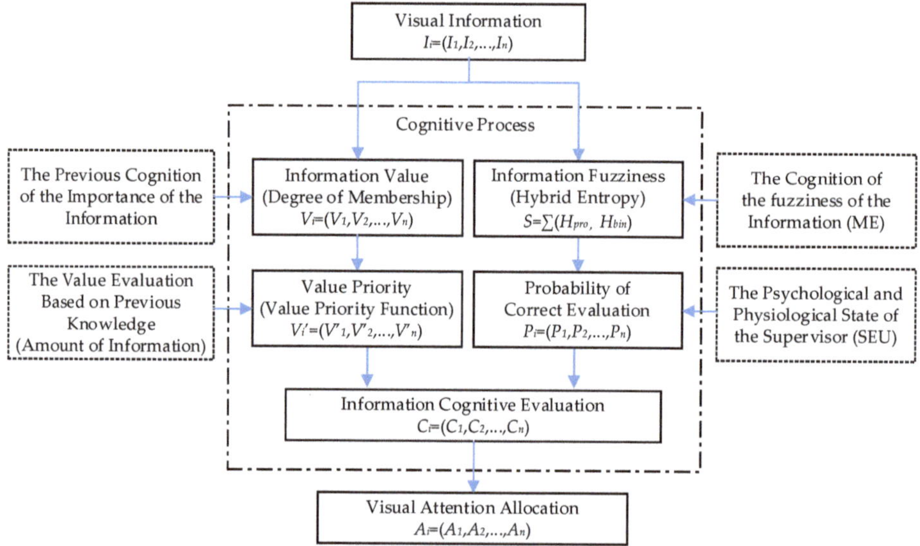


**Figure 1.** The framework of the proposed visual attention allocation model for the supervisors.

## 4. Experiment

### 4.1. Apparatus

The experiment interface was a simulator running the BAS system showing the statuses of the main air exchange fans in the subway system (Figure 2a). The system information was shown on a 22-inch digital screen with a resolution ratio of 1680 × 1050. Based on capturing the reflected infrared lights with the eyes, the SMI RED500 (Silicon Microstructures Inc., California, CA, USA) tracked the participant's eye movements with a 60Hz infrared-based camera. We used it to record the participant's visual behaviors including the gaze points on the screen, the fixation distribution. The experiment environment is shown in Figure 2b.

### 4.2. Participants

Fourteen students from the Beijing Jiaotong University, Beijing, China participated in the study (seven men, seven women, 25.3 ± 2.6 years old). All participants were familiar with the operation of a computer keyboard and had background knowledge of the subway operation. All participants were right-handed with normal vision.

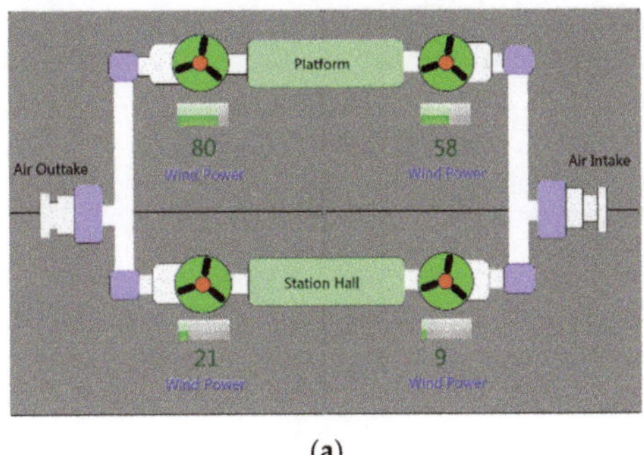


(a)

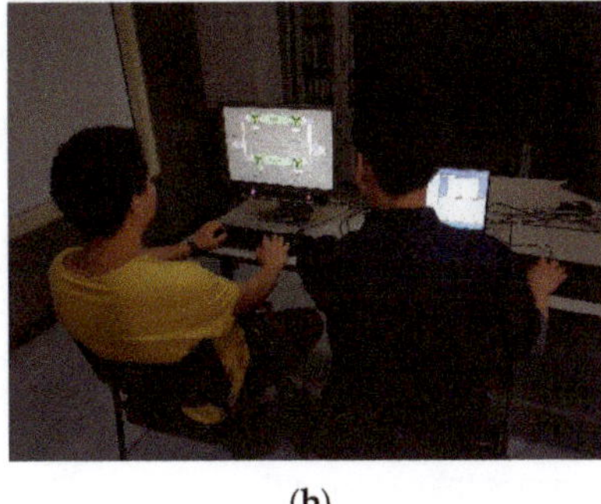

(b)

**Figure 2.** (a) The human-machine interface of the Building Automatic System (BAS). (b) The experiment environment.

### 4.3. Experimental Task

The BAS interface showed four main sections for four air fans in a fire scene. During the task, the participants needed to monitor the four speed indicators of the air fans and allocated their attention resources based on the pre-given membership degrees of the importance of the four sections. The speeds of the air fans continuously changed every second which was shown in the indicators. When the indicators showed an excess speed of the fans (>80% rated), the participants had to press the corresponding key (Insert, Delete, Home, or End for the four sections) on the keyboard to control its speed for overload protection. The abnormal excess speed would remain for one second. If the participants missed it or entered the wrong response to the overload air fans, they would be considered as not having paid attention to the corresponding section on the screen. The accuracy rates and eye behaviors were recorded during the whole task. We used the keys Insert, Delete, Home, and End, because the layout of these four keys is similar to the HMI of the BAS simulator.

The correct response to the abnormal section results in a corresponding score point based on the membership degrees of the importance, e.g., a correct response to areas of interest (AOI) 0.9 will get 0.9 points. It is obvious that response to the section which has higher information value and responses to more abnormal sections will get a higher total score point. The goal of the participants is to achieve the highest total score points.

### 4.4. Experimental Procedure

The operation of the BAS interface was explained to the participants. At first, the membership degrees of the importance of the four air fans were set based on their relative priorities in a fire scene. The participants were instructed to remember and understand the membership degrees given the possibility that the system would encounter a serious failure if the supervisor missed the overload control. Participants were asked to practice task operations twice to simulate the supervisor's experience and previously acquired knowledge. Through practice, the participants became familiar with the operation of the BAS and the functioning of the system. They would not need to look at the keyboard when they pressed the keys.

During the formal experiment, the participants were asked to complete the calibration process for the eye tracking devices first. Then, they were asked to freely allocate their attention to the four sections. They need to try their best to response to all the abnormal sections in the HMI. The test continued for five minutes and during the whole test eye behaviors were recorded.

### 4.5. Data An alysis

The experimental results of the key-press response showed that the sections had a different correct response ratio, $O_i$, which was calculated by the number of correct responses and total overload occurrences during this section. The correct response to the overload section was considered as selective attention to the corresponding section. Therefore, the fractional attention, $A_{k\_i}$ (key), was quantified by the experimental key-press data as:

$$A_{k\_i} = \frac{O_i}{\sum_{i=1}^{n} O_i}, (i = 1, 2, 3, 4) \tag{11}$$

After the experiment, the participants' eye tracking data were analyzed using the eye behavior analysis software Begaze, which was developed by Silicon Microstructures Inc., California, CA, USA. In Begaze, the four sections were identified by the four areas of interest (AOIs). The fixation behaviors of the different AOIs were extracted from the original data, which meant that the participants paid attention to the corresponding sections. Based on the fixation times, $m_i$, for a certain AOI, the fractional attention, $A_{e\_i}$ (eye), was quantified by the experimental eye tracking data with:

$$A_{e\_i} = \frac{m_i}{\sum_{i=1}^{n} m_i}, (i = 1, 2, 3, 4) \tag{12}$$

Using Equation (10), the theoretical results of the proposed supervisors' visual attention allocation model could be calculated as:

$$A_{p\_i} = \frac{-P_i \ln(1 - V_i)}{\sum_{i=1}^{n} -P_i \ln(1 - V_i)}, (i = 1, 2, 3, 4) \tag{13}$$

Matsui's and Wanyan's model was used as a comparison model; their model was used for aircraft pilots [15,42]. The theoretical results of their model can be calculated using Equation (14). This model is referred to as the Matsui's Model, as he was the first to create the basic method:

$$A_{m\_i} = \frac{P_i V_i}{\sum_{i=1}^{n} P_i \ln V_i}, (i = 1, 2, 3, 4) \tag{14}$$

The experiment aimed to compare $A_{k\_i}$ (Key), $A_{e\_i}$ (Eye), and $A_{p\_i}$ (Proposed) and $A_{m\_i}$ (Matsui's). We adopted the SPSS 25.0 statistics software (developed by IBM, California, CA, USA) to process the data. The results are expressed as the mean ± standard deviation (m ± s). Bivariate Pearson correlation analysis was used to analyze the relationship between the theoretical results and the experimental models. Considering the main difference between the Matsui's Model and the proposed model, the one-sample T test was used to analyze the difference between the two experimental results and the two theoretical results at the sections that had a high membership degree of importance.

## 5. Results

### 5.1. Theoretical and Experimental Results

Through the information value, $V_i$, pre-given by the experts for the four sections, in one scene the section of the air intake fan in the station hall (intake@hall) had 0.1 membership degrees of information importance, the section of the air outtake fan in the station hall (outtake@hall) had 0.3; and the section of the air outtake fan in the platform (outtake@platform) had 0.7. The section of the air intake fan in the platform (intake@platform) had 0.9 membership degrees of information importance.

The fractional attention, $A_i$ (%), of each section can be predicted by both Matsui's Model, $A_{m\_i}$, and the proposed model, $A_{p\_i}$. The theoretical values are shown in Table 1. There was a significant difference between the two models in the section that had a high membership degree of importance. The proposed model, $A_{p\_i}$, monotonically increased with the information value, $V_i$, while Matsui's Model, $A_{m\_i}$, did not.

**Table 1.** Information value $V_i$ and theoretical values of Matsui's model $A_{m\_i}$, proposed model $A_{p\_i}$.

| Sections on the HMI | Intake@hall | Outtake@hall | Outtake@platform | Intake@platform |
| --- | --- | --- | --- | --- |
| $V_i$ | 0.1 | 0.3 | 0.7 | 0.9 |
| $Am\_i$ | 4.29 | 17.13 | 39.97 | 38.61 |
| $Ap\_i$ | 2.35 | 10.58 | 35.73 | 51.34 |

The experimental results of the key-press response are shown in Table 2. The key press results showed that a higher information value, $V_i$, led to a higher correct response ratio, $O_i$. This indicted that supervisors paid more attention to the information that had a higher information value, $V_i$, and obtained a higher ratio of correct responses, $O_i$.

**Table 2.** Experimental values based on the key press response data.

| Sections on the HMI | Intake@hall | Outtake@hall | Outtake@platform | Intake@platform |
| --- | --- | --- | --- | --- |
| $V_i$ | 0.1 | 0.3 | 0.7 | 0.9 |
| $O_i$ | $0.12 \pm 0.08$ | $0.23 \pm 0.09$ | $0.67 \pm 0.08$ | $0.89 \pm 0.06$ |
| $A_{k\_i}$ | $6.29 \pm 3.92$ | $11.56 \pm 4.22$ | $35.34 \pm 5.23$ | $46.81 \pm 3.79$ |

The experimental results of the eye tracking are shown in Table 3. The results showed a similar attention tendency as the key-press results. A higher information value, $V_i$, led to more fixation points on the higher-value sections.

**Table 3.** Experimental values based on the eye tracking data.

| Sections on the HMI | Intake@hall | Outtake@hall | Outtake@platform | Intake@platform |
| --- | --- | --- | --- | --- |
| $V_i$ | 0.1 | 0.3 | 0.7 | 0.9 |
| $A_{e\_i}$ | $5.77 \pm 4.50$ | $8.47 \pm 3.91$ | $34.78 \pm 6.20$ | $50.98 \pm 7.80$ |

The eye tracking results provided the most practical evidence of the supervisors' attention allocation strategy. Figure 3 shows the fixation points of one participant. The figure shows that the participant paid more attention to the section that had a higher information value, $Vi$ (AOI 0.9 > AOI 0.7 > AOI 0.3 > AOI 0.1).

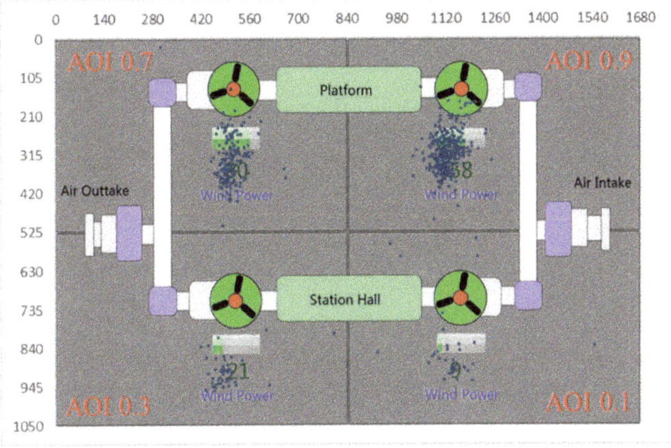


**Figure 3.** Fixation points of the eye tracking data on the screen.

*5.2. Comparison of Theoretical and Experimental Results*

The fractional attention values of the key-press response experiment and the eye movement tracking experiment as well as the two theoretical values are shown in Figure 4.

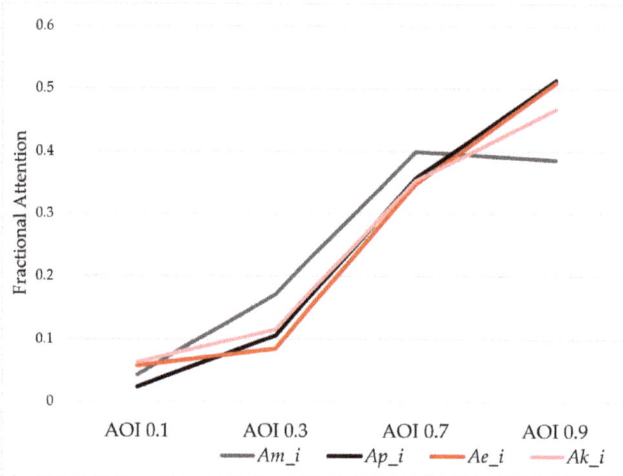

**Figure 4.** Comparison of the theoretical and experimental results.

At Figure 4 shows, the experimental results better supported the proposed model compared to Matsui's Model. The correlation analysis between the four results were processed and the results are shown in Table 4, which shows that the proposed model was significantly associated with the participants' experimental behaviors in both Key Press and Eye Tracking ($P < 0.01$). The two experimental behaviors, Key Press and Eye Tracking, were significantly correlated ($P < 0.01$), the two experimental results showed coincident behaviors, confirming that the data analysis method is effective. We also found that the correlation between Matsui's Model and the proposed model was 0.939, which means that these two models were close but different. The proposed model was more effective.

**Table 4.** The correlation between models.

| | | Matsui's Model | Proposed Model | Key Press | Eye Tracking |
|---|---|---|---|---|---|
| **Matsui's Model** | Pearson Correlation | 1 | 0.939 | 0.912 | 0.944 |
| | Sig. (2-tailed) | | 0.061 | 0.088 | 0.056 |
| **Proposed Model** | Pearson Correlation | 0.939 | 1 | 0.995 * | 0.998 * |
| | Sig. (2-tailed) | 0.061 | | 0.005 | 0.002 |
| **Key Press** | Pearson Correlation | 0.912 | 0.995 * | 1 | 0.996 * |
| | Sig. (2-tailed) | 0.088 | 0.005 | | 0.004 |
| **Eye Tracking** | Pearson Correlation | 0.944 | 0.998 * | 0.996 * | 1 |
| | Sig. (2-tailed) | 0.056 | 0.002 | 0.004 | |

* Correlation was significant at the 0.01 level (two-tailed).

Based on the method used in the proposed model, the significant difference between the two theoretical models were observed for AOI 0.7 and 0.9. The T-test was used to analyze the difference. The results of the statistics are shown in Table 5.

**Table 5.** The one-sample T-test between the models at areas of interest (AOI) 0.7 and 0.9.

| Models | Experimental Results | t | Sig. (2-tailed) | Mean Difference | 95% Confidence Interval of the Difference | |
|---|---|---|---|---|---|---|
| | | | | | Lower | Upper |
| Proposed Model @0.7 | EyeTracking@0.7 | −0.551 | 0.592 * | −0.00948 | −0.0470 | 0.0280 |
| Test Value = 0.3573 | KeyPress@0.7 | −0.270 | 0.792 * | −0.00392 | −0.0356 | 0.0277 |
| Proposed Model @0.9 | EyeTracking@0.9 | −0.168 | 0.869 * | −0.00365 | −0.0508 | 0.0435 |
| Test Value = 0.5134 | KeyPress@0.9 | −4.317 | 0.001 | −0.04534 | −0.0682 | −0.0225 |
| Matsui's Model @0.7 | EyeTracking@0.7 | −3.015 | 0.011 | −0.05188 | −0.0894 | −0.0144 |
| Test Value = 0.3997 | KeyPress@0.7 | −3.189 | 0.008 | −0.04632 | −0.0780 | −0.0147 |
| Matsui's Model @0.9 | EyeTracking@0.9 | 5.712 | 0.000 | 0.12365 | 0.0765 | 0.1708 |
| Test Value = 0.3861 | KeyPress@0.9 | 7.803 | 0.000 | 0.08196 | 0.0591 | 0.1048 |

* Significance level is at the 0.05 level (two-tailed).

The statistics showed that the experimental Eye Tracking and Key Press results were not significantly different ($P > 0.05$) from the proposed model at AOI 0.7, but were different from Matsui's Model at AOI 0.7.

For AOI 0.9, the experimental Key Press result showed a significant difference with the proposed model because the participants may not respond to the AOI 0.9 section, even if the participants focused on the section while the overload scene for AOI 0.9 was random. However, the eye tracking results showed no significant difference ($P > 0.05$) with the proposed model, which is more practical.

For Matsui's Model, the experimental results showed a significant difference for AOI 0.7 and AOI 0.9.

## 6. Discussion

### 6.1. Discussion of the Value Priority Function

The experimental results showed that the proposed model predicts supervisors' visual attention allocation more accurately than Matsui's Model. The improvement in the results from the proposed model was in the high information value, $V_i$, which was due to the proposed value priority function, $F(V_i)$, in Equation (5). The role of this function is discussed in depth below.

The proposed value priority function, $F(V_i)$, processes the information value, $V_i$, and the processed value is $V_i'$. The proposed model used $V_i'$ to present the value priority, whereas Matsui's Model uses the original information value, $V_i$. This finally affected the information cognitive evaluation, $C_i$, process. Therefore, the two theoretical models are based on a different information cognitive evaluation, $C_i$. The fractional cognitive evaluation in Matsui's Model, $C_{m\_i}$, and the proposed model, $C_{p\_i}$, can be calculated using Equations (15) and (16), respectively:

$$C_{m\_i} = P_i V_i \tag{15}$$

$$C_{p\_i} = P_i V_i' = P_i F(V_i) = -P_i \ln(1 - V_i) \tag{16}$$

Assume that the number of the independent information sections, $i$, reaches infinity. Assuming that the corresponding information value, $V_i$ (membership degree of the importance), ranges from 0 to 1, the probability of the correct evaluation, $P_i$, can be calculated using Equation (17) based on Equation (8):

$$P_i = \frac{\exp h(V_i)}{\int_{V_i=0,i=0}^{V_i=1,i=\infty} \exp h(V_i)} \tag{17}$$

Along with the information value, $V_i$, the information cognitive evaluation, $C_i$, values based on Matsui's Model, $C_{m\_i}$, and the proposed model, $C_{p\_i}$, are shown in Figure 5.

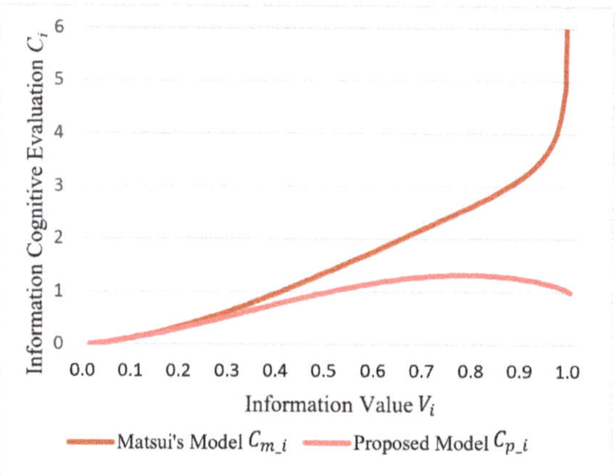

**Figure 5.** The information cognitive evaluation $C_i$ based on Matsui's Model, $C_{m\_i}$, and the proposed model, $C_{p\_i}$, along with information value, $V_i$.

At the figure shows, the proposed model is more reasonable than Matsui's Model in the following aspects:

(1)  The proposed $C_{p\_i}$ monotonically increases along with the information value, $V_i$, which is more reasonable in practice. The supervisor pays more attention to the section that has a higher information value; the supervisor is afraid of missing the most important information that may have a great probability to cause a system failure.

(2)  When the information value, $V_i$, tends to be 1, the information cognitive evaluation of the proposed $C_{p\_i}$ tends to be infinity, but Matsui's $C_{m\_i}$ showed a convergent value (Equation (18)). In practice, the highest information value ($V_i = 1$) (the membership degree of the information importance is 100%) means that the model is absolutely important and the supervisor cannot miss it. For this point, the proposed model has rationality:

$$C_{m\_\infty} = \lim_{V_i \to 1, i \to \infty} C_{m_i} = 0.006 C_{p\_\infty} = \lim_{V_i \to 1, i \to \infty} C_{p\_i} = \infty \tag{18}$$

(3)  The proposed $C_{p\_i}$ increases after 0.7822 along with the information value, $V_i$. However, Matsui's $C_{m\_i}$ decreases after 0.7822, which means that a high information value above 0.7822 will not lead to a higher information cognitive evaluation status (Equation (19)), which is not realistic. Therefore, our proposed value priority function, $F(V_i)$, is an improvement that corrects the unreasonable part of Matsui's Model:

$$C_{m\_i}' = \exp((V_i - 1)\ln(1 - V_i) - V_i \ln(V_i))(\ln(1 - V_i) - \ln(V_i)) C_{m\_i}' = 0 \big| V_i = 0.7822 \tag{19}$$

(4)  At the overall curve of the proposed $C_{p\_i}$ becomes steeper, the attention allocation of the supervisor tends to be more concentrated, and the adjustment of the supervisors' attention allocation is more reasonable.

*6.2. Discussion of Attention Allocation Models*

The proposed value priority function, $F(V_i)$, affects the information cognitive evaluation, $C_i$; $C_i$ affects the whole visual attention allocation model, $A_i$. The difference between Matsui's and the proposed model in theory is discussed in depth below.

Based on the different information cognitive evaluation models, $C_i$, the fractional attention in Matsui's Model, $A_{m\_i}$, and the proposed allocation model, $A_{p\_i}$, can be calculated using Equations (20) and (21) based on Equation (10), respectively:

$$A_{m\_i} = \frac{C_{m\_i}}{\int_{V_i=0,i=0}^{V_i=1,i=\infty} C_{m\_i}} = \frac{P_i V_i}{\int_{V_i=0,i=0}^{V_i=1,i=\infty} P_i V_i} \tag{20}$$

$$A_{p\_i} = \frac{C_{p\_i}}{\int_{V_i=0,i=0}^{V_i=1,i=\infty} C_{p\_i}} = \frac{P_i V_i'}{\int_{V_i=0,i=0}^{V_i=1,i=\infty} P_i V_i'} = \frac{P_i F(V_i)}{\int_{V_i=0,i=0}^{V_i=1,i=\infty} P_i F(V_i)} = \frac{-P_i \ln(1-V_i)}{\int_{V_i=0,i=0}^{V_i=1,i=\infty} -P_i \ln(1-V_i)} \tag{21}$$

Along with the information value, $V_i$, the attention allocation based on Matsui's Model, $A_{m\_i}$, and the proposed model, $A_{p\_i}$, is shown in Figure 6. Based on Equation (17), we added the probability of the correct evaluation $P_i$ into the figure. $P_i$ is a factor of information fuzziness tendency, which affects the model.

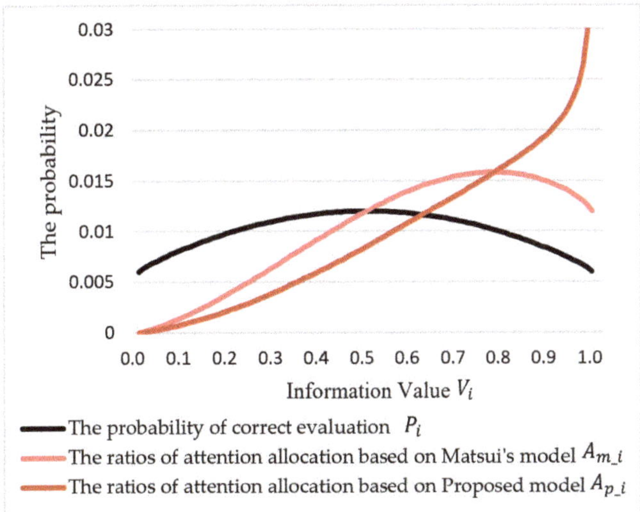


**Figure 6.** The ratios of attention allocation based on Matsui's Model, $A_{m\_i}$, and the proposed model, $A_{p\_i}$, and the probability of the correct evaluation, $P_i$, along with the information value, $V_i$.

At the figure shows, the proposed model, $A_{p\_i}$, and Matsui's Model, $A_{m\_i}$, are significantly different:

(1)  The proposed model, $A_{p\_i}$, monotonically increased along with the information value, $V_i$; but the Matsui's model did not. The supervisor pays more attention to the section that has a higher information value $V_i$;

(2)  The Matsui's attention allocation model has a critical value at 0.7822, calculated using Equation (22). This means that a higher information value, $V_i$, may not lead to a higher attention allocation ratio, $A_i$. The highest information value ($V_i \rightarrow 1$) will not acquire the supervisors' entire visual attention ($A_i \rightarrow 100\%$):

$$A_{m\_i}' = \exp((V_i-1)\ln(1-V_i)-V_i \ln V_i)+V_i \exp((V_i-1)\ln(1-V_i)-V_i \ln V_i)(\ln(1-V_i)-\ln V_i)$$
$$A_{m\_i} = 0 | V_i = 0.7822 \tag{22}$$

(3) The probability of the correct evaluation, $P_i$, reaches the highest value when the information value, $V_i$, = 0.5 (Equation (23)), which means that the supervisor has a higher successful probability to process the information in the visual section that has medium information value, $V_i$:

$$P_0 = \lim_{V_i \to 0, i \to 0} P_i = 0.006 P_m = \lim_{V_i \to 0.5, i \to mid} P_i = 0.012 P_\infty = \lim_{V_i \to 1, i \to \infty} P_i = 0.006 \quad (23)$$

(4) The proposed attention allocation model is not significantly different from Matsui's Model before the intersection near the critical point in Matsui's Model. After the intersection, the ratio of the attention allocation tended to be a steep curve. This means that the participants focused on the highest value information.

In summary, the proposed model is more reasonable and effective, as shown through the above analysis. The experimental results supported the above theoretical discussion. The proposed model is an accurate quantitative method that can be used to analyze the attention allocation strategy of supervisors.

The proposed model can basically quantify attention allocation using hybrid entropy. The other current models based on Matsui's Model, which consider the fatigue, effort, salience, and information detection efficiency [15–17], can replace the basic Matsui Model with the proposed model to improve results. The above factors were weakened in the experiment in this article on purpose to highlight the research achievement that prevented it from being overwhelmed by the above factors.

## 7. Conclusions

By referencing the definition of the information amounts, the value priority function was proposed in this paper. Considering supervisors as information processors, the information fuzziness was quantified based on hybrid entropy theory. Supervisors tend to pay more attention to important and fuzzy information. Combining these two aspects, a quantitative visual attention allocation model for supervisors was built. The experiment showed that the proposed model was more effective than the current model. The difference between the proposed theory and the current theory was further discussed, which showed that the proposed model has mathematical specialties that coincide more with practical applications and compensated for the deficiency in the current model.

Further Application: Using the proposed model, visual attention behavior can be predicted before the task. This will help researchers analyze supervisors' behaviors and evaluate the ergonomics of the HMI. The risk of cognitive deficits can be detected early, and targeted attention training can help supervisors schedule limited behavioral resources. Optimizing the HMI design with human behavior will make the system safer and more efficient.

**Author Contributions:** Conceptualization, W.F. and H.B.; Data Curation, P.W.; Funding Acquisition, W.F. and B.G.; Methodology, H.B.; Project Administration, W.F.; Software, P.W.; Validation, B.G. and P.W.; Writing–Original Draft, H.B.; Writing–Review & Editing, H.B.

**Funding:** This research was funded by the National Natural Science Foundation of China (grant number 51575037) and the Research Foundation of State Key Laboratory of Rail Traffic Control and Safety (grant number RCS2018ZT009).

**Conflicts of Interest:** The authors declare no conflict of interest.

## References

1. Wickens, C.; Lee, J.; Liu, Y.; Becker, S.G. *An Introduction to Human Factors Engineering*; Person Prentice Hall: Upper Saddle River, NJ, USA, 2009.
2. Tabai, B.H.; Bagheri, M.; Sadeghi-Firoozabadi, V.; Shahidi, V. The Relationship between Train Drivers' Attention and Accident Involvement. In Proceedings of the 4th International Conference on Transportation Information and Safety (ICTIS), Alberta, Canada, 8 August 2017; pp. 1034–1039.

3. Li, W.-C.; Kearney, P.; Braithwaite, G.; Lin, J.J. How much is too much on monitoring tasks? Visual scan patterns of single air traffic controller performing multiple remote tower operations. *Int. J. Ind. Ergonom.* **2018**, *67*, 135–144. [CrossRef]

4. Liu, C.-L. Countering the loss of extended vigilance in supervisory control using a fuzzy logic model. *Int. J. Ind. Ergonom.* **2009**, *39*, 924–933. [CrossRef]

5. Wixted, F.; O'Riordan, C.; O'Sullivan, L. Inhibiting the Physiological Stress Effects of a Sustained Attention Task on Shoulder Muscle Activity. *Int. J. Environ. Res. public health* **2018**, *15*, 115. [CrossRef] [PubMed]

6. Itti, L.; Koch, C. Computational modelling of visual attention. *Nat. Rev. Neurosci.* **2001**, *2*, 194. [CrossRef] [PubMed]

7. Burnett, K.; d'Avossa, G.; Sapir, A. Dimensionally Specific Capture of Attention: Implications for Saliency Computation. *Vision* **2018**, *2*, 9. [CrossRef]

8. Actis-Grosso, R.; Ricciardelli, P. Gaze and arrows: The effect of element orientation on apparent motion is modulated by attention. *Vision* **2017**, *1*, 21. [CrossRef]

9. Sharma, P. Modeling Bottom-Up Visual Attention Using Dihedral Group D4. *Symmetry* **2016**, *8*, 79. [CrossRef]

10. Wickens, C.D.; Hellenberg, J.; Xu, X. Pilot maneuver choice and workload in free flight. *Human factors* **2002**, *44*, 171–188. [CrossRef]

11. Wickens, C.D.; Goh, J.; Helleberg, J.; Horrey, W.J.; Talleur, D.A. Attentional models of multitask pilot performance using advanced display technology. *Human factors* **2003**, *45*, 360–380. [CrossRef]

12. Wickens, C.D.; Helleberg, J.; Goh, J.; Xu, X.; Horrey, W.J. Pilot task management: Testing an attentional expected value model of visual scanning. *Savoy, IL, UIUC Institute of Aviation Technical Report* **2001**.

13. Albonico, A.; Malaspina, M.; Daini, R. Target Type Modulates the Effect of Task Demand on Reflexive Focal Attention. *Vision* **2017**, *1*, 13. [CrossRef]

14. Matsuka, T.; Corter, J.E. Observed attention allocation processes in category learning. *Q. J. Exp. Psychol.* **2008**, *61*, 1067–1097. [CrossRef]

15. Wanyan, X.; Zhuang, D.; Wei, H.; Song, J. Pilot attention allocation model based on fuzzy theory. *Comput. Math. Appl.* **2011**, *62*, 2727–2735. [CrossRef]

16. Wu, X.; Wanyan, X.; Zhuang, D. Pilot's visual attention allocation modeling under fatigue. *Technol. Health Care* **2015**, *23*, S373–S381. [CrossRef]

17. Wu, X.; Wanyan, X.; Zhuang, D. Attention allocation modeling under multifactor condition. *J. Beijing Univ. Aeronaut. Astronaut.* **2013**, *8*, 1086.

18. Neokleous, K.C.; Avraamides, M.N.; Neocleous, C.K.; Schizas, C.N. A neurocomputational model of visual selective attention for human computer interface applications. In Proceedings of the 3rd International Conference on Human Computer Interaction, Bangalore, India, 7–10 April 2011; pp. 107–110.

19. Göbel, F.; Giannopoulos, I.; Raubal, M. The Importance of Visual Attention for Adaptive Interfaces. In Proceedings of the 18th International Conference on Human-Computer Interaction with Mobile Devices and Services Adjunct (MobileHCI '16), At Florence, Italy, 6–9 September 2016.

20. Fitts, P.M.; Jones, R.E.; Milton, J.L. Eye movements of aircraft pilots during instrument-landing approaches. *Aeronaut. Engineering Rev.* **1950**, *9*, 24–29.

21. Pradhan, A.K.; Divekar, G.; Masserang, K.; Romoser, M.; Zafian, T.; Blomberg, R.D.; Thomas, F.D.; Reagan, I.; Knodler, M.; Pollatsek, A. The effects of focused attention training on the duration of novice drivers' glances inside the vehicle. *Ergonomics* **2011**, *54*, 917–931. [CrossRef]

22. Wickens, C.D.; McCarley, J.S.; Alexander, A.L.; Thomas, L.C.; Ambinder, M.; Zheng, S. Attention-situation awareness (A-SA) model of pilot error. *Human perform. Model. aviation* **2008**, 213–239.

23. Xie, B.; Salvendy, G. Prediction of mental workload in single and multiple tasks environments. *Int. J. Cogn. Ergonom.* **2000**, *4*, 213–242. [CrossRef]

24. Bao, H.; Fang, W.; Guo, B.; Wang, J. Real-time wide-view eye tracking based on resolving the spatial depth. *Multimed Tools Appl.* **2018**, 1–23. [CrossRef]

25. Itti, L.; Koch, C.; Niebur, E. A Model of Saliency-Based Visual Attention for Rapid Scene An alysis. *IEEE T. Pattern. An al.* **2002**, *20*, 1254–1259. [CrossRef]

26. Bruce, N.D.; Tsotsos, J.K. Saliency, attention, and visual search: An information theoretic approach. *J. Vis.* **2009**, *9*, 5. [CrossRef]

27. Avraham, T.; Lindenbaum, M. Esaliency (extended saliency): Meaningful attention using stochastic image modeling. *IEEE Trans. Pami.* **2010**, *32*, 693–708. [CrossRef]

28. Mnih, V.; Heess, N.; Graves, A. Recurrent models of visual attention. In *Advances in Neural Information Processing Systems*; MIT Press: Cambridge, MA, USA, 2014; pp. 2204–2212.

29. Wu, Q.; McGinnity, T.M.; Maguire, L.; Cai, R.; Chen, M. A visual attention model based on hierarchical spiking neural networks. *Neurocomputing* **2013**, *116*, 3–12. [CrossRef]

30. Wickens, C.; McCarley, J.; Steelman-Allen, K. NT-SEEV: A model of attention capture and noticing on the flight deck. In Proceedings of the human factors and ergonomics society annual meeting, Sage Publications Sage CA, Los An geles, CA, USA, 1 October 2009; pp. 769–773.

31. Cassavaugh, N.D.; Bos, A.; McDonlad, C.; Gunaratne, P.; Backs, R.W. Assessment of the SEEV model to predict attention allocation at intersections during simulated driving. In Proceedings of the 7th International Driving Symposium on Human Factors in Driver Assessment, Training, and Vehicle Design, New York, NY, USA, 17–20 June 2013.

32. Bos, A.J.; Ruscio, D.; Cassavaugh, N.D.; Lach, J.; Gunaratne, P.; Backs, R.W. Comparison of novice and experienced drivers using the SEEV model to predict attention allocation at intersections during simulated driving. In Proceedings of the 8th International Driving Symposium on Human Factors in Driver Assessment, Training, and Vehicle Design, Salt Lake, UT, USA, 22–25 June 2015.

33. Bai, J.; Yao, B.; Yang, K. Quantitative research on impact factors of pilot's attention allocation on HUD. *J. Civil Aviation Univer. China* **2015**.

34. Lin, Y.; Zhang, W.-J.; Wu, C.; Yang, G.; Dy, J. A fuzzy logics clustering approach to computing human attention allocation using eyegaze movement cue. *Int. J. Hum-Comput. St.* **2009**, *67*, 455–463. [CrossRef]

35. Frutos-Pascual, M.; Garcia-Zapirain, B. Assessing visual attention using eye tracking sensors in intelligent cognitive therapies based on serious games. *Sensors* **2015**, *15*, 11092–11117. [CrossRef]

36. Senders, J.W. The human operator as a monitor and controller of multidegree of freedom systems. *IEEE T. Hum. Factors Electron.* **1964**, *5*, 2–5. [CrossRef]

37. Sheridan, T.B. On how often the supervisor should sample. *IEEE T. Systems Sci. Cybernetics* **1970**, *6*, 140–145. [CrossRef]

38. Lin, X.; Zhou, S.; Tang, H.; Qi, Y.; Xie, X. A novel fractional-order chaotic phase synchronization model for visual selection and shifting. *Entropy* **2018**, *20*, 251. [CrossRef]

39. Pan, J.; Hu, H.; Liu, X.; Hu, Y. Multiscale entropy analysis on human operating behavior. *Entropy* **2016**, *18*, 3. [CrossRef]

40. Pan, W.; Dong, W.; Cebrian, M.; Kim, T.; Fowler, J.H.; Pentland, A.S. Modeling dynamical influence in human interaction: Using data to make better inferences about influence within social systems. *IEEE Signal Processing Magazine* **2012**, *29*, 77–86. [CrossRef]

41. Matsui, N.; Bamba, E. Consideration of the attention allocation problem on the basis of fuzzy entropy. *T. Soc. Instru. Control Engineers* **1986**, *22*, 623–628. [CrossRef]

42. Matsui, N.; Bamba, E. Evaluative cognition and attention allocation in human interface. *Systems Comput. Jpn.* **1988**, *19*, 79–86. [CrossRef]

43. Karni, E. Subjective expected utility theory without states of the world. *J. Mathem. Econom.* **2006**, *42*, 325–342. [CrossRef]

44. Kleinman, D. Solving the optimal attention allocation problem in manual control. *IEEE T. Automatic Control* **1976**, *21*, 813–822. [CrossRef]

*Article*

# Saliency Detection Based on the Combination of High-Level Knowledge and Low-Level Cues in Foggy Images

**Xin Zhu [1], Xin Xu [1,2,3,*] and Nan Mu [1]**

[1]  School of Computer Science and Technology, Wuhan University of Science and Technology, Wuhan 430065, China; zhuxin1002@gmail.com (X.Z.); munan528@gmail.com (N.M.)

[2]  Hubei Province Key Laboratory of Intelligent Information Processing and Real-time Industrial System, Wuhan 430065, China

[3]  School of Electronic Information and Electrical Engineering, Shanghai Jiao Tong University, Shanghai 200240, China

*  Correspondence: xuxin@wust.edu.cn; Tel.: +86-27-68893531

Received: 28 February 2019; Accepted: 3 April 2019; Published: 6 April 2019

**Abstract:** A key issue in saliency detection of the foggy images in the wild for human tracking is how to effectively define the less obvious salient objects, and the leading cause is that the contrast and resolution is reduced by the light scattering through fog particles. In this paper, to suppress the interference of the fog and acquire boundaries of salient objects more precisely, we present a novel saliency detection method for human tracking in the wild. In our method, a combination of object contour detection and salient object detection is introduced. The proposed model can not only maintain the object edge more precisely via object contour detection, but also ensure the integrity of salient objects, and finally obtain accurate saliency maps of objects. Firstly, the input image is transformed into HSV color space, and the amplitude spectrum (AS) of each color channel is adjusted to obtain the frequency domain (FD) saliency map. Then, the contrast of the local-global superpixel is calculated, and the saliency map of the spatial domain (SD) is obtained. We use Discrete Stationary Wavelet Transform (DSWT) to fuse the cues of the FD and SD. Finally, a fully convolutional encoder–decoder model is utilized to refine the contour of the salient objects. Experimental results demonstrate that the presented model can remove the influence of fog efficiently, and the performance is better than 16 state-of-the-art saliency models.

**Keywords:** saliency detection; foggy image; spatial domain; frequency domain; object contour detection; discrete stationary wavelet transform

## 1. Introduction

There is great influence on the visibility of the human tracking in the wild under foggy environments on account of how dust particles suspend in the air. Therefore, the foggy images typically have low contrast and faded color features, in which the main objects are difficult to be recognized. Saliency detection is advantageous to this task, and it is a cognitive process that simulates the attention mechanism of human visual system (HVS) [1–3], which has an astonishing capability to rapidly judge the most attractive image region from a scene for further processing in the human brain.

In the past several years, the detection of visual salient objects has drawn much attention to most image processing applications. Saliency detection in foggy images acts a pivotal part in fields such as human tracking in the wild, object recognition, object segmentation, remote sensing, intelligent vehicles, and surveillance. So far, all kinds of defogging techniques [4–7] have been proposed, and they can reach comparatively good performance.

At present, image processing methods in foggy weather can be split into image enhancement and image restoration methods.

Image restoration methods include Dark channel prior algorithm [8], Visual enhancement algorithms for uniform and non-uniform fog [9], and defogging algorithms based on deep learning [10]. The method of image restoration based on the physical model is mainly to explore the physical mechanism of images degraded by fog, and to establish a general foggy weather degradation model. Then, the degradation model is calculated to compensate for the loss of image information caused by the degradation process. Finally, the quality of foggy images can be improved. However, the image restoration algorithm is a physical model based on atmospheric scattering. It requires more priori knowledge. Image enhancement methods can be divided into contrast enhancement and color enhancement. Image enhancement representative algorithms include histogram equalization [11], Retinex [12], and Wavelet based approaches [13,14]. However, the main drawbacks of these algorithms include: (1) High complexity makes their execution time-consuming, thereby making it difficult to guarantee the real-time performance of saliency detection. (2) During the process of dehazing, the visibility of foreground and background is increased simultaneously, so the recognition of salient objects is disturbed to some extent. (3) Image color distortion leads to visual features such as the edge and contour of the target cannot be accurately extracted.

Due to the low-resolution and low-contrast characteristics of foggy images, traditional spatial or frequency-based saliency models have a poor performance under fog environment. In view of this problem, this paper presents a frequency-spatial saliency model based on the atmospheric scattering distribution of foggy images, which can obtain effective information under foggy weather. Since the traditional machine learning method leads to the loss of boundary information, the object contour detection method of deep learning is added to enrich the edge information of the saliency map. As illustrated in Figure 1, the object contour detection method obviously improves the quality of the saliency map.

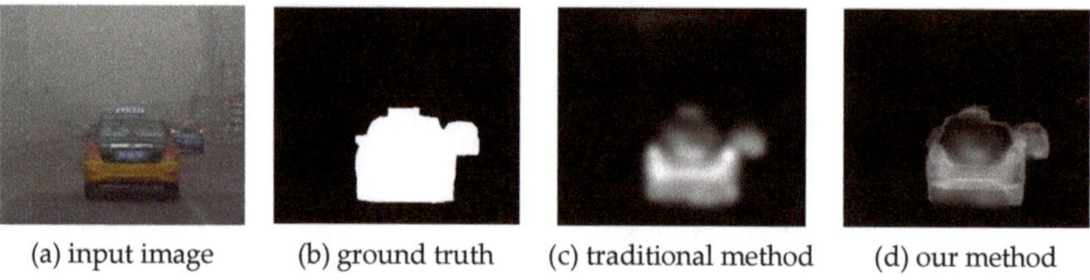

(a) input image    (b) ground truth    (c) traditional method    (d) our method

**Figure 1.** Example of salient object detection in foggy images.

In this paper, traditional methods and deep learning methods are combined to effectively detect salient objects for human tracking in the wild. In step one, the frequency domain (FD) and the spatial information are fused by DSWT. We utilize the object contour detection method of deep learning to obtain the map of the edge of the object at step two. Last, we obtain the final saliency map of the foggy image by fusing the two maps. Specifically, in step one, the foggy image is transformed into HSV color space first, and the amplitude information of FD is utilized to obtain feature maps in each channel. Then, segmenting the image into superpixels and computing the saliency of each superpixel by the local-global spatial contrast. Finally, the DSWT is applied to fuse the FD and spatial domain (SD) saliency maps, and the Gaussian filter is employed to refine the results. The flow diagram of the presented method is shown in Figure 2. The experimental results show that the proposed method can effectively detect salient objects under fog conditions.

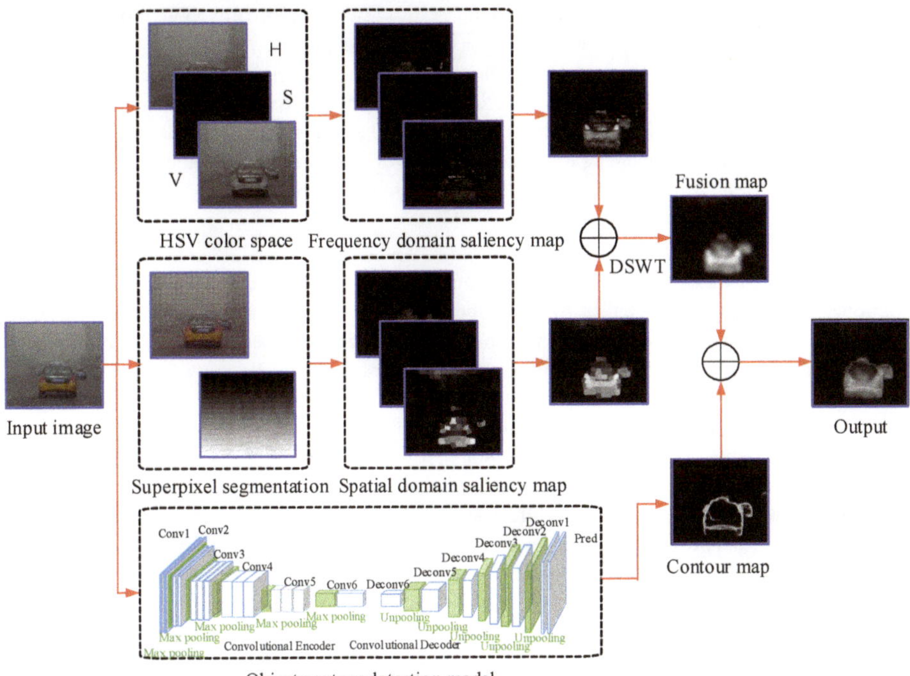


**Figure 2.** Flowchart of the proposed salient object detection model in single foggy image.

## 2. Related Works

Saliency detection is generally driven by low-level knowledge and high-level cues. Therefore, visual saliency computation under foggy environments for human tracking in the wild can be typically categorized into two classes: Saliency computational models and object contour detection approaches. Traditional saliency computational models are data-driven and primarily utilize low-level image features; while top-down object contour detection models are task-driven and usually utilize cognitive visual features.

### 2.1. Saliency Computational Models

From the perspective of information processing, traditional saliency models can be divided into two categories: SD and FD based models.

The SD saliency models are usually based on the contrast analysis to establish the algorithms. Itti et al. [15] presented a famous saliency model by utilizing the center-surround differences of multiple features. Goferman et al. [16] introduced a context-aware saliency approach, which measures the similarity of image patches in a local-global manner. Xu et al. [17] proposed a superpixel-level saliency method through a support vector machine (SVM) to train unique features. Cheng et al. [18] considered the histogram information and spatial relations, and then developed a global contrast-based saliency algorithm. Peng et al. [19] integrated tree-structed sparsity-inducing and Laplacian regularizations to construct a structured matrix decomposition model. However, most of the features used in these spatial models are not ideal for foggy images.

The saliency models of the FD develop an algorithm by converting the to a spectrum. Hou and Zhang [20] employed a spectral residual saliency method, which utilizes the log spectra to represent images. Guo et al. [21] extended the FPT algorithm and denoted four features of image by quaternion. Then, they utilized the Fourier transform of the quaternion to acquire the saliency map. Achanta et al. [22] built a frequency-tuned method, which estimates the contrast of several features. The color and brightness characteristics of each pixel are adopted to calculate the saliency map by Bian and Zhang [23]. Li et al. [24] explored saliency detection by analyzing the scale-space information of the amplitude spectrum (AS). Li et al. [25] studied the image saliency in the FD to design the model. Arya et al. [26] integrated local and global features to propose a biologically feasible FD saliency algorithm. These existing FD saliency models do not work well in foggy images due to the low-frequency information representing salient objects are greatly reduced in foggy weather.

## 2.2. Object Contour Detection

Object contour detection is a traditional computer vision problem with a long history. The traditional computer vision methods include Roberts, Prewitt, Sobel, canny, and other algorithms.

In the process of object contour detection, Roberts' algorithm does not smooth the image, so the image noise is generally not well suppressed, which also affects the loss of a part of the edge when calculating the positioning. However, Roberts' algorithm has higher positioning accuracy and better effect on steep low-noise images. Prewitt algorithm can suppress noise. The principle of noise suppression is pixel average, which is equal to low-pass filtering of the image. Thus, Prewitt's algorithm is inferior to Roberts' algorithm in edge positioning. The practical application of the Sobel edge detection algorithm [27] is when the efficiency requirements are high and the fine texture is not of interest. Sobel is usually directional and can detect only vertical or horizontal edges or both. The Sobel algorithm is improved on the basis of the Prewitt algorithm. Compared with the Prewitt algorithm, the Sobel algorithm can suppress the smoothing noise better. The Canny algorithm [28] pays more attention to the edge information reflected by the pixel gradient change and does not consider the actual object. However, it leads to loss of spatial information of the image at the same time. For some images where the edge color is similar to the background color, the edge information may be lost. The Canny algorithm is one of the best algorithms for detecting edge effects in traditional first-order differentials. It has stronger denoising capabilities than the Prewitt and Sobel algorithms. On the other hand, it is also easy to smooth some edge information, and its checking method is more complicated. However, the traditional edge detection algorithm uses the maximum gradient or the zero-crossing value of the second derivative to obtain the edge of the image. Although these algorithms have better real-time performance, they have poor anti-interference and cannot effectively overcome the influence of noise. In addition, the positioning is not good.

With the development of deep learning, the fast edge algorithm, HED and RCF algorithms are introduced. Fast edge algorithm [29] uses random forests to generate edge information. Ground truth is used to extract the edge of the image patch. This can not only reflect the actual object, but also reflect the spatial information of the picture. HED [30] used the network modified by VGG. Feature information is extracted from the whole image through multi-scale fusion, multi-loss and other methods. Similarly, it can reflect the feature information of the edge. RCF [31] takes advantage of the features of all convolutional layers in each stage compared to HED. The use of more features has also brought about an improvement in results and achieved good results. Inspired but different from these deep learning models, we employ an encoder-decoder network with full convolution to guide better salient object detection.

In our previous work, we trained an encoder-decoder network with full convolution using Caffe to optimize the performance of saliency detection. The proposed fully convolutional encoder-decoder network can learn the object contour to better represent saliency map in low contrast foggy images. The key contributions of this paper are summarized below: (1) We compute the saliency map via a frequency-spatial fusion saliency model based on DSWT. (2) This framework is further refined by a fully convolutional encoder-decoder model based on fully convolutional networks [32] and deconvolutional networks [33]. (3) The presented saliency computational model has better performance in foggy images than traditional models.

## 3. Proposed Saliency Detection Method

In this paper, we propose a frequency and spatial cues based traditional method through DSWT and a deep learning-based edge detection method fused salient object computational model to obtain the saliency map in foggy images effectively.

This section first analyses the features of foggy images, including the imaging model and effect of fog distortion on images in Section 3.1. We describe the FD based algorithm and some important computational formulas in Section 3.2. Then we give the detailed description of the SD based algorithm in Section 3.3. Section 3.4 provides the implementation of the discrete stationary wavelet transform based image fusion, which combines the above-mentioned two algorithms to generate elementary saliency map. Finally, Section 3.5 introduces the object contour detection method to refine the contour of the saliency map. It makes the position of the salient object more precise.

### 3.1. Analysis of Foggy Image Features

#### 3.1.1. Imaging Model of Foggy Image

Under fog conditions, there are a lot of tiny water droplets and aerosols in the atmosphere, which seriously affect the spread of light, resulting in a decrease in image clarity and contrast in foggy days. Especially for color images, it also produces severe color distortion and misalignment. From the respective of the computer vision, there are plentiful models [34,35] which are widely used for describing the information of foggy images. Narasimhan and Nayar [35] proposed imaging model of foggy images as shown following:

$$I_x^c = J_x^c t_x + A^c(1 - t_x) \tag{1}$$

where $c \in \{r, g, b\}$ denotes the color space of the images and $I_x^c$ denotes the foggy image captured by an imaging device. $J_x^c$ and $t_x$ denote the scene reflected light and scene transmissivity, respectively. $A^c$ is a constant and represents the ambient light.

In Equation (1), $J_x^c t_x$ and $A^c(1 - t_x)$ denote the direct attenuation [10] and air light [36], respectively. Direct attenuation is defined as the radiance of the scene and its attenuation in the medium. Air light, on the other hand, is caused by the previous scattering light, resulting in a change in the color of the scene. The transmission $t$ can be indicated as follows in which the atmosphere is homogenous:

$$t_x = e^{-\beta d_x} \tag{2}$$

let $\beta$ denote the scattering coefficient of the atmosphere.

The results show that the scene brightness decays exponentially with the scene depth $d$.

#### 3.1.2. Effect of Foggy Distortion on Images

The degraded effect of fog on the image [37] is called fog distortion. The degraded effect of fog distortion brings great challenges to the saliency computation of images. The effect of fog distortion on image quality is mainly concentrated in three aspects:

(1) The original information of the image is destroyed by the fog, and the structural information of the image is regarded as the high frequency component with enough energy in the image. The generation of fog destroys the structural information of the image and affects the details and texture of the scene object.

(2) The fog adds some information to the image. The existence of fog can be seen as adding the relevant channel information of the image and making the overall brightness of the image rises.

(3) Fog distortion combines with the original information of the image to generate some information. Due to the interaction between fog particles and the information of the image itself, the foggy image adds some multiplicative information, such as fog noise. It blurs the image, which reduces the contrast of the image.

*3.2. FD Based Algorithm*

Given a foggy image, it is transformed into HSV color space firstly, which has shown strong stimuli to human visual cortex in foggy image [38], thus the hue, saturation, value features of H, S, and V channels can be considered as the important indicators for detecting saliency.

Then, the H, S, and V channels are converted into FD respectively by conducting the Fast Fourier Transform (FFT) as:

$$F(u,v) = \sum_{x=0}^{M-1}\sum_{y=0}^{N-1} f(x,y)e^{-j2\pi(\frac{ux}{M}+\frac{vy}{N})},\tag{3}$$

where $M$ and $N$ denote the image's width and height. $f(x,y)$ and $F(u,v)$ denote image pixels in SD and FD, respectively.

$A(u,v)$ and $P(u,v)$ represent the AS and the *phase spectrum* (PS), respectively. And they can be computed via:

$$P(u,v) = \text{angle}(F(u,v)),\tag{4}$$

$$A(u,v) = \text{abs}(F(u,v)),\tag{5}$$

where the AS function and the PS function are denoted as $\text{abs}(\cdot)$ and $\text{angle}(\cdot)$, respectively. In PS function, each element of the complex array $F(u,v)$ returns the phase angle (in radians). This angle is between $\pm\pi$. Amplitude spectrum $A(u,v) = \text{abs}(F(u,v))$ means the absolute value of image pixels in frequency domain.

For foggy images, the low amplitude in FD can be regarded as a cue of the object, and the high amplitude can represent the fog background. Therefore, restraining the high amplitude information to highlight the object region in other words, the salient object can be extracted by removing the peaks of the AS via:

$$A(u,v) = \text{medfilt2}(A(u,v)),\tag{6}$$

where the median filter function is represented as $\text{medfilt2}(\cdot)$, which can effectively eliminate the peaks of $A(u,v)$. $\text{medfilt2}(I)$ performs median filtering of the image I in two dimensions. Each output pixel contains the median value in a 3-by-3 neighborhood around the corresponding pixel in the input image.

Next, it can compute a new FD map via:

$$F(u,v) = \left|A(u,v)\right|e^{-jP(u,v)},\tag{7}$$

where the absolute value is represented as $|\cdot|$.

The FD map is then transformed back to SD by performing the Inverse Fast Fourier Transform (IFFT) via:

$$f(x,y) = \frac{1}{MN}\sum_{u=0}^{M-1}\sum_{v=0}^{N-1} F(u,v)e^{j2\pi(\frac{ux}{M}+\frac{vy}{N})}.\tag{8}$$

The saliency maps (denoted as *Hmap*, *Smap*, and *Vmap*) of each channel in HSV color space can be acquired by (3)–(8).

Finally, we calculate the sum of *Hmap*, *Smap*, and *Vmap*, and obtain the map of FD saliency (represented as $S1$).

*3.3. SD Based Algorithm*

To reduce the amount of computation and guarantee the integrity of the object, the input foggy image is first divided into superpixels (presented as $SP(i)$, $i = 1,\cdots,Num$, $Num = 300$) through the simple linear iterative clustering (SLIC) algorithm [39]. Then, the obtained $H_{map}$, $S_{map}$, and $V_{map}$ of H, S, and V channels are regarded as the features of saliency.

The local-global saliency of every superpixel $SP(i)$ in $H_{map}$ can be obtained through:

$$S_{H_{map}}(i) = 1 - exp\left\{-\frac{1}{Num-1}\sum_{j=1,j\neq i}^{Num}\frac{d_{H_{map}}(SP(i),SP(j))}{1+E(SP(i),SP(j))}\right\},\qquad(9)$$

where $d_{H_{map}}(SP(i),SP(j))$ is the difference in the mean of $SP(i)$ and $SP(j)$ in $H_{map}$. The mean Euclidean distance between $SP(i)$ and $SP(j)$ is represented as $E(SP(i),SP(j))$.

Through (9), saliency values $S_{S_{map}}(i)$ and $S_{V_{map}}(i)$ of superpixels $SP(i)$ in $S_{map}$ and $V_{map}$ can be figured out.

In the end, the saliency value of each pixel $SP(i)$ is acquired by the sum of $S_{H_{map}}(i)$, $S_{S_{map}}(i)$, and $S_{S_{map}}(i)$. And $S_2$ is the saliency map of SD.

### 3.4. DSWT Based Image Fusion

The presented model mainly employs 2-levels DSWT to remove the noise of the saliency map and to accomplish the wavelet decomposition on it.

Low-pass filter and high-pass filter of the 1-level conversion are represented as $h_1[n]$ and $g_1[n]$. Up sample of the 1-level can calculate the 2-levels filters $h_2[n]$ and $g_2[n]$. Next, we can obtain the horizontal high-frequency subband $H_2$, the approximation low-pass subband $A_2$, and the diagonal high-frequency subband $D_2$, the vertical high-frequency subband $V_2$. The high-pass and low-pass subband has the same size as the initial image. Therefore, the information of detail can be preserved adequately. Thereby, it makes DSWT have translation invariance.

According to above steps, the saliency map based on the FD $S_1$ and the saliency map based on the SD $S_2$ is obtained. Then, we fuse the two maps through the 2-levels DSWT as:

$$[A_1S_1, H_1S_1, V_1S_1, D_1S_1] = swt2(S_1,\ 1,\ 'sym2'),\qquad(10)$$

$$[A_1S_2, H_1S_2, V_1S_2, D_1S_2] = swt2(S_2,\ 1,\ 'sym2'),\qquad(11)$$

$$[A_2S_1, H_2S_1, V_2S_1, D_2S_1] = swt2(A_1S_1,\ 1,\ 'sym2'),\qquad(12)$$

$$[A_2S_2, H_2S_2, V_2S_2, D_2S_2] = swt2(A_1S_2,\ 1,\ 'sym2'),\qquad(13)$$

where the multilevel DSWT is represented as $swt2(\cdot)$. $swt2(\cdot)$ performs a multilevel 2-D stationary wavelet decomposition using either an orthogonal or a biorthogonal wavelet. Equations (10)–(13) compute the stationary wavelet decomposition of the real-valued 2-D or 3-D matrix at 1-level by using 'sym2'. The output three-dimensional array $A_iS_j$ is represented as the result of the i-level low frequency approximation coefficients of saliency map $S_j$ employing 'sym2' filter, and $D_iS_j$, $H_iS_j$, $V_iS_j$ represent the high frequency coefficients of the diagonal, vertical and horizontal directions, respectively.

Next, the 2-level fusion is calculated using the following formulas:

$$A_2S_f = 0.5 \times (A_2S_1 + A_2S_2),\qquad(14)$$

$$H_2S_f = D\cdot H_2S_1 + \widetilde{D}\cdot H_2S_2,\ D = (|H_2S_1| - |H_2S_2|) \geq 0,\qquad(15)$$

$$V_2S_f = D\cdot V_2S_1 + \widetilde{D}\cdot V_2S_2,\ D = (|V_2S_1| - |V_2S_2|) \geq 0,\qquad(16)$$

$$D_2S_f = D\cdot D_2S_1 + \widetilde{D}\cdot D_2S_2,\ D = (|D_2S_1| - |D_2S_2|) \geq 0.\qquad(17)$$

The 1-level fusion is calculated using the following formulas:

$$A_1S_f = iswt2(A_2S_f, H_2S_f, V_2S_f, D_2S_f,\ 'sym2'),\qquad(18)$$

$$H_1S_f = D\cdot H_1S_1 + \widetilde{D}\cdot H_1S_2,\ D = (|H_1S_1| - |H_1S_2|) \geq 0,\qquad(19)$$

$$V_1S_f = D\cdot V_1S_1 + \widetilde{D}\cdot V_1S_2,\ D = (|V_1S_1| - |V_1S_2|) \geq 0,\qquad(20)$$

$$D_1 S_f = D \cdot D_1 S_1 + \widetilde{D} \cdot D_1 S_2, \quad D = (|D_1 S_1| - |D_1 S_2|) \geq 0, \tag{21}$$

where the inverse DSWT function is represented as $\mathrm{iswt2}(\cdot)$. For example, $X = \mathrm{iswt2}(A, H, V, D,'\,\mathrm{sym2}')$ reconstructs the matrix $X$ based on the multilevel stationary wavelet decomposition structure $[A, H, V, D]$ in Equation (18) and Equation (22).

Then, the fusion image can be calculated using the following formulas:

$$Salmap = \mathrm{iswt2}\big(A_1 S_f, H_1 S_f, V_1 S_f, D_1 S_f,'\,\mathrm{sym2}'\big). \tag{22}$$

In the end, the proposed method utilizes a Gaussian filter to generate a smoothed saliency map.

### 3.5. Object Contour Detection

Object contour detection model [40] can filter and ignore the edge information in the background and obtain the contour detection result by centering the object in the foreground. Inspired by the fully convolutional networks and deconvolutional networks [33], an object contour detection model is introduced to extract the target contour and suppress background boundaries.

The layers up to 'fc6' from VGG-16 [41] are used in the edge detection model as the encoder of the network. The deconv6 decoder convolutional layer uses $1 \times 1$ kernel, and all remaining decoder convolutional layers use $5 \times 5$ kernel. Except for the decoder convolutional layer next to the output layer which uses the sigmoid activation function, all other decoder convolutional layers are followed by the relu activation function.

We trained the network using Caffe. The parameters of the encoder are fixed when training the network, while only the parameters of the decoder are optimized. This maintains the generalization of the ability of the encoder and enables the decoder network to be easily combined with other tasks.

## 4. Experimental Results

### 4.1. Experiment Setup

**Datasets:** Abundant experiments are executed on two datasets to assess the performance of the proposed saliency model.

A foggy image dataset (FI) was collected from the Internet, which contained 200 foggy images. We also provide the corresponding manual labeled ground truths. The FI dataset can be downloaded at https://drive.google.com/file/d/1aqro3U2lU8iRylyfJP1WRKxTWrrFzizh/view?usp=sharing. The other one is the BSDS500 Dataset. It includes 500 natural images with carefully annotated boundaries by different users. The dataset is divided into three parts: 200 for training, 100 for validation and the other 200 for testing. Object contour detection is utilized to optimize the saliency map which was obtained by traditional machine learning methods of salient object detection. Due to the use of traditional methods, the edge information of the saliency map is incomplete.

**Evaluation Criteria:** For quantitative evaluation, the average computation time, the mean absolute error (MAE) score, the overlapping ratio (OR) score, the precision-recall (PR) curve, the true positive rates (TPRs) *and false positive rates* (FPRs) curve, the area under the curve (AUC) score, the F-measure curve, the weighted F-measure (WF) score, and various saliency models are computed, respectively.

The precision, recall, TPR and FPR values are generated by converting the saliency map into binary map via thresholding to compare the difference of each pixel with ground truth. $\beta^2$ is the parameter to weigh the precision and recall, which is set to 0.3 in our experiments [18,22].

The ratio of the number of salient pixels correctly labeled to all salient pixels in this binary map is defined as the precision. In other words, precision refers to how many of the samples that are positively judged by the model that are true positive samples. The recall rate refers to how many positive samples are judged as positive samples by the model in the ground-truth map:

$$precision = \frac{|TS \cap DS|}{|DS|}, recall = \frac{|TS \cap DS|}{|TS|}, \tag{23}$$

where $TS$ and $DS$ denote true salient pixels and detected salient pixels by the binary map, respectively.

The TPRs represents the probability that have a right classification of positive examples, and the FPRs represents the probability of splitting a negative sample into a positive sample.

$$\text{TPR} = \frac{TP}{(TP+FN)}, \text{FPR} = \frac{FP}{(FP+TN)}, \tag{24}$$

F-measure value, denoted as $F_\beta$, is obtained by computing the weighted harmonic mean of precision and recall.

$$F_\beta = \frac{(1+\beta^2) \times \text{Precision} \times \text{Recall}}{\beta^2 \times \text{Precision} + \text{Recall}}, \tag{25}$$

where $\beta^2$ is set to 0.3 to weight precision more than recall as suggested in [42].

Given a ground truth main subject region G and a detected main-subject region D. The OR score is the ratio between two times the correctly detected main-subject region to the sum of detected and ground truth main subject region.

$$\text{OR} = \frac{2 \times A(D \cap G)}{A(D) + A(G)}, \tag{26}$$

The percentage of area under the TPRs-FPRs curve is called as the AUC score. It intuitively reflects the classification ability of ROC curve.

$$\text{AUC} = \frac{\sum_{i \in positiveClass} rank_i - \frac{M(1+M)}{2}}{M \times N}, \tag{27}$$

The MAE score to calculate the average difference of each pixel between the saliency map which is predicted and ground truth. It is acquired by:

$$\text{MAE} = \frac{1}{W \times H} \sum_{x-1}^{W} \sum_{y-1}^{H} |S(x,y) - G(x,y)|, \tag{28}$$

where $S$ is predicted saliency map and $G$ is ground truth, the width and height of saliency map $S$ are presented as $W$ and $H$.

## 4.2. Comparison and Analysis

The presented method is compared with 16 well-known saliency detection methods including: IT [15], CA [16], SMD [19], SR [20], FT [22], MR [41], NP [43], IS [44], LR [45], PD [46], SO [47], BSCA [48], BL [49], GP [50], SC [51], and MIL [52]. The source code provided by others was used to test on our foggy dataset. Each foggy image in our dataset was tested on 16 methods of others to produce the corresponding saliency map.

Figure 3 shows the PR, TPRs-FPRs, and F-measure curves of various saliency models to evaluate the proposed model quantitatively. The larger the area under the curve is, the better the performance of the saliency model will be.

It can be seen from the three figures that the proposed model is superior to other saliency models, which validates that our saliency result is robust in foggy images.

The greatest three results in Table 1 are emphasized in red fonts, blue fonts and green fonts when comparing performance with other methods. Table 1 shows that the presented model yields the greatest performance in terms of AUC and OR scores and obtains the second best in MAE and WF. These results indicate that the presented saliency model reaches the better performance under fog conditions. Moreover, our proposed method has a shorter running time than most, ranking fifth out of other 16 methods.

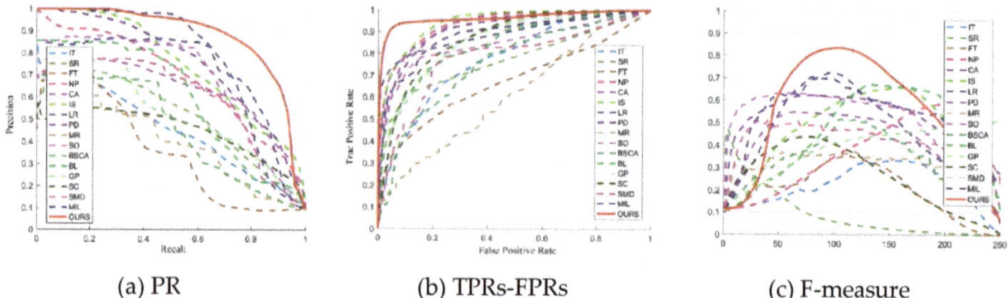

(a) PR        (b) TPRs-FPRs        (c) F-measure

**Figure 3.** The quantitative comparisons of the proposed saliency model with 16 state-of-the-art models in foggy images.

**Table 1.** The performance comparisons of various saliency models in foggy images.

| Saliency Models | AUC | MAE | WF | OR | TIME(s) |
|---|---|---|---|---|---|
| IT | 0.7916 | 0.3434 | 0.1250 | 0.1629 | 5.7661 |
| SR | 0.5602 | 0.1118 | 0.0730 | 0.3253 | 10.7458 |
| FT | 0.6809 | 0.1724 | 0.1268 | 0.1703 | 0.8717 |
| NP | 0.9156 | 0.2881 | 0.1879 | 0.4357 | 4.6347 |
| CA | 0.8718 | 0.1328 | 0.2729 | 0.4145 | 59.2286 |
| IS | 0.9077 | 0.1736 | 0.2378 | 0.4115 | 1.2987 |
| LR | 0.8687 | 0.1174 | 0.2661 | 0.4274 | 146.0651 |
| PD | 0.8277 | 0.1073 | 0.3602 | 0.4449 | 28.5625 |
| GBMR | 0.5658 | 0.2219 | 0.2058 | 0.1809 | 2.4929 |
| SO | 0.7705 | 0.0913 | 0.4431 | 0.4557 | 2.5251 |
| BSCA | 0.7327 | 0.1850 | 0.2028 | 0.2420 | 6.7254 |
| BL | 0.8053 | 0.2220 | 0.2159 | 0.5007 | 53.3103 |
| GP | 0.8241 | 0.2235 | 0.2556 | 0.3133 | 20.2649 |
| SC | 0.8077 | 0.1491 | 0.2005 | 0.3215 | 39.0530 |
| SMD | 0.7311 | 0.1418 | 0.2976 | 0.3455 | 7.6693 |
| MIL | 0.8636 | 0.1365 | 0.3127 | 0.5009 | 341.9829 |
| Proposed | 0.9177 | 0.0995 | 0.4060 | 0.6050 | 5.0756 |

Figure 4 shows the visual comparisons of varieties of saliency detection models on the foggy image dataset, which demonstrates that the saliency maps obtained by our method are much closer to the ground truths. Compared to the baselines, our method yields a better performance, which means that it suppresses background clutters well and generates visually good contour maps. Based on the saliency maps compared with other models, this paper makes a few basic observations:

The IT, NP, IS and BL models find it difficult to suppress the fog background. The map did not highlight salient objects but detected the fog background together. It is treated with fog as the foreground. As can be seen from Figure 4, there has a very poor effect.

Saliency maps of the MR, BSCA and GP models show that the fog background areas are too bright, and background and foreground are marked as salient regions at the same time. Therefore, saliency maps are blurred. However, the results are relatively better than IT, NP, IS and BL.

The FT, LR, PD, SC, CA models detect salient objects in the foreground while also clearly detecting non-salient objects such as trees, streetlights, and roads in the back-ground. Such an algorithm cannot achieve the purpose of saliency detection and is meaningless for tracking humans in the wild.

Although the fog background has less interferential in the SR than others, the salient objects are also not detected. It is the worst model on test dataset. Due to the features they used are ineffective in foggy images.

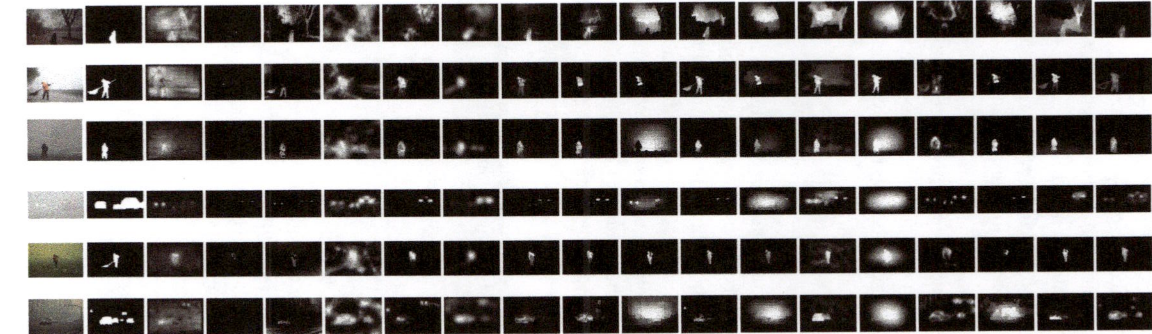

(a)Input (b)GT (c) IT (d) SR (e) FT (f) NP (g) CA (h) IS (i) LR (j) PD (k)MR (l) SO (m)BSCA (n) BL (o) GP (p) SC (q)SMD (r)MIL (s)OURS

**Figure 4.** The saliency maps of the proposed model in comparison with 16 models in foggy images. (**a**) testing foggy images, (**b**) ground truth binary masks, (**c–r**) saliency maps obtained by 16 state-of-the-art saliency models, (**s**) saliency maps obtained by the proposed model.

The SO, SMD and MIL have poor performance for foggy images with a slightly complex background. Although salient objects in the foreground are detected, the brightness of the fog in the image affects the final detection results. In other words, these models are not robust in fog environment.

The experiment results show that other models cannot detect the salient objects well under foggy weather. It is evident that the proposed method can better detect the salient objects in foggy images and more effective than other models. The reasons are summarized as follows:

(1)   The local and global information of the images are utilized, so that the salient objects can complement the information in the FD and the SD. However, the traditional machine learning method leads to the loss of edge information. Thus, causes the boundary of the saliency map to be blurred.

(2)   The object contour detection method of deep learning is added to enrich the edge information of the saliency map. It can suppress the interference from the fog background and acquire the edge of salient objects more precisely.

(3)   In the meantime, traditional based method and deep learning-based method are combined to effectively detect salient objects. The proposed method can not only retain the edge more accurately via object contour detection, but also ensure the salient objects' integrity. By this means, can obtain a more precise and clearer saliency map.

## 5. Conclusions

In our study, we present a high-efficiency model to handle the salient object detection of foggy images. The proposed model combines traditional machine learning based frequency-spatial saliency detection algorithm and deep learning-based object contour detection algorithm to cope with the matter of salient object detection under fog environments. In traditional saliency detection method, the saliency map is acquired by fusing the frequency and spatial saliency maps via DSWT. Then, a fully convolutional encoder–decoder model is utilized to improve the contour of the salient objects. Experimental results on foggy image dataset demonstrate that the proposed saliency detection model performs obviously better against other 16 well-known models.

**Author Contributions:** X.Z. analyzed the data and wrote the paper; X.X. guided the algorithm design and provided the funding support; X.Z. and N.M. designed the algorithm and conducted the experiments with technical assistance.

**Funding:** This research was funded by the Natural Science Foundation of China, grant number 61602349.

**Conflicts of Interest:** The authors declare no conflict of interest.

## References

1.   Zhang, L.; Li, A.; Zhang, Z.; Yang, K. Global and local saliency analysis for the extraction of residential areas in high-spatial-resolution remote sensing image. *IEEE Trans. Geosci. Remote Sens.* **2016**, *54*, 3750–3763. [CrossRef]

2.   Zhang, L.; Yang, K. Region-of-interest extraction based on frequency domain analysis and salient region detection for remote sensing image. *IEEE Geosci. Remote Sens. Lett.* **2014**, *11*, 916–920. [CrossRef]

3.   Zhang, L.; Yang, K.; Li, H. Regions of interest detection in panchromatic remote sensing images based on multiscale feature fusion. *IEEE J. Sel. Top. Appl. Earth Obs. Remote Sens.* **2014**, *7*, 4704–4716. [CrossRef]

4.   Schaul, L.; Fredembach, C.; Susstrunk, S. Color image dehazing using the near-infrared. In Proceedings of the 16th IEEE International Conference on Image Processing, Cairo, Egypt, 7–10 November 2009; pp. 1629–1632.

5.   Ancuti, C.O.; Ancuti, C. Single image dehazing by multi-scale fusion. *IEEE Trans. Image Process.* **2013**, *22*, 3271–3282. [CrossRef] [PubMed]

6.   Lee, C.; Shao, L. Learning-based single image dehazing via genetic programming. In Proceedings of the 23th International Conference on Pattern Recognition, Cancun, Mexico, 4–8 December 2016; pp. 745–750.

7.   Bui, T.M.; Kim, W. Single image dehazing using color ellipsoid prior. *IEEE Trans. Image Process.* **2018**, *27*, 999–1009. [CrossRef]

8.   He, K.; Sun, J.; Tang, X. Single image haze removal using dark channel prior. *IEEE Trans. Pattern Anal. Mach. Intell.* **2011**, *33*, 2341–2353.

9. Fattal, R. Single image dehazing. *ACM Trans. Graph.* **2008**, *27*, 1–9. [CrossRef]

10. Tan, R.T. Visibility in bad weather from a single image. In Proceedings of the 2008 IEEE Conference on Computer Vision and Pattern Recognition, Anchorage, AK, USA, 23–28 June 2008; pp. 1–8.

11. Kim, J.Y.; Kim, L.S.; Hwang, S.H. An advanced contrast enhancement using partially overlapped sub-block histogram equalization. *IEEE Trans. Circuits Syst. Video Technol.* **2001**, *11*, 475–484.

12. Jang, J.H.; Bae, Y.; Ra, J.B. Contrast-enhanced fusion of multi sensor images using subband-decomposed multiscale retinex. *IEEE Trans. Image Process.* **2012**, *21*, 3479. [CrossRef]

13. Ganeshnagasai, P.V. Image enhancement using Wavelet transforms and SVD. *Int. J. Eng. Sci. Technol.* **2012**, *4*, 1080–1087.

14. Neena, K.A.; Aiswariya, R.; Rajesh, C.R. Image Enhancement based on Stationary Wavelet Transform, Integer Wavelet Transform and Singular Value Decomposition. *Int. J. Comput. Appl.* **2012**, *58*, 30–35.

15. Itti, L.; Koch, C.; Niebur, E. A model of saliency-based visual attention for rapid scene analysis. *IEEE Trans. Pattern Anal. Mach. Intell.* **1998**, *20*, 1254–1259. [CrossRef]

16. Goferman, S.; Zelnik-Manor, L.; Tal, A. Context-aware saliency detection. *IEEE Trans. Pattern Anal. Mach. Intell.* **2012**, *34*, 1915–1926. [CrossRef]

17. Xu, X.; Mu, N.; Zhang, H.; Fu, X. Salient object detection from distinctive features in low contrast images. In Proceedings of the 2015 IEEE International Conference on Image Processing, Quebec City, QC, Canada, 27–30 September 2015; pp. 3126–3130.

18. Cheng, M.-M.; Mitra, N.J.; Huang, X.; Torr, P.H.S.; Hu, S.-M. Global contrast based salient region detection. *IEEE Trans. Pattern Anal. Mach. Intell.* **2015**, *37*, 569–582. [CrossRef]

19. Peng, H.; Li, B.; Ling, H.; Hu, W.; Xiong, W.; Maybank, S.J. Salient object detection via structured matrix decomposition. *IEEE Trans. Pattern Anal. Mach. Intell.* **2017**, *39*, 818–832. [CrossRef]

20. Hou, X.; Zhang, L. Saliency detection: A spectral residual approach. In Proceedings of the 2007 IEEE Conference on Computer Vision and Pattern Recognition, Minneapolis, MN, USA, 17–22 June 2007; pp. 1–8.

21. Guo, C.; Ma, Q.; Zhang, L. Spatio-temporal saliency detection using phase spectrum of quaternion fourier transform. In Proceedings of the 2008 IEEE Conference on Computer Vision and Pattern Recognition, Anchorage, AK, USA, 23–28 June 2008; pp. 1–8.

22. Achanta, R.; Hemami, S.; Estrada, F.; Susstrunk, S. Frequency-tuned salient region detection. In Proceedings of the 2009 IEEE Conference on Computer Vision and Pattern Recognition, Miami, FL, USA, 20–25 June 2009; pp. 1597–1604.

23. Bian, P.; Zhang, L. Visual saliency: A biologically plausible contourlet-like frequency domain approach. *Cogn. Neurodyn.* **2010**, *4*, 189–198. [CrossRef]

24. Li, J.; Levine, M.D.; An, X.; Xu, X.; He, H. Visual saliency based on scale-space analysis in the frequency domain. *IEEE Trans. Pattern Anal. Mach. Intell.* **2013**, *35*, 996–1010. [CrossRef]

25. Li, J.; Duan, L.-Y.; Chen, X.; Huang, T.; Tian, Y. Finding the secret of image saliency in the frequency domain. *IEEE Trans. Pattern Anal. Mach. Intell.* **2015**, *37*, 2428–2440. [CrossRef]

26. Arya, R.; Singh, N.; Agrawal, R.K. A novel hybrid approach for salient object detection using local and global saliency in frequency domain. *Multimed. Tools Appl.* **2016**, *75*, 8267–8287. [CrossRef]

27. Forsyth, D.A.; Ponce, J. *Computer Vision: A Modern Approach*; Pearson Education: Upper Saddle River, NJ, USA, 2003; Volume 17, pp. 21–48.

28. Canny, J. A computational approach to edge detection. *IEEE Trans. Pattern Anal. Mach. Intell.* **1986**, *6*, 679–698. [CrossRef]

29. Dollár, P.; Zitnick, C.L. Fast edge detection using structured forests. *IEEE Trans. Pattern Anal. Mach. Intell.* **2015**, *37*, 1558–1570. [CrossRef]

30. Xie, S.; Tu, Z. Holistically-nested edge detection. In Proceedings of the 2015 IEEE international conference on computer vision, Santiago, Chile, 7–13 December 2015; pp. 1395–1403.

31. Liu, Y.; Cheng, M.M.; Hu, X.; Wang, K.; Bai, X. Richer convolutional features for edge detection. *IEEE Trans. Pattern Anal. Mach. Intell.* **2017**. [CrossRef]

32. Long, J.; Shelhamer, E.; Darrell, T. Fully convolutional networks for semantic segmentation. In Proceedings of the 2015 IEEE Conference on Computer Vision and Pattern Recognition, Boston, MA, USA, 7–12 June 2015; pp. 3431–3440.

33. Noh, H.; Hong, S.; Han, B. Learning deconvolution network for semantic segmentation. In Proceedings of the 2015 IEEE International Conference on Computer Vision, Santiago, Chile, 7–13 December 2015; pp. 1520–1528.

34. Narasimhan, S.G.; Nayar, S.K. Chromatic framework for vision in bad weather. In Proceedings of the 2000 IEEE Computer Conference on Computer Vision and Pattern Recognition, Hilton Head Island, NC, USA, 15 June 2000; pp. 598–605.
35. Narasimhan, S.G.; Nayar, S.K. Vision and the atmosphere. *Int. J. Comput. Vis.* **2002**, *48*, 233–254. [CrossRef]
36. Koschmieder, H. Theorie der horizontalen sichtweite. *Beitr. Phys. Freien Atm.* **1924**, *12*, 171–181.
37. Li, C.; Lu, W.; Xue, S.; Shi, Y.; Sun, X. Quality assessment of polarization analysis images in foggy conditions. In Proceedings of the 2014 IEEE International Conference on Image Processing, Paris, France, 27–30 October 2014; pp. 551–555.
38. Chen, W.; Shi, Y.Q.; Xuan, G. Identifying Computer Graphics using HSV Color Model and Statistical Moments of Characteristic Functions. In Proceedings of the 2007 IEEE International Conference on Multimedia and Expo, Beijing, China, 2–5 July 2007; pp. 1123–1126.
39. Achanta, R.; Shaji, A.; Smith, K.; Lucchi, A.; Fua, P.; Susstrunk, S. SLIC superpixels compared to state-of-the-art superpixel methods. *IEEE Trans. Pattern Anal. Mach. Intell.* **2012**, *34*, 2274–2282. [CrossRef]
40. Yang, J.; Price, B.; Cohen, S.; Lee, H.; Yang, M. Object contour detection with a fully convolutional encoder-decoder network. In Proceedings of the 2016 IEEE Conference on Computer Vision and Pattern Recognition, Las Vegas, NV, USA, 27–30 June 2016; pp. 193–202.
41. Simonyan, K.; Zisserman, A. Very deep convolutional networks for large-scale image recognition. *arXiv* **2014**, arXiv:1409.1556.
42. Yang, C.; Zhang, L.; Lu, H.; Ruan, X.; Yang, M.-H. Saliency detection via graph-based manifold ranking. In Proceedings of the 2013 IEEE Conference on Computer Vision and Pattern Recognition, Portland, OR, USA, 23–28 June 2013.
43. Murray, N.; Vanrell, M.; Otazu, X.; Parraga, C.A. Saliency estimation using a non-parametric low-level vision model. In Proceedings of the 2011 IEEE Computer Conference on Computer Vision and Pattern Recognition, Colorado Springs, CO, USA, 20–25 June 2011; pp. 433–440.
44. Hou, X.; Harel, J.; Koch, C. Image signature: Highlighting sparse salient regions. *IEEE Trans. Pattern Anal. Mach. Intell.* **2012**, *34*, 194–201.
45. Shen, X.; Wu, Y. A unified approach to salient object detection via low rank matrix recovery. In Proceedings of the 2012 IEEE Conference on Computer Vision and Pattern Recognition, Providence, RI, USA, 16–21 June 2012; pp. 853–860.
46. Margolin, R.; Zelnik-Manor, L.; Tal, A. What makes a patch distinct? In Proceedings of the 2013 IEEE Conference on Computer Vision and Pattern Recognition, Portland, OR, USA, 23–28 June 2013; pp. 1139–1146.
47. Zhu, W.; Liang, S.; Wei, Y.; Sun, J. Saliency optimization from robust background detection. In Proceedings of the 2014 IEEE Conference on Computer Vision and Pattern Recognition, Columbus, OH, USA, 23–28 June 2014; pp. 2814–2821.
48. Qin, Y.; Lu, H.; Xu, Y.; Wang, H. Saliency detection via cellular automata. In Proceedings of the 2015 IEEE Conference on Computer Vision and Pattern Recognition, Boston, MA, USA, 7–12 June 2015; pp. 110–119.
49. Tong, N.; Lu, H.; Yang, M. Salient object detection via bootstrap learning. In Proceedings of the 2015 IEEE Conference on Computer Vision and Pattern Recognition, Boston, MA, USA, 7–12 June 2015; pp. 1884–1892.
50. Jiang, P.; Vasconcelos, N.; Peng, J. Generic promotion of diffusion-based salient object detection. In Proceedings of the 2015 IEEE International Conference on Computer Vision, Santiago, Chile, 7–13 December 2015; pp. 217–225.
51. Zhang, J.; Wang, M.; Zhang, S.; Li, X.; Wu, X. Spatiochromatic context modeling for color saliency analysis. *IEEE Trans. Neural Netw. Learn. Syst.* **2016**, *2*, 1177–1189. [CrossRef]
52. Huang, F.; Qi, J.; Lu, H.; Zhang, L.; Ruan, X. Salient object detection via multiple in-stance learning. *IEEE Trans. Image Process.* **2017**, *26*, 1911–1922. [CrossRef]

*Article*

# Detecting Toe-Off Events Utilizing a Vision-Based Method

**Yunqi Tang [1], Zhuorong Li [1], Huawei Tian [2], Jianwei Ding [3,*] and Bingxian Lin [4,5,*]**

[1]  School of Forensic Science, People's Public Security University of China, Beijing 100000, China;
tangyunqi@ppsuc.edu.cn (Y.T.); lizhuorong98@126.com (Z.L.)

[2]  School of Criminal Investigation and Counter Terrorism, People's Public Security University of China,
Beijing 100000, China; hwtian@ppsuc.edu.cn

[3]  School of Information Engineering and Network Security, People's Public Security University of China,
Beijing 100000, China

[4]  Jiangsu Center for Collaborative Innovation in Geographical Information Resource Development and
Application, Nanjing 210000, China

[5]  Key Laboratory of Virtual Geographic Environment, Nanjing Normal University, Ministry of Education,
Nanjing 210000, China

*  Correspondence: jwding@foxmail.com (J.D.); 09345@njnu.edu.cn (B.L.)

Received: 15 February 2019; Accepted: 24 March 2019; Published: 27 March 2019

**Abstract:** Detecting gait events from video data accurately would be a challenging problem. However, most detection methods for gait events are currently based on wearable sensors, which need high cooperation from users and power consumption restriction. This study presents a novel algorithm for achieving accurate detection of toe-off events using a single 2D vision camera without the cooperation of participants. First, a set of novel feature, namely consecutive silhouettes difference maps (CSD-maps), is proposed to represent gait pattern. A CSD-map can encode several consecutive pedestrian silhouettes extracted from video frames into a map. And different number of consecutive pedestrian silhouettes will result in different types of CSD-maps, which can provide significant features for toe-off events detection. Convolutional neural network is then employed to reduce feature dimensions and classify toe-off events. Experiments on a public database demonstrate that the proposed method achieves good detection accuracy.

**Keywords:** toe-off detection; gait event; silhouettes difference; convolutional neural network

---

## 1. Introduction

Gait is the periodic motion pattern of human walking or running. Different people owns different gait patterns, due to the reason that gait pattern is uniquely decided by the personal factors, such as personal habits, injury, and disease. Base on this character, researchers in pattern recognition area employ gait pattern to recognition the identity of walkers, namely gait recognition. And gait pattern is also used for disease diagnosing by the researchers in the field of medicine, namely gait analysis. No matter gait recognition or gait analysis, gait events detection is the basic problem of the both applications. Automatic detection of gait events is desirable for artificial intelligence applications, such as gait recognition and medicine abnormal gait analysis [1].

A gait cycle is the minimum periodic movement of human walking. Usually, a gait cycle is defined as a period from a heel strikes on the ground to the same heel strikes on the ground again the next time. According to the swing character of legs, a gait cycle can be divided into two phases, which are stance phase and swing phase. And there are also important six gait events within each gait cycle (shown as Figure 1), which are right heel strike, left toe-off, mid stance, left heel strike, right toe-off and mid swing. Accurate detection of the six gait events would raises the accuracy of gait recognition and analysis. In this paper, we focus on automatic detection of toe-off events using vision methods.

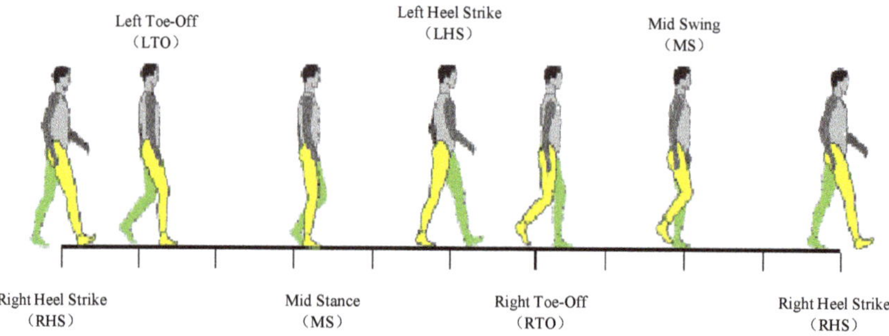


**Figure 1.** Graphic demonstration of the gait events within a gait cycle.

Currently, gait events detection methods can be mainly classified into two types: wearable sensors-based and vision-based methods [2]. The wearable sensors-based methods can accurately detect gait events by collecting motion data from the joints and segments of human lower limb with wearable devices. This type of method is widely used in the medicine area for evaluating abnormal gait due to its high accuracy performance. However, wearable sensors-based methods rely on high cooperation of participants. The participants have to first wear particular devices and then walk around the given area.

Conversely, vision-based methods detect gait event directly from video data captured by a single or several cameras without the aid of any other special sensors. Various cameras including structured light camera [3], stereo camera [4] and 2D vision camera [5] have been applied within these methods. Compared with the wearable sensors, cameras would be cheaper and easier to use. Detecting gait events from 2D video data is a challenging problem due to variations of illumination, perspective, and clothing. Previously, researchers attached markers to the joints of the human limb as participants walked on a clearly marked walkway. This setup requires the cooperation from participants.

In this paper, a new method of toe-off events detection based on a single 2D vision camera system is proposed. Consecutive pedestrian silhouettes extracted from video frames are combined to generate consecutive silhouettes difference maps (CSD-maps). Different number of consecutive silhouettes would result in different CSD-maps, namely $n$-CSD-maps, while $n$ represents the number of consecutive silhouettes. Convolutional neural network is finally employed to learn the toe-off events detection features from CSD-maps. The main contribution of this paper is designing of a set of novel features, namely, consecutive silhouettes difference maps, for toe-off event detection. This method can be used to accurately detect gait event from video data captured from a single 2D vision camera under different viewing angles. If gait events can be accurately detected from 2D video data without participants cooperation, it would be greatly benefit to gait recognition and gait analysis.

The remainder of this study is organized as follows. In Section 2, the advancements of gait events detection methods are reviewed. In Section 3, the proposed method is discussed in detail. Section 4 reports the experimental results on publicly available databases. Finally, Section 5 concludes this study.

## 2. Related Work

In this section, we review the recent progress of gait event detection, which can be coarsely classified into two categories: wearable sensors-based methods and vision-based methods.

### 2.1. Wearable Sensors-Based Methods

Wearable sensors-based methods employ various wearable sensors placed on joints or segments of human limbs (such as feet, knees, thighs or waist) to collect their motion data. Accelerometers and gyroscopes are desirable sensors for gait event detection, which have drawn much attention from researchers. Rueterbories et al. [6] placed accelerometers on the foot to detect gait events. Aung et al. [7] placed tri-axial accelerometers on the foot, ankle, shank or waist to detect heel strike and toe off events. Formento et al. [8] placed a gyroscope on the shank to determine initial contact and foot-off events. Mannini et al. [9] used a uniaxial gyroscope to measure the angular velocity of foot instep in a sagittal plane. Anoop et al. [10] utilized force myography signals from thighs to determine the heel strike (HS) and toe-off (TO) events. Jiang et al. [11] proposed a gait phase detection method based on force myography technique.

The inertial measurement unit (IMU), which is composed of gyroscope and accelerometer, is also a powerful sensor for capturing human limb motion data. Bejarano et al. [12] employed two inertial and magnetic sensors placed on the shanks to detect gait events. Olsen et al. [13] accurately and precisely detected gait events using the features from trunk- and distal limb-mounted IMUs. And latter, Trojaniello et al. [14] mounted a single IMU at the waist level to detect gait events. Ledoux [15] presented a method for walking gait event detection using a single inertial measurement unit (IMU) mounted on the shank.

These sensors can accurately capture motion signals of the points where sensors are placed. Thus, these methods can accurately detect gait events and have been widely used for gait analysis in the medicine area. The disadvantages of these type of methods mainly lie in power consumption restriction, high cost and user cooperation restriction.

A smartphone would contain a 3-dimensional accelerometer, a 3-dimensional gyroscope, and a digital compass. Thus, smartphones are new convenient sensors for gait analysis. Pepa et al. [16] utilized smartphones to detection gait events (such as heel strike) by securing them to an individual's lower back or sternum. Manor et al. [17] proposed a method to detect the heel strike and toe off events by placing a smartphone in the user's pants pocket. Ellis et al. [18] presented a smartphone-based mobile application to quantify gait variability for Parkinson's disease diagnosing. Smartphones are also powerful sensors for gait recognition. Fernandez-Lopez et al. [19] compared the performance of four state-of-art algorithms on a smartphone before 2016. Muaaz et al. [20] evaluated the security strength of a smartphone-based gait recognition system against zero-effort and live minimal-effort impersonation attacks under realistic scenarios. Gadaleta et al. [21] proposed a user authentication framework from smartphone-acquired motion signals. The goal of this work is to recognize a target user from their way of walking, using the accelerometer and gyroscope (inertial) signals provided by a commercial smartphone worn in the front pocket of the user's trousers.

### 2.2. Vision-Based Methods

Vision-based methods can be also divided into two sub-categories: marker-based and no marker-based methods.

Marker-based methods calculate human limb motion parameters by tracking the markers attached to the joints of human limb. Ugbolue et al. [22] employed an augmented-video-based-portable-system (AVPS) to achieve gait analysis. In this study, bull's eye markers and retroreflective markers are attached to human lower limb. In [23], Yang et al. proposed an alternative, inexpensive, and portable gait kinematics analysis system using a single 2D vision camera. Markers are also attached on the hip, knee, and ankle joints for motion data capture. And three years later, the authors enhanced the initial single-camera system by designing a novel autonomous gait event detection method [5]. These methods achieve good accuracy of gait event detection. However, a calibration step is needed, where the participant has to walk on a clearly marked walkway, thus indicating user cooperation is required.

No marker-based methods can achieve gait event detection without user cooperation. With respect to this type of method very few research studies have worked on gait event detection techniques. The directly related work is Auvinet's work [3], in which a depth camera (Kinect) is employed to achieve gait analysis on a treadmill for routine outpatient clinics. In [3], a heel-strike event detection algorithm is presented by searching for extreme values of the distance between knee joints along the walking longitudinal axis. Although it achieves accurate detection results, Kinect used in [3] is also a special camera compared with a widely used web camera. In this study, we attempt to detect toe-off events using a web camera. As far as we know, this paper would be the first effort to detect gait events utilizing video data without the cooperation of participants.

Some research works about gait cycle detection algorithm have been presented in gait recognition methods. These methods can detect whole gait cycle or gait phase from video data without the help of markers. In [24], a gait periodicity detection method is presented based on dual-ellipse fitting (DEF). The periodicity is defined as the internal between the first extreme point and the third extreme point of DEF signals. Kale et al. [25] employed the norm of the width vector to show a periodic variation. Sarkar et al. [26] estimated gait cycle by counting the number of foreground pixels in the silhouette in each frame overtime. Mori et al. [27] detected the gait period by maximizing the normalized autocorrelation of the gait silhouette sequence for the temporal axis. These methods mentioned above can achieve gait cycle detection, but cannot obtain accurate gait event detection results.

## 3. Toe-Off Events Detection Based on CSD-Maps

In this section, we present the technique detail of the toe-off events detection method. The framework of the proposed method is graphically presented in Figure 2. Several consecutive silhouettes of a pedestrian are first combined to generate a consecutive silhouettes difference map. Convolutional neural networks are then employed to learn the features for toe-off events classification.

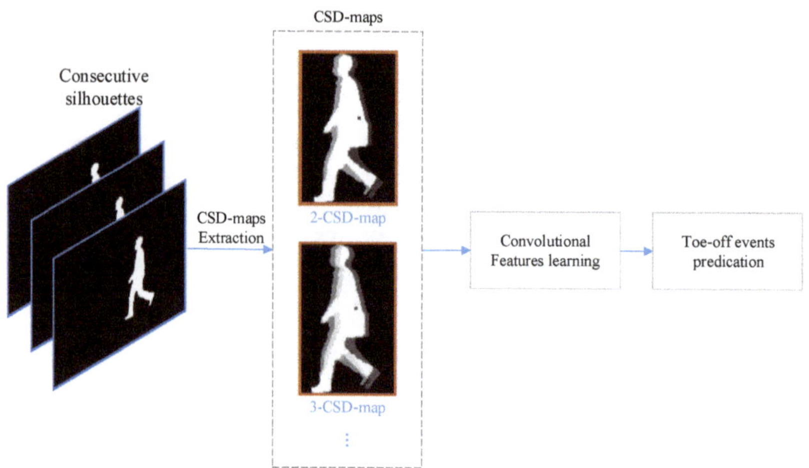


**Figure 2.** The framework of the proposed method.

### 3.1. Consecutive Silhouettes Difference Maps

There are rich temporal and spatial information contained in video data. Mining and fusing temporal and spatial information is currently an interest in computer vision. Inspired by the principle of the exclusive OR operation, we employ a frame difference method to encode the temporal and spatial information contained in several consecutive frames into a map. The difference map generated from $n$ consecutive silhouettes is named as a $n$-CSD-map. We first take a 2-CSD-map as an example to explain how consecutive silhouette frames are encoded into a map.

### 3.1.1. 2-CSD-Maps

The main idea of 2-CSD-maps is graphically presented in Figure 3. The 2-CSD-map of the $i^{th}$ frame is generated from two consecutive silhouette frames. Let $\Gamma_i^2$ present the 2-CSD-map of the $i^{th}$ frame, $I_{i-1}$ and $I_i$ present the binary silhouette images of the $(i-1)^{th}$ and $i^{th}$ frame. For any pixel $P_{j,k}^2$ in $\Gamma_i^2$, it's pixel value can be formulated as following:

$$\Gamma_i^2(j,k) = \begin{cases} 1 & \text{if } (P_{j,k}^2 \notin \Omega_{i-1}) \cap (P_{j,k} \in \Omega_i) \\ 2 & \text{if } (P_{j,k}^2 \in \Omega_{i-1}) \cap (P_{j,k} \notin \Omega_i) \\ 3 & \text{if } (P_{j,k}^2 \in \Omega_{i-1}) \cap (P_{j,k} \in \Omega_i) \end{cases} \tag{1}$$

while, $\Omega_{i-1}$ represents the pixel set of the silhouette area in $I_{i-1}$, and $\Omega_i$ represents the pixel set of the silhouette in $I_i$. In order to achieve a good visual effect, the pixel values in Figure 3c are normalized to [0,1].

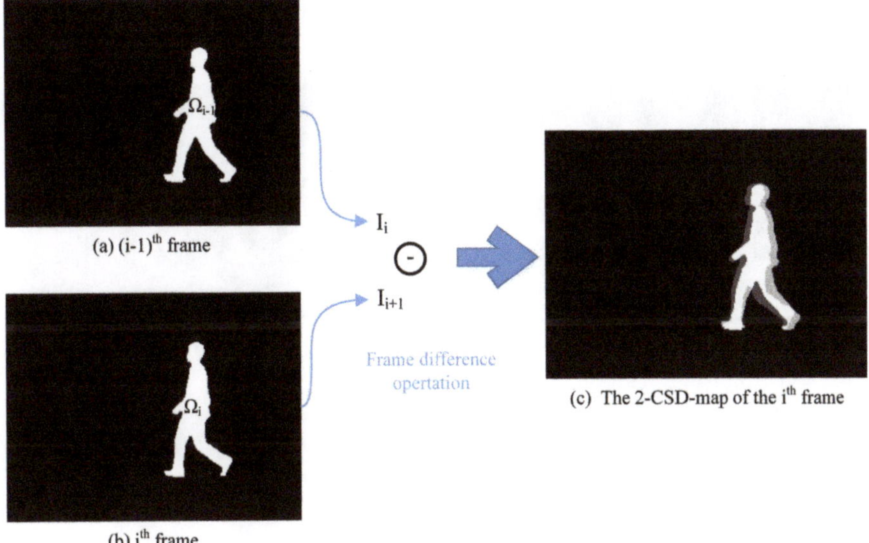


**Figure 3.** The basic idea of the 2-CSD-map. The pixel values in (c) are are normalized to [0,1] for good visual effect.

In practice, a pedestrian silhouette is presented as a binary image. Thus, a 2-CSD-map of two consecutive silhouettes can be computed using following three steps to achieve fast extraction of 2-CSD-maps.

First, copy gray value of pixels from $I_{i-1}$ to $\Gamma_i^2$. A temporary matrix $I$ is then computed as:

$$I = I_i - I_{i-1} \tag{2}$$

Secondly, modify the pixel value of $\Gamma_i^2$ according to the value of matrix $I$:

$$\Gamma_i^2(j,k) = \begin{cases} 1 & \text{if } I(j,k) > 0 \\ 2 & \text{if } I(j,k) < 0 \end{cases} \tag{3}$$

Finally, the pixel value of $\Gamma_i^2$ is modified as follows:

$$\Gamma_i^2(j,k) = \begin{cases} 3 & \text{if } \Gamma_i^2(j,k) == 255 \\ \Gamma_i^2(j,k) & \text{else} \end{cases} \tag{4}$$

Some samples of 2-CSD-maps are graphically presented in Figure 4. We can see that 2-CSD-maps are distinctive features for toe-off events detection compared with original silhouette images.

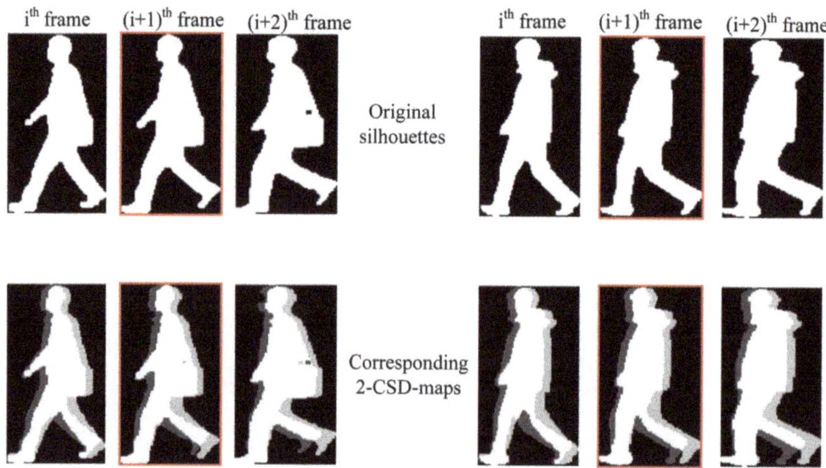


**Figure 4.** Samples of 2-CSD-maps compared with original silhouettes. The images presented in the first row are original silhouettes of two different persons, and the corresponding 2-CSD-maps are presented in the second row. The images with red edging are the toe-off frames. The pixel values in 2-CSD-maps are normalized to [0,1] for good visual effect.

### 3.1.2. *n*-CSD-Maps

Suppose that there are $n$ consecutive silhouettes images $I_1$, $I_2$, ..., and $I_n$. The $n$-CSD-maps $\Gamma_i^n$ can be formulated as following:

$$\Gamma_i^n(j,k) = \begin{cases} 1 & \text{if } (P_{j,k}^n \in \Omega_1) \cap (P_{j,k} \notin \Omega_2) \cap (P_{j,k} \notin \Omega_3) \cap ... \cap (P_{j,k} \notin \Omega_n) \\ 2 & \text{if } (P_{j,k}^n \notin \Omega_1) \cap (P_{j,k} \in \Omega_2) \cap (P_{j,k} \notin \Omega_3) \cap ... \cap (P_{j,k} \notin \Omega_n) \\ 3 & \text{if } (P_{j,k}^n \notin \Omega_1) \cap (P_{j,k} \notin \Omega_2) \cap (P_{j,k} \in \Omega_3) \cap ... \cap (P_{j,k} \notin \Omega_n) \\ .... \\ 2^n - 1 & \text{if } (P_{j,k}^n \in \Omega_1) \cap (P_{j,k} \in \Omega_2) \cap (P_{j,k} \in \Omega_3) \cap ... \cap (P_{j,k} \in \Omega_n) \end{cases} \tag{5}$$

while, $\Gamma_i^n(j,k)$ stands for the pixel value of the pixel $P_{j,k}^n$ in the generated $n$-CSD-map. $\Omega_1$, $\Omega_2$, ..., and $\Omega_n$ represent the pixel set of the silhouette areas in frame $I_1$, $I_2$, ..., and $I_n$ respectively.

Given $n$ consecutive silhouette images, the $n$-CSD-maps extraction algorithm can be described as Algorithm 1. With this algorithm, the CSD-map generated from the given consecutive silhouette images is also presented as an image with the same size as silhouette images, shown as Figure 3c. Thus, a further normalization step is necessary. In this paper, CSD-map images are initially normalized to a certain size (such as 90 × 140) using Algorithm 2.

Figure 5 shows some consecutive normalized CSD-maps. we can see that the CSD-maps under toe-off state are obviously different with other CSD-maps.

---

**Algorithm 1** Algorithm for generating $n$-CSD-maps

---

**Require:**

    Consecutive silhouette images: $I[w, h, n]$.

    Parameter $w$ and $h$ represent the width and height of the silhouette images respectively. Parameter $n$ represents the number of consecutive silhouette images.

**Ensure:**

    The CSD-map: $\Gamma$

  1: **for** $i = 1$ to $w$ **do**

  2:     **for** $j = 1$ to $h$ **do**

  3:       $t = I(i, j, :)$;

  4:       $value = 0$;

  5:       **for** $k = 1$ to $n$ **do**

  6:         $value = value + 2^{(k-1)} * t(k)$;

  7:       **end for**

  8:       $\Gamma(i, j) = value$;

  9:     **end for**

10: **end for**

11: **return** $\Gamma$;

---

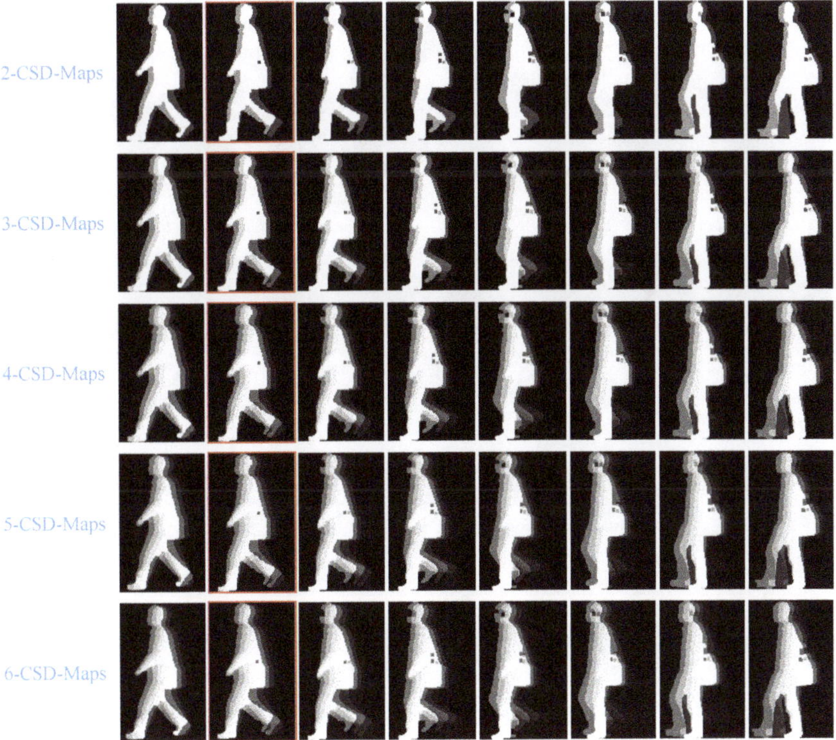

**Figure 5.** Samples of normalized CSD-maps. From the first row to the fifth row, the normalized 2-CSD-maps, 3-CSD-maps, 4-CSD-maps, 5-CSD-maps and 6-CSD-maps are respectively presented. The images with red edging are the toe-off frames. The pixel values in all CSD-maps are normalized to [0,1] for good visual effect.

---

**Algorithm 2** Algorithm for normalizing a CSD-map

---

**Require:**
    The original CSD-map image: $OM$
    The width of the normalized CSD-map:$w$
    The height of the normalized CSD-map:$h$
**Ensure:**
    The normalized CSD-map: $NM$
 1: $[x, y] = find(OM > 0)$;
 2: $segm = OM(min(x) : max(x), min(y) : max(y))$;
 3: $NM = imresize(segm, [h, w])$;
 4: return $NM$;

---

### 3.2. Convolutional Neural Network

Convolutional neural networks have a feed-forward network architecture with multiple interconnected layers which may be of any of the following types: convolution, normalization, pooling and fully connected layers. CNNs have recently achieved many successes in visual recognition tasks, including image classification [28], object detection [29], and scene parsing [30]. CNNs are chosen as a detector for this study because they outperform other traditional methods in many image classification challenges, such as ImageNet [28] and many other image-based recognition problems, e.g., face recognition and digital recognition [31]. Comparing with traditional methods which rely on feature engineering, CNNs are able to learn feature representation through the back propagation algorithm without the need for much intervention and also achieve much higher accuracy.

The aim of this study is not to propose another CNN but use a classic CNN to address the problem of toe-off events detection. In this paper, we employ the CNN architecture presented in Figure 6. It is modified from DeepID [32]. The network includes three convolutional layers and three fully connected layers. The three convolutional layers have 64, 128 and 256 kernels and their sizes are respectively $5 \times 5, 3 \times 3 \times 64$ and $3 \times 3 \times 128$. The first fully connected layer has 1024 neurons and the second fully connected layer has 512 neurons. In the last fully connected layer, there are two neurons, one for toe-off frame output and the other for non-toe-off frame output. The max-pooling with a size of 2 and a stride of 2 follows the three convolutional layers.

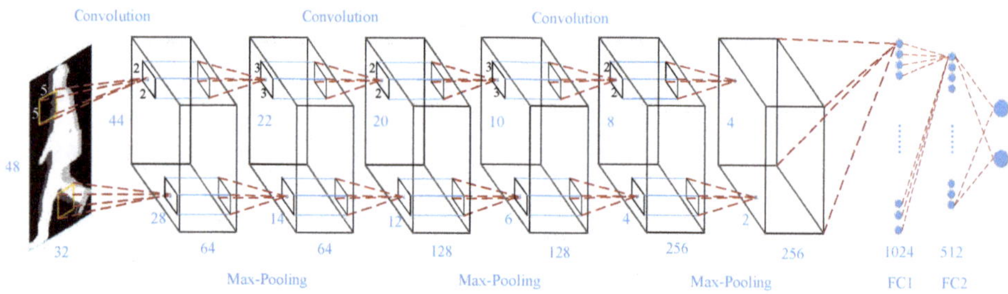


**Figure 6.** The architecture of the CNN employed in this study.

## 4. Experiments and Results Analysis

### 4.1. Database

Experiments are conducted on CASIA gait database (Dataset B) [33] to evaluate the accuracy of the performance of the proposed method. The data contained in this database are collected from 124 subjects (93 males and 31 females) in an indoor environment under 11 different viewing angles. The data from a subject is simultaneously captured by 11 USB cameras (with a resolution of $320 \times 240$, and a frame rate of 25 fps) around the left hand side of the subject when he/she was walking, and the angle between two nearest view directions is 18°. When a subject walked in the scene, he/she was asked to walk naturally along a straight line 6 times first, and 11 × 6 = 66 normal walking video sequences were captured for each subject. After normal walk, the subjects were asked to put on their coats or carried a bag, and then walked twice along the straight line. In each viewing angle, there are totally 10 videos collected from every subject under three different clothing conditions, namely normal condition, coat condition and bag condition. The CASIA Gait Database is provided free of charge at web site http://www.cbsr.ia.ac.cn.

In this study, we considered the data captured under the viewing angles of 36°, 54°, 72°, 90°, 108°, 126° and 144° (approximately 500,000 frames in total) for training and testing. The data captured under the frontal viewing angles of 0°, 18°, 172°, 180°, are not used in the experiments primarily because there is very little difference between two consecutive silhouettes. The CSD-maps generated from the video data captured in the viewing of sagittal plane do not contain much information for gait events detection. This means that the method proposed by this paper cannot deal with the video data captured in the viewing of sagittal plane. Even so, the proposed method can deal with the video data captured from most viewing angles. This makes the proposed method useful in practice.

### 4.2. Toe-Off Frame Definition and Data Preparation

The ground truth of all the silhouette frames should be manually labeled for modal training and testing. Thus, the toe-off frames should be first and clearly defined.

Human gait is a continuous and periodic movement. In medical field, the toe-off event is defined as the moment that the stance limb leaves the ground, shown as in Figure 1. While, the video data is the sampling record of human gait with a certain frame rate $\theta$. Usually, the frame rate $\theta$ would be 30 fps. And the gait cycle of a person is averagely about 1 s time consuming. This means that one gait movement cycle of a person would be recoded as about 30 consecutive frames with an interval of 33 ms. The problem is that the moment the stance limb leaving the ground may not be included in the 30 consecutive sampling frames. In this paper, the first frame after the stance limb leaves the ground is defined as a toe-off frame. For example, as shown in Figure 7, if the moment that the stance limb leaves the ground falls within the period of $t_n < t < t_{n+1}$, then the frame $(n + 1)$ is defined as the toe-off frame.

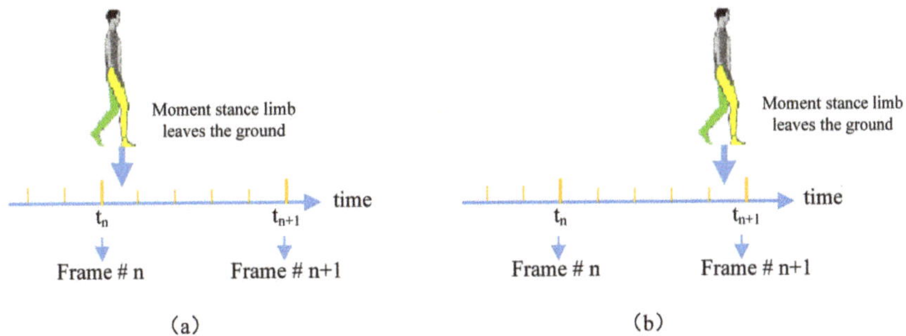


**Figure 7.** Toe-off event definition of video frames.

According to the definition, there would exist error in the labeled groundtruth. Let $\theta$ be the frame rate of the video data. The during time between two continuous frames would be $\frac{1}{\theta}$, which means $t_{n+1} - t_n = \frac{1}{\theta}$. If the toe-off event happens during the period of $(t_n, t_{n+1})$ but nearer to $t_n$ shown as Figure 7a, then at frame $n + 1$, the foot would have swung in the air for about $\frac{1}{\theta}$ seconds. However, if the toe-off event happens during the period of $(t_n, t_{n+1})$ but nearer to $t_{n+1}$ shown as Figure 7b. At frame $n + 1$, the foot would have just left the ground. The frames $n + 1$ in both Figure 7a,b are regarded as toe-off frames. Obviously, the toe-off frames in Figure 7a,b may be different with each other. But this error doesn't change the validity of the proposed method.

*4.3. Experimental Configuration*

The experiments are conducted by using Caffe [34], which is a deep learning framework created by Yangqing Jia during his PhD at UC Berkeley. The experiments are conducted as following.

- *Configuration of n-CSD-maps.* Several pre-tests have been conducted under the viewing angles of $72°$, $90°$ and $108°$ for choosing the size of normalized CSD-maps. As shown in Figure 8, the pre-test results show that different sizes of normalized CSD-maps practically cause almost no change to the detection accuracy. The main reason is that CSD-maps are generated from binary pedestrian silhouettes. The decline of the size of normalized CSD-maps would not result in much change to the detection accuracy of this method. Thus, in the following experiments, the size of normalized CSD-maps is set as 48*32. As to the parameter $n$ of $n$-CSD-maps, it is set as 2, 3, 4, 5, and 6. This means that 2-CSD-maps, 3-CSD-maps, 4-CSD-maps, 5-CSD-maps and 6-CSD-maps are used in the experiments. The reason is that the increase of parameter $n$ brings little increase of detection accuracy, while costs more time for features extraction, shown as Table 1 and Figure 9.
- *Configuration of Training set and test set.* The samples from subject #001 to subject #90 of each viewing angle are selected for model training. The rest of samples (from subject #091 to subject #124) is used for testing.
- *Configuration of CNN Solver.* The initialized learning rate is 0.001, the momentum is 0.9 and the weight decay is 0.0005. The maximum number of iteration in each experiment is 20,000. The weights in the CNN are initialized with a zeromean Gaussian distribution with standard deviation of 0.0001. The bias is set to one.

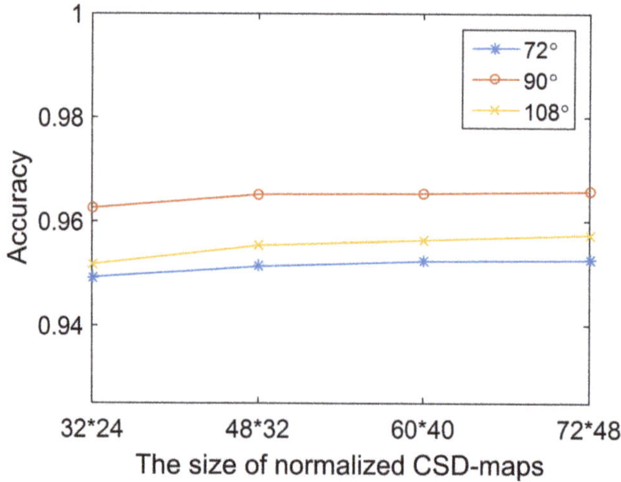


**Figure 8.** The relationship between detection accuracy of the proposed method and the size of normalized $n$-CSD-map.

**Table 1.** Detection accuracy of the proposed method.

| $n$-CSD-Maps | 36 Degree | 54 Degree | 72 Degree | 90 Degree | 108 Degree | 126 Degree | 144 Degree |
|---|---|---|---|---|---|---|---|
| 2-CSD | 93.2% | 94.34% | 94.3% | 96.26% | 94.68% | 94.83% | 92.72% |
| 3-CSD | 93.45% | 94.74% | 95.14% | 96.52% | 95.54% | 95.52% | 93.08% |
| 4-CSD | 93.52% | 95.18% | 95.24% | 96.58% | 95.64% | 95.58% | 93.16% |
| 5-CSD | 93.55% | 95.38% | 95.36% | 96.62% | 95.74% | 95.62% | 93.22% |
| 6-CSD | **93.63%** | **95.4%** | **95.44%** | **96.78%** | **95.78%** | **95.65%** | **93.44%** |

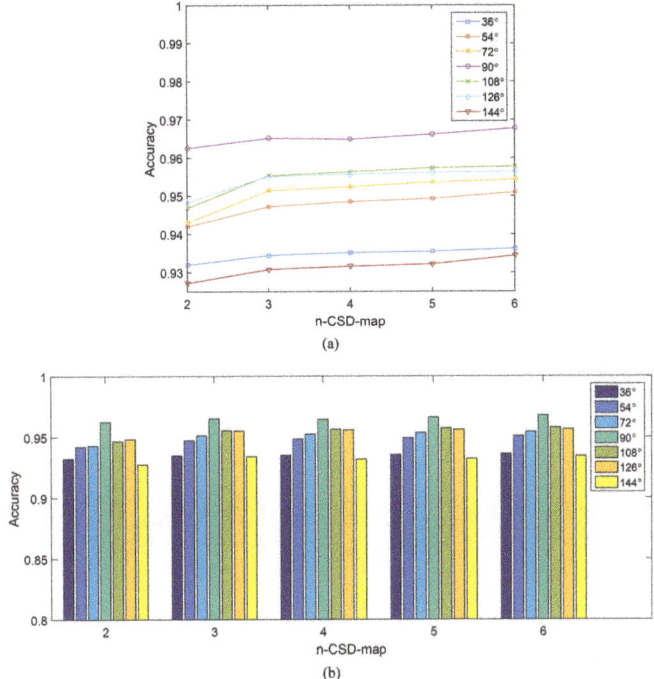

(a)

(b)

**Figure 9.** The relationship between detection accuracy of the proposed method and *n*-CSD-map. (**a**) The detection accuracy as a function of *n*-CSD-map. (**b**) The bars of the detection accuracy VS. *n*-CSD-map.

## 4.4. Experimental Results and Discussion

In this paper, a new evaluation indicator, namely n-frame-error cumulative detection accuracy, is designed to evaluate the performance of the proposed method besides detection accuracy and ROC curve. The n-frame-error cumulative detection accuracy is similar with cumulative match characteristics (CMC) curves [35]. Let's $d$ represents the difference between the sequence number of predicted toe-off frame and the ground truth, shown as Figure 10. *n*-frame-error cumulative detection accuracy indicates the detection accuracy with the condition of $d \leq n$.

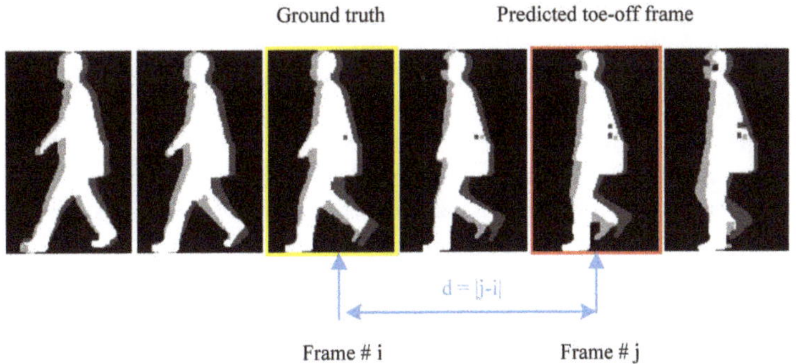


**Figure 10.** Graphical demonstration of the n-frame-error cumulative detection accuracy. The frame difference between the sequence number of predicted toe-off frame and the ground truth is noted as *d*. The image with red edging is the predicted toe-off frame. The image with yellow edging is the ground truth.

Table 1 shows the detection accuracy of the proposed method. We can see that the proposed method achieves good detection accuracy. The proposed method reaches the accuracy around 93% under the viewing angles of 36°, and achieves the peak value of 93.63% by using 6-CSD-maps. Under the viewing angle of 54°, the proposed method reaches the accuracy around 94% and achieves the peak value of 95.4% by using 6-CSD-maps. Under the viewing angle of 72°, the proposed method reaches the accuracy around 95% and achieves the peak value of 95.44% by using 6-CSD-maps. Under the viewing angle of 90°, the proposed method reaches the accuracy around 96% and achieves the peak value of 96.78% by using 6-CSD-maps. Under the viewing angle of 108°, the proposed method reaches the accuracy around 95% and achieves the peak value of 95.78% by using 6-CSD-maps. Under the viewing angle of 126°, the proposed method also reaches the accuracy around 95% and achieves the peak value of 95.65% by using 6-CSD-maps. Under the viewing angle of 144°, the proposed method reaches the accuracy around 93% and achieves the peak value of 93.44% by using 6-CSD-maps.

The relationship between detection accuracy of the proposed method and $n$-CSD-map is graphically presented in Figure 9. Figure 9a demonstrates the detection accuracy of the proposed method as a function of $n$-CSD-map, and the corresponding bars are presented in Figure 9b. Generally, the detection accuracy is slightly improved with the increase of $n$. The reason is that the bigger the parameter $n$ is, the more consecutive silhouettes will be encoded into a CSD-map, and the more information will be contained in the CSD-map. The detection accuracy gets a good promotion when the parameter $n$ changes from 2 to 3. For example, under viewing angle of 108°, the accuracy of the proposed method increase from 94.68% to 95.54% when the parameter $n$ increases from 2 to 3. However, the accuracy gets a few increase when the parameter $n$ goes to 4, 5 and 6. This demonstrates that 3-CSD-map is a good choice for toe-off detection, which can achieve good accuracy with little additional computation cost. Figure 11 shows the ROC curves of the proposed method under different viewing angles. The ROC curves of the proposed method under the viewing angles of 36°, 54°, 72°, 90°, 108°, 126° and 144° are respectively presented in the figures from Figure 11a–g. As shown in Figure 11, under all viewing angles, the proposed method gets higher detection performance by using larger parameter $n$ of $n$-CSD-map.

The ROC curves of the proposed method using 3-CSD-map under different viewing angles are presented in Figure 9h. Generally, we can see that the proposed method obtains higher detection accuracy around coronal plane viewing angles than sagittal plane viewing angles. Especially, the proposed method achieves the accuracy of 96.78% under the viewing angle 90°, which is higher than other viewing angles. This demonstrate that CSD-maps generated from the video data captured in sagittal plane viewing angles contain less useful information for gait events detection than coronal plane viewing angles. The reason is that there is fewer different between two consecutive silhouettes of video frames captured under sagittal plane viewing angles compared with coronal plane viewing angles.

The plots presented in Figure 12a are the $n$-frame-errors cumulative detection accuracy of the proposed method against different viewing angles. The 1-frame-error cumulative detection accuracy of the proposed method reaches the accuracy of 99.3%, 99.86%, 99.9%, 99.9%, 99.9%, 99.8%, and 99.4% for the viewing angles of 36°, 54°, 72°, 90°, 108°, 126°, and 144° respectively. For the 2-frame-error, the cumulative detection accuracy of the proposed method achieves 100% for the viewing angles of 54°, 72°, 90°, 108°, and 126°. This demonstrates that the maximum time error of the proposed method detecting toe-off events in coronal plane viewing angles is less than $\frac{2}{\theta}$, where $\theta$ is the frame rate of the video data. Practically, we can promote the time accuracy of this method by increasing the frame rate of the video.

Figure 12b shows the detection accuracy of the proposed method as a function of viewing angles compared with [24,25,36]. Due to the reason that [24,25] do not provide toe-off event detection results directly, we implemented the both algorithms for toe-off event detection according to the main ideas of [24,25]. Ref. [36] is our previous work based on principal component analysis and support vector machine. In this experiment, all frames are used for training and testing in 5-fold cross validation. We can see that our CNN-based method significantly outperforms Ben's method [24], Kale's method [25] and our previous work [36] in the viewing angles of 36°, 54°, 72°, 90°, 108°, 126°, and 144°.

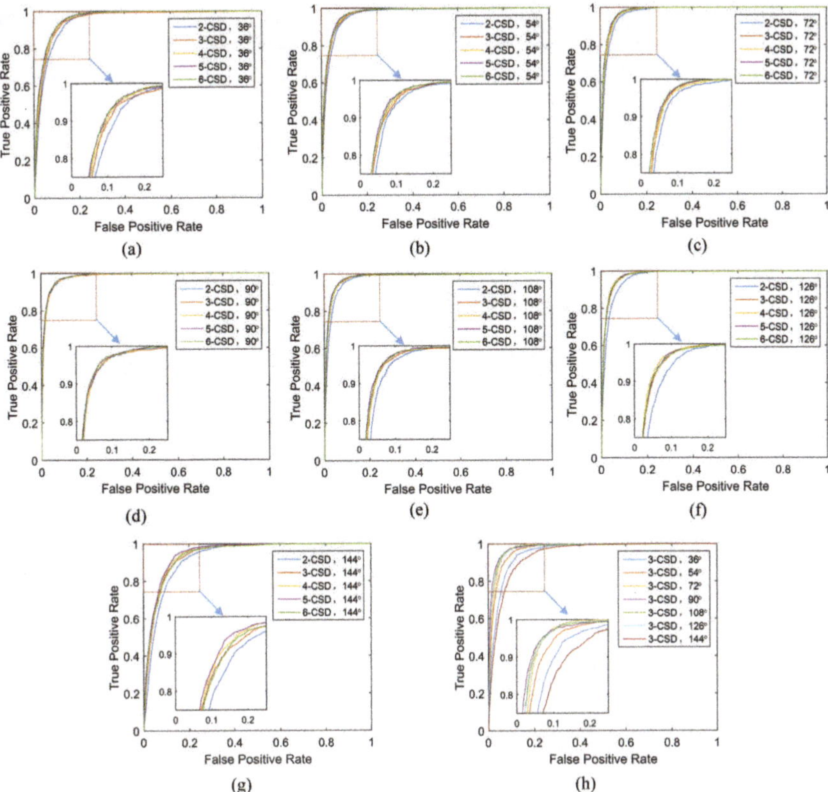

**Figure 11.** The ROC curves of the proposed method. (**a**) The ROC curves under the viewing angle of 36°. (**b**) The ROC curves under the viewing angle of 54°. (**c**) The ROC curves under the viewing angle of 72°. (**d**) The ROC curves under the viewing angle of 90°. (**e**) The ROC curves under the viewing angle of 108°. (**f**) The ROC curves under the viewing angle of 126°. (**g**) The ROC curves under the viewing angle of 144°. (**h**) The ROC curves of the proposed method with 3-CSD-map under different viewing angles of 36°, 54°, 72°, 90°, 108°, 126° and 144°.

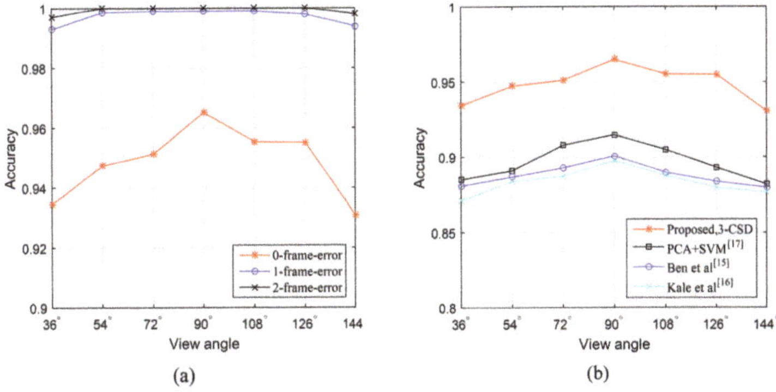

**Figure 12.** The n-frame-error cumulative detection accuracy of the proposed method. (**a**) the detection accuracy of the proposed method against different frame-errors. (**b**) The detection accuracy of the proposed method compared with [24,25,36].

In Figure 13, we use a confusion matrix to evaluate cross viewing angle detection accuracy of this method using 3-CSD-maps. As can be seen in the figure, this method achieves the best accuracy in the counter-diagonal and around 90% in the other areas, which means that this method can get good accuracy for cross view toe-off detection. Figure 14 presents the ROC curves of this method under all viewing angles compared with [24,25,36]. We can see that the proposed method significantly outperforms the comparation methods.

|  | 36° detector | 54° detector | 72° detector | 90° detector | 108° detector | 126° detector | 144° detector |
|---|---|---|---|---|---|---|---|
| 36° probe | 0.9345 | 0.9294 | 0.9145 | 0.9088 | 0.9054 | 0.8986 | 0.8919 |
| 54° probe | 0.9256 | 0.9474 | 0.9328 | 0.9194 | 0.9286 | 0.9208 | 0.9056 |
| 72° probe | 0.9123 | 0.9324 | 0.9514 | 0.9142 | 0.9386 | 0.9248 | 0.9152 |
| 90° probe | 0.9056 | 0.9174 | 0.9265 | 0.9652 | 0.9128 | 0.9258 | 0.9189 |
| 108° probe | 0.8972 | 0.9054 | 0.9266 | 0.9080 | 0.9554 | 0.9390 | 0.9212 |
| 126° probe | 0.8913 | 0.9122 | 0.9074 | 0.9128 | 0.9302 | 0.9552 | 0.9279 |
| 144° probe | 0.8901 | 0.9014 | 0.9124 | 0.9156 | 0.9234 | 0.9244 | 0.9308 |

**Figure 13.** The confusion matrix of cross viewing angle detection accuracy of this method using 3-CSD-maps.

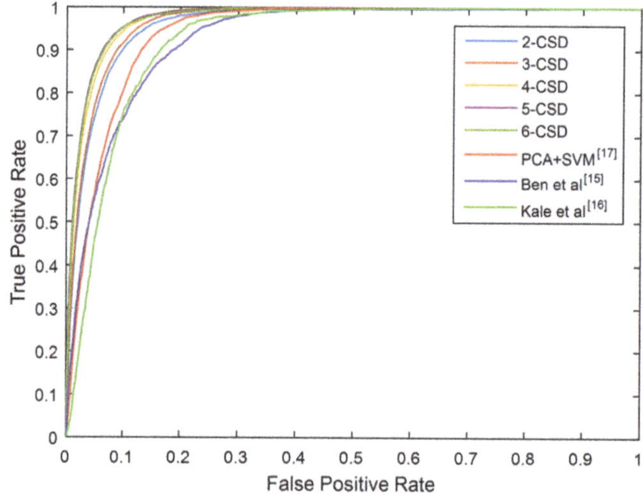


**Figure 14.** The ROC curves of this method compared with [24,25,36] under all viewing angles.

## 5. Conclusions and Future Work

This paper presents a promising vision-based method to detect toe-off events. The main contribution of this paper is the design of consecutive silhouettes difference maps for toe-off event detection. Convolutional neural network is employed for feature dimension reduction and toe-off event classification. Experiments on a public database have demonstrated good performance of our method in terms of detection accuracy. The main advantages of the proposed method can be described as following.

- Comparing with wearable sensors-based methods, this method can detect toe-off event from 2D video data without the cooperation of participants. Usually, in the field of medicine, wearable sensors-based methods are the first choice for gait analysis, due to their high accuracy. However, these methods are suffering the disadvantages of high cooperation from users and power consumption restriction. The method proposed by this paper, which also achieves good accuracy for toe-off event detection by using a web camera, can overcome the disadvantages of wearable sensors-based methods for gait analysis.

- Comparing with other vision-based methods, this method provides a better accuracy for toe-off event detection. Gait cycle detection is a basic step of gait recognition. An accurate toe-off event detection algorithm can produce an accurate gait cycle detection algorithm. Thus, the method proposed by this paper would be beneficial to gait recognition.

Although a promising feature representation method is proposed in this paper for toe-off event detection, more efforts are needed to improve the method of gait events detection from video data in our future work.

- A much larger database is needed to test the practical performance of toe-off event detection under different conditions.

- CSD-map provides a good feature representation for detecting toe-off events from video data. It also would be applicable for other gait events detection, such as heel strike, foot flat, mid-stance, heel-off, and mid-swing.

**Author Contributions:** Data curation, Z.L.; Formal analysis, J.D.; Investigation, Y.T.; Methodology, Y.T. and B.L.; Software, Z.L.; Supervision, J.D. and B.L.; Validation, H.T.; Visualization, H.T.; Writing–original draft, Y.T.; Writing–review & editing, J.D.

**Funding:** This research was funded by National Key Research and Development Program of China (No.2017YFC0803506, 2017YFC0822003), the Fundamental Research Funds for the Central Universities of China (Grant No.2018JKF217), the National Natural Science Foundation of China (No. 61503387, 61772539).

**Conflicts of Interest:** The authors declare no conflict of interest.

## References

1. Muro-de-la-Herran, A.; Garcia-Zapirain, B.; Mendez-Zorrilla, A. Gait analysis methods: An overview of wearable and non-wearable systems, highlighting clinical applications. *Sensors* **2014**, *14*, 3362–3394. [CrossRef]

2. Fraccaro, P.; Walsh, L.; Doyle, J.; O'Sullivan, D. Real-world gyroscope-based gait event detection and gait feature extraction. In Proceedings of the Sixth International Conference on eHealth, Telemedicine, and Social Medicine, Barcelona, Spain, 23–27 March 2014; pp. 247–252.

3. Auvinet, E.; Multon, F.; Aubin, C.E.; Meunier, J.; Raison, M. Detection of gait cycles in treadmill walking using a kinect. *Gait Posture* **2015**, *41*, 722–725. [CrossRef] [PubMed]

4. Richards, J.G. The measurement of human motion: A comparison of commercially available systems. *Hum. Mov. Sci.* **1999**, *18*, 589–602. [CrossRef]

5. Yang, C.; Ugbolue, U.C.; Kerr, A.; Stankovic, V.; Stankovic, L.; Carse, B.; Kaliarntas, K.T.; Rowe, P.J. Autonomous gait event detection with portable single-camera gait kinematics analysis system. *J. Sens.* **2016**, *2016*, 5036857. [CrossRef]

6. Rueterbories, J.; Spaich, E.G.; Andersen, O.K. Gait event detection for use in fes rehabilitation by radial and tangential foot accelerations. *Med. Eng. Phys.* **2014**, *36*, 502–508. [CrossRef]

7. Aung, M.S.H.; Thies, S.B.; Kenney, L.P.; Howard, D.; Selles, R.W.; Findlow, A.H.; Goulermas, J.Y. Automated detection of instantaneous gait events using time frequency analysis and manifold embedding. *IEEE Trans. Neural Syst. Rehabil. Eng.* **2013**, *21*, 908–916. [CrossRef] [PubMed]

8. Formento, P.C.; Acevedo, R.; Ghoussayni, S.; Ewins, D. Gait event detection during stair walking using a rate gyroscope. *Sensors* **2014**, *14*, 5470–5485. [CrossRef] [PubMed]

9.	Mannini, A.; Genovese, V.; Sabatini, A.M. Online decoding of hidden markov models for gait event detection using foot-mounted gyroscopes. *IEEE J. Biomed. Health Inform.* **2014**, *18*, 1122–1130. [CrossRef]

10.	Anoop, K.G.; Hemant, K.V.; Nitin, K.; Deepak, J. A Force Myography-Based System for Gait Event Detection in Overground and Ramp Walking. *IEEE Trans. Instrum. Meas.* **2018**, *67*, 2314–2323.

11.	Jiang, X.; Chu, K.H.T.; Khoshnam, M.; Menon, C. A Wearable Gait Phase Detection System Based on Force Myography Techniques. *Sensors* **2018**, *18*, 1279. [CrossRef] [PubMed]

12.	Chia, B.N.; Ambrosini, E.; Pedrocchi, A.; Ferrigno, G.; Monticone, M.; Ferrante, S. A novel adaptive, real-time algorithm to detect gait events from wearable sensors. *IEEE Trans. Neural Syst. Rehabil. Eng.* **2014**, *23*, 413–422. [CrossRef]

13.	Olsen, E.; Andersen, P.H.; Pfau, T. Accuracy and precision of equine gait event detection during walking with limb and trunk mounted inertial sensors. *Sensors* **2012**, *12*, 8145–8156. [CrossRef]

14.	Trojaniello, D.; Cereatti, A.; Della, C.U. Accuracy, sensitivity and robustness of five different methods for the estimation of gait temporal parameters using a single inertial sensor mounted on the lower trunk. *Gait Posture* **2014**, *40*, 487–492. [CrossRef]

15.	Ledoux, E.D. Inertial Sensing for Gait Event Detection and Transfemoral Prosthesis Control Strategy. *IEEE Trans. Biomed. Eng.* **2018**, *65*, 2704–2712. [CrossRef] [PubMed]

16.	Pepa, L.; Verdini, F.; Spalazzi, L. Gait parameter and event estimation using smartphones. *Gait Posture* **2017**, *57*, 217–223. [CrossRef] [PubMed]

17.	Manor, B.; Yu, W.; Zhu, H.; Harrison, R.; Lo, O.Y.; Lipsitz, L.; Travison, T.; Pascual-Leone, A.; Zhou, J. Smartphone app-based assessment of gait during normal and dual-task walking: demonstration of validity and reliability. *JMIR MHealth UHealth* **2018**, *6*, e36. [CrossRef]

18.	Ellis, R.J.; Ng, Y.S.; Zhu, S.; Tan, D.M.; Anderson, B.; Schlaug, G.; Wang, Y. A validated smartphone-based assessment of gait and gait variability in Parkinson's disease. *PLoS ONE* **2015**, *10*, e0141694. [CrossRef] [PubMed]

19.	Fernandez-Lopez, P.; Liu-Jimenez, J.; Sanchez-Redondo, C.S.; Sanchez-Reillo, R. Gait recognition using smartphone. In Proceedings of the 2016 IEEE International Carnahan Conference on Security Technology (ICCST), Orlando, FL, USA, 24–27 October 2016.

20.	Muaaz, M.; Mayrhofer, R. Smartphone-based gait recognition: From authentication to imitation. *IEEE Trans. Mob. Comput.* **2017**, *16*, 3209–3221. [CrossRef]

21.	Gadaleta, M.; Rossi, M. Idnet: Smartphone-based gait recognition with convolutional neural networks. *Pattern Recognit.* **2018**, *74*, 25–37. [CrossRef]

22.	Ugbolue, U.C.; Papi, E.; Kaliarntas, K.T.; Kerr, A.; Earl, L.; Pomeroy, V.M.; Rowe, P.J. The evaluation of an inexpensive, 2D, video based gait assessment system for clinical use. *Gait Posture* **2013**, *38*, 483–489. [CrossRef] [PubMed]

23.	Yang, C.; Ugbolue, U.; Carse, B.; Stankovic, V.; Stankovic, L.; Rowe, P. Multiple marker tracking in a single-camera system for gait analysis. In Proceedings of the 2013 20th IEEE International Conference on Image Processing (ICIP), Melbourne, Victoria, Australia, 15–18 September 2013; pp. 3128–3131.

24.	Ben, X.; Meng, W.; Yan, R. Dual-ellipse fitting approach for robust gait periodicity detection. *Neurocomputing* **2012**, *79*, 173–178. [CrossRef]

25.	Kale, A.; Sundaresan, A.; Rajagopalan, A.N.; Cuntoor, N.P.; Roy-Chowdhury, A.K.; Krüger, V.; Chellappa, R. Identification of humans using gait. *IEEE Trans. Image Process.* **2004**, *13*, 1163–1173. [CrossRef] [PubMed]

26.	Sarkar, S.; Phillips, P.J.; Liu, Z.; Bowyer, K.W. The humanid gait challenge problem: Data sets, performance, and analysis. *IEEE Trans. Pattern Anal. Mach. Intell.* **2005**, *27*, 162–177. [CrossRef] [PubMed]

27.	Mori, A.; Makihara, Y.; Yagi, Y. Gait recognition using period-based phase synchronization for low frame-rate videos. In Proceedings of the IEEE 20th International Conference on Pattern Recognition (ICPR), Istanbul, Turkey, 23–26 August 2010; pp. 2194–2197.

28.	Krizhevsky, A.; Sutskever, I.; Hinton, G.E. Imagenet classification with deep convolutional neural networks. In Proceedings of the 25th International Conference on Neural Information Processing Systems, Lake Tahoe, NV, USA, 3–6 December 2012; pp. 1097–1105.

29.	Girshick, R.; Donahue, J.; Darrell, T.; Malik, J. Rich feature hierarchies for accurate object detection and semantic segmentation. In Proceedings of the IEEE Conference on Computer Vision and Pattern Recognition, Columbus, OH, USA, 23–28 June 2014; pp. 580–587.

30. Farabet, C.; Couprie, C.; Najman, L.; LeCun, Y. Learning hierarchical features for scene labeling. *IEEE Trans. Pattern Anal. Mach. Intell.* **2013**, *35*, 1915–1929. [CrossRef] [PubMed]
31. Schroff, F.; Kalenichenko, D.; Philin, J. FaceNet: A unified embedding for face recognition and clustering. In Proceedings of the IEEE Conference on Computer Vision and Pattern Recognition (CVPR), Boston, MA, USA, 7–12 June 2015; pp. 815–823.
32. Sun, Y.; Wang, X.; Tang, X. Deep learning face representation from predicting 10,000 classes. In Proceedings of the IEEE Conference on Computer Vision and Pattern Recognition (CVPR), Providence, RI, USA, 24–27 June 2014; pp. 1891–1898.
33. Yu, S.; Tan, D.; Tan, T. A framework for evaluating the effect of view angle, clothing and carrying condition on gait recognition. In Proceedings of the IEEE 18th International Conference on Pattern Recognition (ICPR'06), Hong Kong, China, 20–24 August 2006; pp. 441–444.
34. Jia, Y.; Shelhamer, E.; Donahue, J.; Karayev, S.; Long, J.; Girshick, R.; Guadarrama, S.; Darrell, T. Caffe: Convolutional Architecture for Fast Feature Embedding. *arXiv* **2014**, arXiv:1408.5093.
35. Phillips, P.J.; Moon, H.; Rizvi, S.A.; Rauss, P.J. The feret evaluation methodology for face-recognition algorithms. *IEEE Trans. Pattern Anal. Mach. Intell.* **2000**, *22*, 1090–1104. [CrossRef]
36. Tang, Y.; Xue, A.; Ding, J.; Tian, H.; Guo, W. Gait Cycle Detection by Fusing Temporal and Spatial Features with Frame Difference. *J. Data Acquis. Process.* **2017**, *32*, 533–539.

 *entropy*

*Article*

# Incremental Market Behavior Classification in Presence of Recurring Concepts

**Andrés L. Suárez-Cetrulo [1,2], Alejandro Cervantes [1,*] and David Quintana [1]**

[1] Department of Computer Science, Universidad Carlos III de Madrid, Leganés, 28911 Madrid, Spain; andres.suarez-cetrulo@ucd.ie (A.L.S.-C.); dquintan@inf.uc3m.es (D.Q.)

[2] Centre for Applied Data Analytics Research, University College Dublin, D04 V2N9 Dublin, Ireland

* Correspondence: acervant@inf.uc3m.es; Tel.: +34-91-624-8843

Received: 28 November 2018; Accepted: 20 December 2018; Published: 1 January 2019

**Abstract:** In recent years, the problem of concept drift has gained importance in the financial domain. The succession of manias, panics and crashes have stressed the non-stationary nature and the likelihood of drastic structural or concept changes in the markets. Traditional systems are unable or slow to adapt to these changes. Ensemble-based systems are widely known for their good results predicting both cyclic and non-stationary data such as stock prices. In this work, we propose RCARF (Recurring Concepts Adaptive Random Forests), an ensemble tree-based online classifier that handles recurring concepts explicitly. The algorithm extends the capabilities of a version of Random Forest for evolving data streams, adding on top a mechanism to store and handle a shared collection of inactive trees, called concept history, which holds memories of the way market operators reacted in similar circumstances. This works in conjunction with a decision strategy that reacts to drift by replacing active trees with the best available alternative: either a previously stored tree from the concept history or a newly trained background tree. Both mechanisms are designed to provide fast reaction times and are thus applicable to high-frequency data. The experimental validation of the algorithm is based on the prediction of price movement directions one second ahead in the SPDR (Standard & Poor's Depositary Receipts) S&P 500 Exchange-Traded Fund. RCARF is benchmarked against other popular methods from the incremental online machine learning literature and is able to achieve competitive results.

**Keywords:** ensemble methods; adaptive classifiers; recurrent concepts; concept drift; stock price direction prediction

---

## 1. Introduction

Financial market forecasting is a field characterized by data intensity, noise, non-stationary, unstructured nature, a high degree of uncertainty, and hidden relationships [1], being the financial markets complex, evolutionary, and non-linear dynamical systems [2]. Many approaches try to predict market data using traditional statistical methods. Albeit, these tend to assume that the underlying data have been created by a linear process, trying to make predictions for future values accordingly [3]. However, there is a relatively new line of work based on machine learning, whose success has surprised experts given the theory and evidence from the financial economics literature [4–6]. Many of these algorithms are able to capture nonlinear relationships in the input data with no prior knowledge [7]. For instance, Random Forest [8] has been one of the techniques obtaining better results predicting stock price movements [9–12].

In recent years, the notion of concept drift [13] has gained attention in this domain [14]. The Asian financial crisis in 1997 and, more recently, the great crisis in 2007–2008 have stressed the non-stationary nature and the likelihood of drastic structural or concept changes in financial markets [14–19].

Incremental machine learning techniques deal actively or passively [20] with the non-stationary nature of the data and its concept changes [13]. However, the problem of recurring concepts [21–23], where previous model behaviors may become relevant again in the future, is still a subject of study. As part of the so-called stability–plasticity dilemma, most of the incremental approaches need to re-learn previous knowledge once forgotten, wasting time and resources, and losing accuracy while the model is out-of-date. Although some authors have started to consider recurring concepts [24–28], the number of contributions focused on the financial forecasting domain is still very limited [29,30]. This might be partially explained by the fact that, in this context, the presence of noise and the uncertainties related to the number of market states, their nature, and the transition dynamics have a severe impact on the feasibility of establishing a ground truth.

Our contribution is an algorithm that deals with gradual and abrupt changes in the market structure through the use of an adaptive ensemble model, able to remember recurring market behaviors to predict ups and downs. The algorithm proposed improves a previous algorithm, namely Adaptive Random Forest (ARF) [31], by being able to react more accurately in the case of abrupt changes in the market structure. This is accomplished through the use of a concept history [21,22,32–34], which stores previously learned concept representations. When a structural change is detected, it replaces drifting classifiers with either a new concept model or with a concept extracted from the history, using dynamic time-windows to make the decision. As this concept representation is already trained, our algorithm is able to react faster than its predecessor, which is unable to profit from previous models.

The remainder of the paper is organized as follows. In Section 2, we review related work and approaches. In Section 3, we propose the algorithm RCARF. In Section 4, we describe the experimental design, present our empirical results and discuss their implications. Finally, in Section 5 we conclude with a summary of our findings and future lines of research.

## 2. Related Work

The number of approaches proposed for financial applications is vast. In terms of market price forecasting and trend prediction, these can be approached by looking at fundamental and technical indicators. Even though there is controversy regarding the potential of the latter to produce profitable trading strategies [5,35], the fact is that they are widely used in short-term trading [36]. Kara et al. [37] proposed a set of 10 technical indicators identified by domain experts and previous studies [38–44]. This approach has been used in more recent works (e.g., [4]). Some of them, such as the work of Patel [12], discretize features based on a human approach to investing, deriving the technical indicators using assumptions from the stock market.

Stock markets are non-stationary by nature. Depending on the period, they can show clear trends, cycles, periods where the random component is more prevalent, etc. Furthermore, stock prices are affected by external factors such as the general economic environment and political scenarios that may result in cycles [12]. Under these circumstances, incremental and online machine learning techniques [28,45] that adapt to structural changes, usually referred to as concept drift [13], are gaining traction in the financial domain [14].

In parallel, ensemble techniques are known for their good performance at predicting both cyclic and non-stationary data such as stock prices [9,12,46]. Ensembles are able to cover many different situations by using sets of learners. If a specific type of pattern reappears after a certain time, some of the trained models should be able to deal with it. These techniques, which are commonly used for trend prediction in financial data, are also one of the current trends of research in incremental learning. Lately, several incremental ensembles have been proposed [47] to deal not only with stationary data and recurring drifts but, also with non-stationary data in evolving data streams [20,22,34,48–51].

There are different types of concept drift detection mechanisms for handling gradual or abrupt changes, blips or recurring drifts [24,26–28,52] that can be used to deal with changes in the market behavioral structure [53]. As opposed to stationary data distributions, where the error rate of the learning algorithm will decrease when the number of examples increases, the presence of changes affects the learning model continuously [54]. This creates the need to retrain the models over time when they are no longer relevant for the current state of the market [15].

In the case of repeated cycles, handling of recurring concepts can help reduce the cost of retraining a model if a similar one has already been generated in the past. Fast recognition of a reappearing model may also improve the overall model accuracy as the trained model will provide good predictions immediately.

Gomes et al. [31] proposed an adaptive version of Random Forest that creates new trees when the accuracy of a participant in the ensemble decreases down to a certain threshold. These trees, considered background learners, are trained only with new incoming data and replace the model that raised a warning when this is flagged as drifting. Their Adaptive Random Forest algorithm (ARF) provides a mechanism to update decision trees in the ensemble and keep historical knowledge only when this is still relevant. However, once a tree is discarded, it is completely removed from memory. In presence of recurring concepts, ARF needs to train the trees from scratch.

Gonçalves et al. [23] proposed a recurring concept drift framework (RCD) that raises warnings when the error rate of a given classifier increases. Their approach creates a collection of classifiers and chooses one based on the data distribution. This data distribution is stored in a buffer of a limited length for each of the classifiers. When there is a warning, the newest data distribution is compared to the data distributions of other stored classifiers, to verify whether the new context has already occurred in the past.

Elwell et al. [20] dealt with recurrent concepts in a similar way. Their approach, Learn++.NSE, keeps one concept per batch, not limiting the number of classifiers. The idea, along the lines of Hosseini et al. [48], is to keep all the accumulated knowledge in a pool of classifiers to be used eventually, if needed. However, this approach suffers from scalability bottlenecks in continuous data streams as it does not prune the list of active classifiers. Other approaches have proposed explicit handling of recurring concepts by checking for similarity [21,22,32–34]. These store old models in a concept history for comparison when the current model is flagged as changing.

An alternative approach is the use of Evolving Intelligent Systems (EIS) [55]. These have achieved great results classifying non-stationary time series [19,29,30]. The latest EIS works apply meta-cognitive scaffolding theory for tuning the learned model incrementally in what-to-learn, when-to-learn, and how-to-learn [56]. These have also introduced the ability to deal with recurrent concepts explicitly, beating other methods at predicting the S&P500 [29,30]. In this space, Pratama et al. recently proposed pENsemble [57], an evolving ensemble algorithm inspired by Dynamic Weighted Majority (DWM) [50]. pENsemble counts with explicit drift detection, and it is able to deal with non-stationary environments and handle recurring drifts because of its base classifiers. These have a method that functions as a rule recall scenario, triggering previously pruned rules portraying old concepts to be valid again. However, pENsemble differs from our approach and the rest of the architectures of this work in the fact that it is built upon an evolving classifier. There is still an important gap between EIS and the rest of the literature for data stream classification. Features such as meta-cognition and explicit handling of recurrent concepts are still in an early level of adoption outside EIS. Furthermore, extensive application of EIS to challenging domains as stock market prediction is only starting.

Our proposal, which is described in detail in the next section, applies explicit recurring drift handling for price direction prediction to intra-day market data. The foundations of the algorithm start with the core ideas of ARF [31] as an evolving and incremental ensemble technique. The proposed approach extends these with the capability to store old models in a concept history. These models are subsequently retrieved when they are deemed suitable to improve the predictions of the current ensemble. The approach leverages certain ideas from some of the papers cited above, also including adaptive windowing to compare old and background learners based on buffers of different sizes, depending on the speed of the changes.

## 3. Adaptive Ensemble of Classifiers for Evolving and Recurring Concepts

The idea behind our proposal, Recurring Concepts Adaptive Random Forest (RCARF), is the development of an algorithm able to adapt itself to gradual, abrupt and also recurring drifts in the volatile data streams of the stock market. The main contribution of the approach is the explicit handling of recurring drifts in an incremental ensemble. This process is managed by two key components: the concept history, and the associated Internal Evaluator. Both are represented in Figure 1, which illustrates the overall structure of the algorithm.

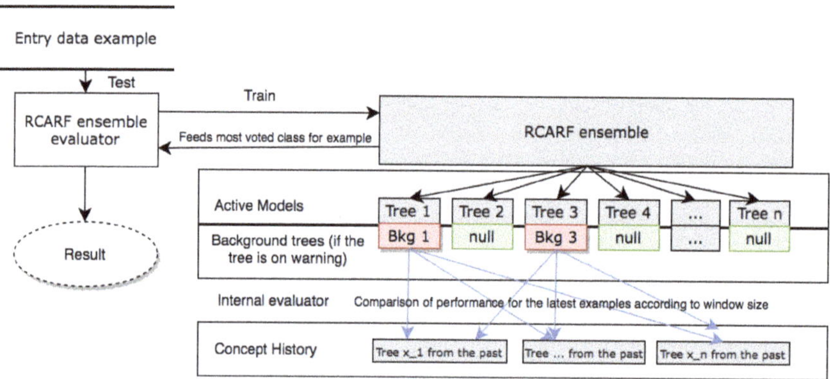


**Figure 1.** RCARF structure.

In Algorithm 1, we show the overall pseudocode for the RCARF algorithm. RCARF inherits several features of the Adaptive Random Forest (ARF) algorithm proposed by Gomes et al. [31].

As mentioned above, RCARF is a Random Forest classifier [8]. These algorithms use a collection (ensemble) of "base" classifiers. Traditionally, the forest is a homogeneous set of tree-based classifiers. The full forest classifier performs a prediction for every example in a data stream. Each example or batch of examples is pushed to all the base classifiers, each of which then casts its vote for the most likely class for the example. Each vote is multiplied by the base classifier's weight, a value that is adapted later depending on whether the related base classifier prediction matches the "real" class of the example. The random component arises from the fact that each of the base classifiers in the ensemble takes into account only a random set of the examples' features. Even though each base classifier is deciding its individual vote based on partial information, the voting mechanism usually provides very accurate predictions, in many circumstances due to the reinforcing process of the voting mechanism.

This general approach requires some adaptations to handle structural changes on the fly. RCARF implements a basic drift handling strategy along these lines inherited from ARF. To be ready to react properly to structural breaks, these algorithms have a mechanism to detect potential drifts in advance and ensure a smooth transition to new trees. A signal (warning) is raised by a very sensitive drift detector. This triggers the creation of a background tree and starts its training. In the case drift is confirmed (drift signal) at a later stage, the background tree replaces the associated one. Otherwise, it is discarded.

Unlike its predecessor, RCARF is also able to spot the recurrence of previously trained trees and retrieve them from a shared collection of inactive classifiers called concept history. Specific mechanisms in the decision process, such as the internal evaluator, are designed to make the best decision under drift conditions by using only the most adequate sample of recent data.

---

**Algorithm 1** RCARF algorithm. Adapted from ARF in [31]. Symbols: $m$, maximum features evaluated per split; $n$: total number of trees ($n = |T|$); $\delta_w$, warning threshold; $\delta_d$, drift threshold; $C(\cdot)$, change detection method; $S$, data stream; $B$, set of background trees; $W(t)$, tree $t$ weight; $P(\cdot)$, learning performance estimation function; $CH$, concept history; $TC$, temporal concept saved at the start of warning window.

---

1: **function** RECURRINGCONCEPTSADAPTIVERANDOMFORESTS(m, n, $\delta_w$, $\delta_d$)
2:
3:     $T \leftarrow CreateTrees(n)$
4:
5:     $W \leftarrow InitWeights(n)$
6:
7:     $B, CH \leftarrow \varnothing$
8:
9:     **while** $HasNext(S)$ **do**
10:
11:         $(x, y) \leftarrow next(S)$
12:
13:         **for all** $t \in T$ **do**
14:
15:             $\hat{y} \leftarrow predict(t, x)$
16:
17:             $W(t) \leftarrow P(W(t), \hat{y}, y)$
18:
19:             $RFTreeTrain(m, t, x, y)$             ▷ Train $t$ on the current instance(x,y)
20:
21:             **if** $C(\delta_w, t, x, y)$ **then**             ▷ Warning detected?
22:
23:                 $lastError \leftarrow evaluate(t)$         ▷ Save overall error of t
24:
25:                 $TC \leftarrow copy(t)$         ▷ Copy current tree at the start of warning window
26:
27:                 $b \leftarrow CreateTree()$         ▷ Init background tree
28:
29:                 $B(t) \leftarrow b$
30:
31:             **end if**
32:
33:             **if** $C(\delta_d, t, x, y)$ **then**             ▷ Drift detected?
34:
35:                 $t \leftarrow bestTransition(t, B(t), CH)$
36:
37:                 $addToConceptHistory(TC)$         ▷ Push current concept to CH
38:
39:             **end if**
40:
41:         **end for**
42:
43:         **for all** $b \in B$ **do**             ▷ Train each background tree
44:
45:             $RFTreeTrain(m, b, x, y)$
46:
47:         **end for**
48:
49:     **end while**
50:
51: **end function**

---

In both adaptive versions of Random Forest, base classifiers are always the Hoeffding Trees used in ARF. That means that, hereafter, we use the term "tree" to refer to each one of these base classifiers. However, it is worth noting that the mechanism we propose does not depend on the type of base classifier, which may be replaced transparently.

For the description of the algorithm, it is important to take into account that every tree generated will be in one of three different states:

- Active trees: Trees currently running in the ensemble for test and train. In the case of drift, an active tree is moved to the concept history.
- Background trees (one per active tree): A new background tree is created when an active tree signals a warning. This starts growing in parallel until a drift is signaled by its active tree (the moment when this gets replaced by either the background tree or a tree from the concept history). We refer to the training time of a background tree as the warning window of its active tree. As in ARF, each background tree inherits warning and drift related parameters values and the maximum subspace size per split from its active tree.
- Concept history trees: These were active trees in the past, but eventually they were replaced because of their low performance in a certain period of time. Throughout this work, when these trees are re-activated, they are called recurring trees.

Code kept from ARF includes the function responsible for inducing each base tree (Algorithm 2) and the warning and drift detection and handling (Lines 1–21, 27–29 and 43–47 in Algorithm 1). The method retains the mechanisms related to the ensemble itself (bagging, weighting and voting). However, in RCARF, we introduce the steps required to manage the concept history and how to perform an informed decision as to how to replace active trees in case of drift (Lines 23–25, and 35–37). These aspects of RCARF are detailed in the sections that follow.

---

**Algorithm 2** Random Forest Tree Train (RFTreeTrain). Symbols: $\lambda$, fixed parameter to Poisson distribution; $GP$, grace period before recalculating heuristics for split test; m: maximum features evaluated per split; t, decision tree selected; (x, y), current training instance. Adapted from [31].

```
 1: function RFTREETRAIN(m, t, x, y)
 2:
 3:     k ← Poisson(λ = 6)
 4:
 5:     if k > 0 then
 6:
 7:         l ← FindLeaf(t, x)
 8:
 9:         UpdateLeafCounts(l, x, k)
10:
11:         if examplesSeen(l) ≥ GP then
12:
13:             AttempSplit(l)
14:
15:             if DidSplit(l) then
16:
17:                 CreateChildren(l, m)
18:
19:             end if
20:
21:         end if
22:
23:     end if
24:
25: end function
26:
```

---

*3.1. Concept History*

As stated previously, one of the core elements of RCARF is the addition of a concept history to the ARF schema. The concept history ($CH$) is a collection of trees shared by all trees in the ensemble. This collection is created during the execution of the algorithm, and is stored for future use when an episode of concept drift impacts the performance of active trees. If an active tree is inserted in the concept history, it becomes available for the whole ensemble. If a tree from the concept history is "promoted" to be an active tree, it is immediately removed from the concept history.

RCARF relies on the assumption that, particularly in the case of abrupt drift, the background tree learned from scratch from the beginning of the warning window may be at a disadvantage compared to an old tree adapted to obtain good results but subsequently discarded. This situation, which would be affected by the speed of the concept drift, is especially likely if we can expect episodes of recurring drift in the data. In that case, the concept history already contains trained trees well-adapted to the recurring concept. Thus, instead of discarding useful trees, the objective would be storing them and then recovering them whenever they become relevant again.

Figure 1 illustrates the structure of RCARF. First, incoming data examples are tested using the ensemble evaluator. Only then, the example is used also for training the active tree.

As stated in the algorithm in Algorithm 1 and by Gomes et al. in [31], when the error statistics increase over time up to a certain threshold, a warning is activated and a background tree is created to replace the active model in the case of drift. After performing these steps, a change detector decides if the algorithm must be prepared for the occurrence of concept drift (warning detection, Line 21) or if a drift has really happened (drift detection, Line 33).

In both ARF and RCARF, the "warning window" is defined as the period of time that starts when an active tree raises a warning and finishes when the same tree detects a drift. Each warning window is specific to an active tree, and resets in the case of false alarm; that is, if a new warning is raised by the same tree before the drift is confirmed. In ARF, if a drift is detected (Line 33), the warning window is finished and the background tree replaces the active tree. In RCARF, during the warning window, there is also an online evaluation on the background tree (the one linked to the active tree that has raised the warning) and all trees in the concept history to compare their performance. This is the task of an "internal evaluator", described below. Only when a drift is detected, the tree with the lowest error according to the internal evaluator is promoted to active (Line 35). The previously stored copy of the active tree is then moved to the concept history (Line 37).

### 3.2. Internal Evaluator

RCARF has two types of evaluators: the ensemble one and the internal one.

- Ensemble evaluator (global, Line 15): It is in charge of the predicted accuracy and results of the algorithm.
- Internal evaluator (tree-specific, Algorithm 3): This is one of the main components of our proposal. During the warning window, we must collect information to be able to take the best decision of which is the best classifier in case drift happens. Given that we are estimating performance when drift is happening, only the latest examples are relevant. This evaluator is updated every time that a new example is tested. As we can see in Algorithm 4, the internal sliding window changes size dynamically during the warning window only for the background trees. The window size is fixed for trees in the concept history, and it is provided as an input parameter.

---

**Algorithm 3** Internal evaluator. It computes the best transition in the case of drift. Symbols: $t$, active tree; $b$, background tree; $CH$, concept history; $c$, tree from $CH$; $WS(CH)$, fixed window size in $CH$; $WS(b)$, current window size in $b$, $W(c)$, error statistics in $c$ for the latest examples in $WS(CH)$; $W(b)$, error statistics in $b$ according to $WS(b)$.

---

1:  **function** BestTransition$(t, b, CH)$
2:
3:    **for all** $c \in CH$ **do**                                     ▷ Rank of errors of each tree in CH
4:
5:      $addToRank(c, countErrors(W(c))/WS(CH))$
6:
7:    **end for**
8:    **if** $minError(rank) \leq (countErrors(W(b))/WS(b))$ **then**
9:
10:      $R \leftarrow extractClassifier(CH, minErrorKey(rank))$      ▷ Get and remove tree from the concept history
11:
12:
13:    **else**
14:
15:      $R \leftarrow b$
16:
17:    **end if**
18:
19:    **return** R
20:
21:  **end function**
22:

---

---

**Algorithm 4** Internal evaluator with dynamic windows for background trees. Symbols: $WS$, evaluator window size; $W$, evaluator window; $SI$, size increments; $MS$, minimum size of window.

---

1:  **function** AddEvaluationResults(value = correctlyClassifies ? 0 : 1)
2:
3:    $removeFirstElement(W)$
4:
5:    $add(W, value)$                                  ▷ Add result [1 (error) or 0 (success)] to window
6:
7:    $updateWindowSize()$
8:    **if** $(countOfErrors(W)/WS) < getErrorBeforeWarning$ **then**
9:
10:      $WS = WS + SI$
11:
12:
13:    **else if** $WS > MS$ **then**
14:
15:      $WS = WS - SI$
16:
17:    **end if**
18:
19:  **end function**
20:

---

The adaptation mechanism for the window size in Algorithm 4 is as follows: if the error obtained by a background tree for its internal evaluator window size ($WS$) in the latest testing examples is lower than the error obtained by the active tree when it raised the warning signal, then $WS$ decreases down to a minimum size. Otherwise, it increases once per iteration (that is, per example evaluated). Increments and decrements of $WS$ are performed according to an input parameter that defines "size increments" ($SI$).

The logic of the resizing mechanism relies on the interpretation of the error obtained by the background trees. In cases where it is greater than the error obtained by the active tree before warning, we believe that the underlying reason must be either because the background tree has not been trained with enough samples yet, or because the structure of the data stream is continuously changing (in a period of transition). In the second scenario, a smaller sample of the latest examples could be more accurate in estimating which is the best classifier for the latest concept ($WS$ decreases). Otherwise, a larger sample would be desirable, as it would provide a more representative set of data ($WS$ increases).

### 3.3. Training of the Available Trees

The addition of the concept history and the differences in the replacement strategy used in RCARF entail the need to discuss the way data are used to train the trees. As in ARF, both the active and background trees are trained with new examples as soon as they are available (Lines 19 and 45 in Algorithm 1). However, trees in the concept history are adapted to data that correspond to a different concept. Therefore, they are not retrained unless they are promoted to active.

As mentioned above, in the case of drift, the active tree is replaced by either the best tree from the concept history or the background tree (Lines 35–37 in Algorithm 1) following Algorithm 3. In the case that the background tree was selected for promotion, the training examples from the warning window would already have been used for its training. Conversely, if a concept history tree were selected for promotion, these training examples would be discarded.

As stated by Alippi et al. [21], there is always a delay from the start of a concept drift to the start of the warning window. During this lag, it is not possible to warrant the isolation of a given concept. In this paper, for simplicity, we avoid taking into consideration this delay as a part of our analysis. Therefore, for the purpose of this work, we assumed that the start of every warning window that ends with the trigger of the drift (thus, when this is not a false alarm), matches the start of a concept drift. For this reason, even though active trees are being updated during warning windows, we consider that the moment in which they are best adapted to a given concept is just before the warning window. Hence, the tree that is pushed to the concept history is a snapshot of the active tree at the start of the warning window (see Lines 25 and 37 in Algorithm 1).

## 4. Experimentation: Predicting the S&P500 Price Trend Direction

### 4.1. Data

Data for this work were produced in the following way. First, we downloaded Exchange-Traded Fund (ETF) SPY prices for the entire first quarter of 2017 at second level from QuantQuote (Data source: https://www.quantquote.com). This ETF, one of most heavily-traded ones, tracks the popular US index S&P 500. Secondly, we selected 10 different technical indicators as feature subsets based on the work by Kara et al. [37]. The default value of 10 s that we set for the number of periods, $n$, was extended in the case of the two moving averages. Once we considered the additional possibilities, 5 and 20 s, we ended up with the 14 features described in Table 1. These were computed with the TA-lib technical analysis library (Technical Analysis library: http://ta-lib.org/) using its default values for all parameters other than the time period.

**Table 1.** Selected technical indicators. Formulas as reported in Kara et al. [37] applied to second-level. Exponential and simple moving averages for 5 and 20 s added as extra features.

| Name of Indicators | Formulas |
|---|---|
| Simple n-second moving average (5, 10, 20) | $\frac{C_t+C_{t-1}+...+C_{t-n+1}}{n}$ |
| Weighted n-second moving average (5, 10, 20) | $\frac{n \times C_t+(n-1) \times C_{t-1}+...+C_{t-n+1}}{n+(n-1)+...+1}$ |
| Momentum | $C_t - C_{t-n}$ |
| Stochastic K% | $\frac{C_t-LL_{t-n}}{HH_{t-n}-LL_{t-n}} \times 100$ |
| Stochastic D% | $\frac{\sum_{i=0}^{n-1} K_{t-i}\%}{n}$ |
| RSI (Relative Strength Index) | $100 - \frac{100}{1+(\sum_{i=0}^{n-1} Up_{t-i}/n)/(\sum_{i=0}^{n-1} Dw_{t-i}/n)}$ |
| MACD (Moving average convergence divergence) | $EMA(12)_t - EMA(26)_t$ |
| Larry William's R% | $\frac{H_n-C_t}{H_n-L_n} \times 100$ |
| A/D (Accumulation/Distribution) Oscillator | $\frac{H_t-C_{t-1}}{H_t-L_t}$ |
| CCI (Commodity Channel Index) | $\frac{M_t-SM_t}{0.015D_t}$ |

$C_t$ is the closing price; $L_t$ the low price; $H_t$ the high price at time $t$; EMA exponential moving average, $EMA(k)_t$: $EMA(k)_{t-1} + \alpha \times (c_t - EMA(k)_{t-1})$; $\alpha$ smoothing factor: $2/1 + k$; $k$ is time period of $k$ second exponential moving average; $LL_t$ and $HH_t$ mean lowest low and highest high in the last $t$ seconds, respectively; $M_t$ : $H_t + L_t + C_t/3$; $SM_t$ : $\sum_{i=1}^{n} M_{t-i+1})/n$; $D_t$ : $(\sum_{i=1}^{n} |M_{t-i+1} - SM_t|)/n$; $Up_t$ means the upward price change; $Dw_t$ means the downward price change at time $t$. $n$ is the period used to compute the technical indicator in seconds.

The label, categorized as "0" or "1", indicates the direction of the next change in the EFT. If the SPY closing price at time $t$ is higher than that at time $t - 1$, direction $t$ is "1". If the SPY closing price at time $t$ is lower or equal than that at time $t - 1$, direction $t$ is "0". Furthermore, as part of the labeling process, a lag of 1 s has been applied over the feature set. Thus, if the technical indicators belong to the instant $t - 1$, the label reflects the price change from $t - 1$ to $t$.

Short sellers are usually averse to holding positions over non-market hours and try to close them at the end of the day [58]. The price may jump, or the market can behave very differently in the next morning. Therefore, only prices during market hours are considered in this work. In addition, as the technical indicators selected depend on the 35 previous seconds of data, the first 35 s are discarded for each day after processing the technical indicators. This filtering aims to avoid the influence of the previous day trends, and prices before market hours.

## 4.2. Experimental Setting

We designed the experiments presented in this section with two separate purposes in mind.

First, were compared the utility of the recurring drift detection implemented in RCARF vs. the basic ARF approach. To perform a fair comparison between RCARF and ARF, both algorithms used the same ADaptive WINdowing (ADWIN) [59] change detector for warnings and drifts. Furthermore, both learners used the same adapted version of Hoeffding Trees as base classifier, and the same number of trees in their configuration.

Secondly, we aimed to prove that RCARF is a suitable candidate for this task compared to other state-of-the-art learners for data stream classification. For this comparison, we selected the following learners, all of them from the literature of online classification of non-stationary data streams: DWM [50] using Hoeffding Trees as base classifiers, a RCD learner [23] of Hoeffding Trees and Hoeffding Adaptive Tree (AHOEFT) [60]. All of the experiments were performed using the MOA framework [61], which provides implementations of the aforementioned algorithms, in a Microsoft Azure Virtual Machine "DS3 v2" with the Intel(R) Xeon(R) CPU E5-2673 v4 @ 2.30 GHz processor and 14 GB RAM.

ARF is able to train decision trees in parallel with multithreading to reduce the running time. However, in RCARF, multithreading would impact the results because of the addition of the concept history as a shared storage space used by all trees. Thus, in this work, all experiments were run on a single thread. The impact of multithreading is out of the scope of our proposal.

The dataset was modeled as a finite, ordered stream of data, and evaluation and training were performed simultaneously using *Interleaved-Test-Then-Train* evaluation [61]. In this method, a prediction is obtained for each example, and success or failure is recorded before the example is used for training and adjusting the model.

We evaluated each algorithm using the accumulated classification error at the end of the period. This error was calculated by dividing the number of misclassified examples by the total number of examples. However, this accumulated error was not adequate to compare how different algorithms behave in particular moments of time. Thus, we also calculated at regular intervals the error of each algorithm calculated over a fixed window of time (500 examples). This sequence could then be compared graphically.

Given the stochastic nature of the algorithms based on Random Forests (RCARF and ARF), in these cases, we performed 20 experiments and averaged the results. The statistical significance of the differences of performance among the algorithmic approaches was formally tested using a protocol that starts verifying the normality of the distribution of prediction errors over the mentioned experiments using the Lilliefors test. In the case that the null hypothesis of normality was rejected, we relied on the Wilcoxon test [62]. Otherwise, we tested for homoscedasticity using Levene's test and, depending on whether we could reject the null hypothesis, the process ends testing for equality of means using either a Welch test or a *t*-test. The significance levels considered in the tests for normality and homoscedasticity were set at 5%. For the rest, we considered both 5% and 1%.

It is worth emphasizing that the approach that we describe predicts short-term market trends, but it does not generate any trading signals (that would require further processing). All the tested algorithms, including our proposed method, used the values of the raw technical indicators at specific points in time to generate a binary class prediction of the price trend (up or stable/down) for new data patterns. Ensemble based approaches have a number of internal classifiers whose predictions are subsequently combined to provide this prediction for the whole ensemble. We emphasize this idea at the end of Section 5.

*4.3. Parameter Selection and Sensitivity*

The parameterization effort was not subject to systematic optimization and, therefore, the performance of the algorithm might be understated. The algorithms in the experiments held most of their respective default parameters or recommended setups according to their authors. Nonetheless, there are certain points common to most of the algorithms that deserve a mention.

- Base learner: All ensembles in our experiments used Hoeffding Trees as base classifier.
- Batch size: The algorithms ARF, RCARF and AHOEFT processed examples one at a time. For RCD, we processed data in batches of 600 examples, that is, 10-min intervals. In DWM, we used its default setup.
- Change detector: All ensembles in our experiments used ADWIN as explicit change detection mechanism. Apart from being the default change detector in ARF, its performance has already been proven in [31].
- Ensemble size: An ensemble size of 40 classifiers was applied by default to RCARF and ARF as this value performed well according to the study in [31]. We used 40 classifiers in all ensembles used in this work.

As mentioned above, in the experiments for ARF and RCARF, as change detector, we used the ADWIN algorithm proposed in [60]. The detailed procedure is described, for instance, in [63]. ADWIN has a variable sized sliding window of performance values. If a drift is detected, the window size is reduced; otherwise, it increases, becoming larger with longer concepts. When used as a drift detector, ADWIN stores two sub-windows to represent older and recent data. When the difference in the averages between these sub-windows surpasses a given threshold, a drift is detected.

The ADWIN change detector uses a parameter, $\delta$, the value of which is directly related to its sensitivity. A large value sets a smaller threshold in the number of changes in the monitored statistic (error rate in our case) that triggers the detection event. Specific values for parameters of this sort are dependent on the signal-to-noise ratio of the specific data stream and may impact the overall performance of the algorithm.

The RCARF algorithm assumes that a background tree is created and starts learning as soon as a change is reported by the ADWIN detector with $\delta_w$ sensitivity. This background tree is only upgraded to "active" status when drift is confirmed by a second change detector triggered with $\delta_d$ sensitivity. Thus, value for warning ($\delta_w$) has to be greater than the value for drift ($\delta_d$). Large values were selected for the RCARF change detector ($\delta_w$ and $\delta_d$) to ensure that concept history trees are given a chance to replace the active tree often enough to detect abrupt changes. These were set to $\delta_w = 0.3$ and $\delta_d = 0.15$.

The starting size of the dynamic windows for the internal evaluator was 10 examples, with increments or decrements of 1 example in the background trees, and a minimum size of 5 examples.

Although this should be confirmed by further analysis, our experiments suggest that ARF is in general more sensitive to the values of $\delta$ than RCARF. We believe that this can be explained by the fact that, in the case of early detection of a drift or abrupt changes, when the background tree is not yet ready to replace the active model, RCARF can still transition to a recurring decision tree that outperforms the incompletely trained background tree. Because of the sensitivity of ARF to these parameters, we tested three configurations for ARF (two of them, "*moderate*" and "*fast*", recommended by the authors in [31]) that are summarized in Table 2. Regarding RCD, given that it uses a single ADWIN change detector, we selected the same value that was used for drift detection in RCARF.

**Table 2.** Sensitivity parameters for the ADWIN change detector in ARF and RCARF.

| Configuration | $\delta_w$ | $\delta_d$ |
|---|---|---|
| ARF$_{moderate}$ | 0.0001 | 0.00001 |
| ARF$_{fast}$ | 0.01 | 0.001 |
| RCARF, ARF$_{ultra}$ | 0.3 | 0.15 |
| RCD | | 0.15 |

### 4.4. Global Performance Comparison

Table 3 summarizes the results of the experimental work providing the main descriptive statistics for the accumulated error (%) in predicting the market trend, for all the algorithms on the whole dataset over 20 runs. As can be seen, RCARF obtains the most competitive results. The reported differences were formally tested using the previously described protocol, and all of them were statistically significant at 1%.

**Table 3.** Global comparison. Accumulated error (%) for all algorithms on the whole dataset, sorted from best to worst result. Main descriptive statistics over 20 runs. Differences are significant at 1%.

| | Mean | Median | Var. | Max. | Min. |
|---|---|---|---|---|---|
| RCARF | 34.7533 | 34.7538 | 0.0002 | 34.7791 | 34.7285 |
| ARF$_{moderate}$ | 34.8008 | 34.8007 | 0.0002 | 34.8362 | 34.7769 |
| ARF$_{fast}$ | 34.8309 | 34.8335 | 0.0003 | 34.8591 | 34.7902 |
| RCD$_{HOEF}$ | 35.0469 | 35.0469 | 0.0000 | 35.0469 | 35.0469 |
| ARF$_{ultra}$ | 35.1104 | 35.1114 | 0.0002 | 35.1392 | 35.0881 |
| DWM | 35.2364 | 35.2364 | 0.0000 | 35.2364 | 35.2364 |
| AHOEFT | 35.4661 | 35.4661 | 0.0000 | 35.4661 | 35.4661 |

The differences between RCARF and ARF are due to the fact that, in some of the abrupt changes, RCARF is able to replace the active tree with a trained tree from its concept history, the performance of which is better than the performance of the background tree used by ARF under the same circumstances. When these gains are over the whole period (including stable periods without concept drift), the final average difference is small. Because of the low signal to noise ratio in this domain, we believe that these small gains in predictive accuracy may create a competitive advantage for any trading system that might use the prediction of RCARF as part of its decision process.

Two configurations of ARF, $ARF_{moderate}$ and $ARF_{fast}$, obtained the second- and third-best results, followed by RCD. AHOEFT obtained the worst result, which was expected, as this algorithm maintains a single tree (not an ensemble of trees). It is well-known that ensemble methods can be used for improving prediction performance [64]. We can also conclude that configurations for ARF suggested by the authors are better than $ARF_{ultra}$, which used the same parameters as RCARF. This may be explained by the fact that this configuration may be too sensitive to noise. It produces too many switches to background trees that are not yet accurately trained when they must be promoted to be active trees. RCARF, instead, is able to switch to pre-trained trees stored in the concept history, thus avoiding the corresponding decrease in performance.

As can be seen in Table 4, RCARF performed an average of 85 drifts per decision tree, for a total of 3411 drifts on average per experiment. However, the final number of decision trees in the concept history was in average 118 trees. As each recurring drift pushes one tree but also deletes one tree from the concept history, the table shows that there were only 118 background drifts in an average experiment, while there were more than 3000 recurring drifts on every experiment. This, together with the obtained results for RCARF, shows that the recurring drift mechanism was used to resolve most of the drift situations.

**Table 4.** Internal statistics for RCARF on the whole dataset over 20 runs. # Drifts, number of total drifts during the execution (both recurring and background); Drifts per tree, number of total drifts during the execution (both recurring and background) divided by the ensemble size; # F. Warnings, number of active warnings at the end of the execution; # *CH* Trees, number of decision trees in the concept history at the end of the execution.

|  | Mean | Median | Var. | Max. | Min. |
|---|---|---|---|---|---|
| # Drifts | 3411.1500 | 3411.5000 | 3120.2395 | 3518 | 3279 |
| Drifts per tree | 85.2788 | 85.2875 | 1.9501 | 88 | 82 |
| # F. Warnings | 13.6000 | 14.0000 | 9.2000 | 19 | 9 |
| # *CH* Trees | 118.2500 | 119.0000 | 46.0921 | 130 | 106 |

Another issue of interest is that the final number of active warnings (at the end of the experiments) was between 9 and 19; that is, with 40 trees, a percentage between 22.5% and 50% of the total were in warning at this point of time. Obviously, this fraction changed continuously during the experiment and was different on every run and for each base classifier. This number depends on the sensitivity parameter in RCARF $\delta_w$, and may be taken as a measure of the number of "open" warning windows in a given experiment. A lower value for $\delta_w$ may be chosen to reduce the number of warning windows opened simultaneously.

In terms of efficiency, the average full running time of RCARF on the whole dataset over 20 experiments was 35,263 s. This is less than the 10 computing hours for an entire quarter of market data at 1-s level. Hence, although the experiments were not run against the market in real time, RCARF demonstrates the ability to operate in an online setting at 1-s level on the server used.

*4.5. Evolution of the Ensemble over Time*

To show the overall behavior of RCARF, we have included Figure 2. It shows the evolution of error in RCARF for a short period of time (the first trading day of the year). Vertical lines are used to signal moments where a drift occurred and an active tree was replaced with one of the trees in the ensemble. Red dotted lines indicate times where a background tree became active, while blue dashed lines indicate times where a concept history tree was re-activated (recurrent drift). As we can observe in Figure 2, drifts are detected throughout the whole period of time.

At the beginning of the experiment, the error was higher because the models had not yet had the opportunity to adjust to the data; therefore, drifts occurred quite often and sometimes with very short intervals among them. Later on, drifts were more sparse. Most of the transitions were to trees that were stored in the concept history (blue dashed drifts in Figure 2), and not very often to background trees (red dotted drifts). That is, concept history trees were used most of the time instead of background trees, which proves that storing information from the past helped the RCARF algorithm in this particular dataset.

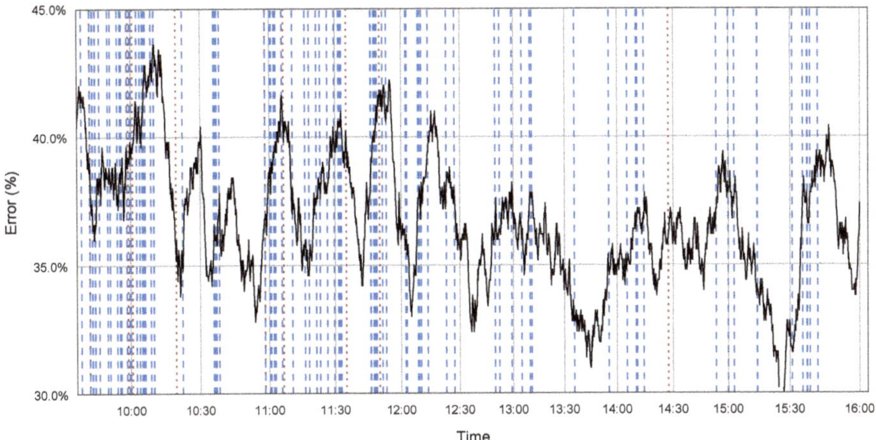

**Figure 2.** Sample run of RCARF on a single test for the trading first day. Error measured on windows of 500 examples. Red dotted vertical lines mark drifts to background trees, and blued dashed vertical lines mark drifts to recurring trees.

Figure 3 compares the results of all of the algorithms over a portion of the training set. Due to the sampling frequency of seconds, we have smoothed the plots averaging error on 1000 examples. The first 1000 examples are excluded from the chart due to this fact. The aim of the figure is to illustrate the performance of the algorithms over a specific period of time. Given the length of the time series used in the experimental analysis and the fact that the algorithms were run a number of times, it is hard to extract clear conclusions out of it. The performance comparison should be made based on the global performance indicators and statistical tests reported Table 3.

Having said that, the figure is consistent with the mentioned results. ARF and RCARF show a similar behavior, and their average error over time tended to be below the one found for the other algorithms. This is interesting because it suggests that these algorithms might indeed be superior under most circumstances, and not under some specific market conditions that might be difficult to capture with the AHOEFT, RCD, and DWM. RCARF and $ARF_{moderate}$, the closest competitor, often overlapped. However, RCARF was often dominant for short periods of time. This would be consistent with the notion that RCARF should benefit from the use of its concept history to adjust faster to drifts than ARF, which would eventually accumulate enough evidence to converge to a similar model.

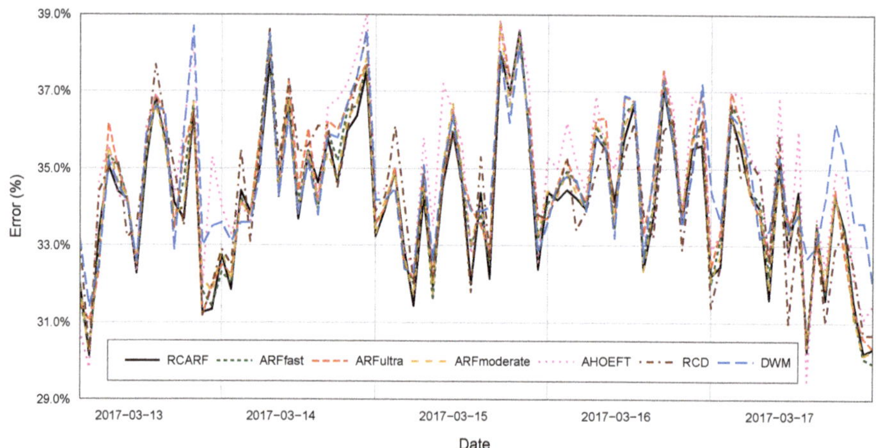

**Figure 3.** Algorithm comparison. Average error measured on windows of 1000 examples for a example period of time. For RCARF, ARF$_{ultra}$, ARF$_{fast}$ and ARF$_{moderate}$, we show the average result of 20 runs.

## 5. Summary and Conclusions

In this paper, we introduce RCARF, an ensemble tree-based online classifier that handles recurring concepts explicitly. The algorithm extends the capabilities of Adaptive Random Forests (ARF) adding a mechanism to store and handle a shared collection of inactive trees, called concept history. This works in conjunction with a decision strategy that reacts to drift by replacing active trees with the best available alternative: either a previously stored tree from the concept history or a newly trained background tree. Both mechanisms are designed to provide fast reaction times and are thus applicable to high-frequency data.

The experimentation was conducted on data from a full quarter of both price and trade volumes for the SPY Exchange-Traded Fund. This ETF, one of most heavily-traded ones, tracks the S&P 500 index. Both series were downloaded with a resolution of 1-s. We defined a classification problem where the objective was to predict whether the price will rise or decrease in the price change. For this classification task, we used as attributes a list of technical indicators commonly found in the literature. These indicators were labeled with the predicted behavior (class) and the result was fed as a data stream to our test bench of online stream classifiers, including our proposal, RCARF.

The experimental results show that RCARF offers a statistically significant improvement over the comparable methods. Given that the main difference between ARF and RCARF is the fact that the second one uses recurring concepts, the new evidence would support the hypothesis that keeping a memory of market models adds value versus a mere continuous adaptation. The idea that old models might end up eventually being more useful than the ones that are being fitted at the time, mostly due to faster adaptation to the market state, has interesting implications from a financial point of view. The reported results would support the idea of history repeating in terms of the price generation process. The market would not always transition to completely new market states, but also switch back to previous (or similar) ones. Recognition of the previous aspect is an extra insight for financial experts that might be used to obtain excess returns. This, however, is something to be analyzed in the future.

This work was focused on trend prediction with adaptation to concept drift, but we did not intend to derive any trading system. Actually, the implementation of such system might require reframing the classification problem to include a larger number of alternatives that could discriminate not only the direction of price changes, but also their magnitude. The current version of the algorithm predicts to a certain point short-term market trends, whether there is a way to exploit profitably market regularities is yet to be determined. For that reason, while it is clear that our the results are compatible with arguments against the efficient-market hypothesis, we cannot claim that we can beat consistently buy and hold and, therefore, we cannot reject it.

Future extensions of this work might include optimization of the algorithm for ultra-high frequencies and the development of further methods to adapt and resize the internal evaluator, such as the possibility of saving the window size as part of the concept to be inherited in case of recurring drifts and new window resizing politics for the historical models. All this might contribute to the optimization of the process that currently selects between recurrent or new decision trees. Finally, another possibility would be the addition of meta-cognition to evaluate recurring behaviors from the history by looking at previous transitions of the model.

**Author Contributions:** A.L.S.-C. and A.C. conceived the algorithm; A.L.S.-C. implemented the solution; A.L.S.-C. and A.C. D.Q designed the experiments; A.L.S.-C. ran the experiments; and A.L.S.-C., A.C., and D.Q analyzed the data and wrote the paper.

**Funding:** This research was funded by the Spanish Ministry of Economy and Competitiveness under grant number ENE2014-56126-C2-2-R.

**Conflicts of Interest:** The authors declare no conflict of interest.

## References


1. Huang, W.; Nakamori, Y.; Wang, S.Y. Forecasting stock market movement direction with support vector machine. *Comput. Oper. Res.* **2005**, *32*, 2513–2522. [CrossRef]
2. Abu-Mostafa, Y.S.; Atiya, A.F. Introduction to financial forecasting. *Appl. Intell.* **1996**, *6*, 205–213. [CrossRef]
3. Cavalcante, R.C.; Brasileiro, R.C.; Souza, V.L.F.; Nobrega, J.P.; Oliveira, A.L.I. Computational Intelligence and Financial Markets: A Survey and Future Directions. *Expert Syst. Appl.* **2016**, *55*, 194–211. [CrossRef]
4. Hsu, M.W.; Lessmann, S.; Sung, M.C.; Ma, T.; Johnson, J.E. Bridging the divide in financial market forecasting: Machine learners vs. financial economists. *Expert Syst. Appl.* **2016**, *61*, 215–234. [CrossRef]
5. Fama, E.F. Efficient Capital Markets: A Review of Theory and Empirical Work. *J. Financ.* **1970**, *25*, 383. [CrossRef]
6. Tsaih, R.; Hsu, Y.; Lai, C.C. Forecasting S&P 500 stock index futures with a hybrid AI system. *Decis. Support Syst.* **1998**, *23*, 161–174.
7. Atsalakis, G.S.; Valavanis, K.P. Surveying stock market forecasting techniques—Part II: Soft computing methods. *Expert Syst. Appl.* **2009**, *36*, 5932–5941. [CrossRef]
8. Breiman, L. Random Forests. *Mach. Learn.* **2001**, *45*, 5–32. [CrossRef]
9. Ballings, M.; Van Den Poel, D.; Hespeels, N.; Gryp, R. Evaluating multiple classifiers for stock price direction prediction. *Expert Syst. Appl.* **2015**, *42*, 7046–7056. [CrossRef]
10. Booth, A.; Gerding, E.; McGroarty, F. Automated trading with performance weighted Random Forests and seasonality. *Expert Syst. Appl.* **2014**, *41*, 3651–3661. [CrossRef]
11. Ładyżyński, P.; Żbikowski, K.; Grzegorzewski, P. Stock Trading with Random Forests, Trend Detection Tests and Force Index Volume Indicators. In Proceedings of the 12th International Conference on Artificial Intelligence and Soft Computing Part II (ICAISC 2013), Zakopane, Poland, 9–13 June 2013; Rutkowski, L., Korytkowski, M., Scherer, R., Tadeusiewicz, R., Zadeh, L.A., Zurada, J.M., Eds.; Springer: Berlin/Heidelberg, Germany, 2013; pp. 441–452.
12. Patel, J.; Shah, S.; Thakkar, P.; Kotecha, K. Predicting stock and stock price index movement using Trend Deterministic Data Preparation and machine learning techniques. *Expert Syst. Appl.* **2015**, *42*, 259–268. [CrossRef]
13. Tsymbal, A. *The Problem of concept drift: Definitions and Related Work*; Technical Report: TCD-CS-2004-15; Department of Computer Science Trinity College: Dublin, Ireland, 2004.
14. Das, R.T.; Ang, K.K.; Quek, C. IeRSPOP: A novel incremental rough set-based pseudo outer-product with ensemble learning. *Appl. Soft Comput. J.* **2016**, *46*, 170–186. [CrossRef]
15. Münnix, M.C.; Shimada, T.; Schäfer, R.; Leyvraz, F.; Seligman, T.H.; Guhr, T.; Stanley, H.E. Identifying States of a Financial Market. *Sci. Rep.* **2012**, *2*, 644. [CrossRef] [PubMed]
16. Vella, V.; Ng, W.L. Enhancing risk-adjusted performance of stock market intraday trading with Neuro-Fuzzy systems. *Neurocomputing* **2014**, *141*, 170–187. [CrossRef]

17. Hu, Y.; Liu, K.; Zhang, X.; Xie, K.; Chen, W.; Zeng, Y.; Liu, M. Concept drift mining of portfolio selection factors in stock market. *Electron. Commer. Res. Appl.* **2015**, *14*, 444–455. [CrossRef]

18. Silva, B.; Marques, N.; Panosso, G. Applying neural networks for concept drift detection in financial markets. In Proceedings of the CEUR Workshop Proceedings, Boston, MA, USA, 11–15 November 2012; Volume 960, pp. 43–47.

19. Gu, X.; Angelov, P.P.; Ali, A.M.; Gruver, W.A.; Gaydadjiev, G. Online evolving fuzzy rule-based prediction model for high frequency trading financial data stream. In Proceedings of the 2016 IEEE Conference on Evolving and Adaptive Intelligent Systems (EAIS), Natal, Brazil, 23–25 May 2016; pp. 169–175.

20. Elwell, R.; Polikar, R. Incremental learning of concept drift in nonstationary environments. *IEEE Trans. Neural Netw.* **2011**, *22*, 1517–1531. [CrossRef]

21. Alippi, C.; Boracchi, G.; Roveri, M. Just-in-time classifiers for recurrent concepts. *IEEE Trans. Neural Netw. Learn. Syst.* **2013**, *24*, 620–634. [CrossRef] [PubMed]

22. Gomes, J.B.; Gaber, M.M.; Sousa, P.A.C.; Menasalvas, E. Mining recurring concepts in a dynamic feature space. *IEEE Trans. Neural Netw. Learn. Syst.* **2014**, *25*, 95–110. [CrossRef]

23. Gonçalves, P.M., Jr.; Souto, R.; De Barros, M. RCD: A recurring concept drift framework. *Pattern Recognit. Lett.* **2013**, *34*, 1018–1025. [CrossRef]

24. Webb, G.I.; Hyde, R.; Cao, H.; Nguyen, H.L.; Petitjean, F. Characterizing concept drift. *Data Min. Knowl. Discov.* **2016**, *30*, 964–994. [CrossRef]

25. Gomes, H.M.; Barddal, J.P.; Enembreck, F.I.; Bifet, A. A Survey on Ensemble Learning for Data Stream Classification. *ACM Comput. Surv.* **2017**, *50*, 1–36. [CrossRef]

26. Ramírez-Gallego, S.; Krawczyk, B.; García, S.; Woźniak, M.; Herrera, F. A survey on data preprocessing for data stream mining: Current status and future directions. *Neurocomputing* **2017**, *239*, 39–57. [CrossRef]

27. Gama, J.A. A Survey on concept drift Adaptation. *ACM Comput. Surv* **2013**, *46*, 44. [CrossRef]

28. Ditzler, G.; Roveri, M.; Alippi, C.; Polikar, R. Learning in Nonstationary Environments: A Survey. *IEEE Comput. Intell. Mag.* **2015**, *10*, 12–25. [CrossRef]

29. Pratama, M.; Lughofer, E.; Er, J.; Anavatti, S.; Lim, C.P. Data driven modelling based on Recurrent Interval-Valued Metacognitive Scaffolding Fuzzy Neural Network. *Neurocomputing* **2017**, *262*, 4–27. [CrossRef]

30. Pratama, M.; Lu, J.; Lughofer, E.; Zhang, G.; Er, M.J. Incremental Learning of concept drift Using Evolving Type-2 Recurrent Fuzzy Neural Network. *IEEE Trans. Fuzzy Syst.* **2016**, *25*, 1175–1192. [CrossRef]

31. Gomes, H.M.; Bifet, A.; Read, J.; Barddal, J.P.; Enembreck, F.; Pfharinger, B.; Holmes, G.; Abdessalem, T. Adaptive Random Forests for evolving data stream classification. *Mach. Learn.* **2017**, *106*, 1469–1495. [CrossRef]

32. Yang, Y.; Wu, X.; Zhu, X. Mining in anticipation for concept change: Proactive-reactive prediction in data streams. *Data Min. Knowl. Discov.* **2006**, *13*, 261–289. [CrossRef]

33. Gomes, J.B.; Menasalvas, E.; Sousa, P.A.C. Tracking Recurrent Concepts Using Context. In *Rough 557 Sets and Current Trends in Computing, Proceedings of the 7th International Conference, RSCTC 2010, Warsaw, Poland, 28–30 June 2010*; Szczuka, M., Kryszkiewicz, M., Ramanna, S., Jensen, R., Hu, Q., Eds.; Springer: Berlin/Heidelberg, Germany, 2010; pp. 168–177.

34. Li, P.; Wu, X.; Hu, X. Mining Recurring concept drifts with Limited Labeled Streaming Data. *ACM Trans. Intell. Syst. Technol.* **2012**, *3*, 1–32. [CrossRef]

35. Lo, A.W.; Mamaysky, H.; Wang, J. Foundations of Technical Analysis: Computational Algorithms, Statistical Inference, and Empirical Implementation. *J. Financ.* **2000**, *55*, 1705–1765. [CrossRef]

36. Tay, F.E.; Cao, L. Application of support vector machines in financial time series forecasting. *Omega* **2001**, *29*, 309–317. [CrossRef]

37. Kara, Y.; Acar Boyacioglu, M.; Baykan, O.K. Predicting direction of stock price index movement using artificial neural networks and support vector machines: The sample of the Istanbul Stock Exchange. *Expert Syst. Appl.* **2011**, *38*, 5311–5319. [CrossRef]

38. Diler, A. Predicting direction of ISE national-100 index with back propagation trained neural network. *J. Istanb. Stock Exch.* **2003**, *7*, 65–81.

39. Armano, G.; Marchesi, M.; Murru, A. A hybrid genetic-neural architecture for stock indexes forecasting. *Inf. Sci.* **2005**, *170*, 3–33. [CrossRef]

40. Huang, C.L.; Tsai, C.Y. A hybrid SOFM-SVR with a filter-based feature selection for stock market forecasting. *Expert Syst. Appl.* **2009**, *36*, 1529–1539. [CrossRef]

41. Kim, K.J. Financial time series forecasting using support vector machines. *Neurocomputing* **2003**, *55*, 307–319. [CrossRef]

42. Kim, K.J.; Han, I. Genetic algorithms approach to feature discretization in artificial neural networks for the prediction of stock price index. *Expert Syst. Appl.* **2000**, *19*, 125–132. [CrossRef]

43. Yao, J.; Tan, C.L.; Poh, H.L. Neural Networks for Technical Analysis: A Study on Klci. *Int. J. Theor. Appl. Financ.* **1999**, *02*, 221–241. [CrossRef]

44. Kumar, M.; Thenmozhi, M. Forecasting Stock Index Movement: A Comparision of Support Vector Machines and Random Forest. In Proceedings of the 9th Capital Markets Conference on Indian Institute of Capital Markets, Mumbai, India, 24 January 2006; pp. 1–16.

45. Lughofer, E.; Angelov, P. Handling drifts and shifts in on-line data streams with evolving fuzzy systems. *Appl. Soft Comput. J.* **2011**, *11*, 2057–2068. [CrossRef]

46. Patel, J.; Shah, S.; Thakkar, P.; Kotecha, K. Predicting stock market index using fusion of machine learning techniques. *Expert Syst. Appl.* **2015**, *42*, 2162–2172. [CrossRef]

47. Krawczyk, B.; Minku, L.L.; Gama, J.; Stefanowski, J.; Woźniak, M.; Wó Zniak, M. Ensemble learning for data stream analysis: A survey. *Inf. Fusion* **2017**, *37*, 132–156. [CrossRef]

48. Hosseini, M.J.; Ahmadi, Z.; Beigy, H. New management operations on classifiers pool to track recurring concepts. In *Lecture Notes in Computer Science*; Springer: Berlin/Heidelberg, Germany, 2012; Volume 7448, pp. 327–339.

49. Hosseini, M.J.; Ahmadi, Z.; Beigy, H. Pool and Accuracy Based Stream Classification: A New Ensemble Algorithm on Data Stream Classification Using Recurring Concepts Detection. In Proceedings of the 2011 IEEE 11th International Conference on Data Mining Workshops, Vancouver, BC, Canada, 11 December 2011; pp. 588–595. [CrossRef]

50. Kolter, J.Z.; Maloof, M.A. Dynamic Weighted Majority: An Ensemble Method for Drifting Concepts. *J. Mach. Learn. Res.* **2007**, *8*, 2755–2790.

51. Karnick, M.; Ahiskali, M.; Muhlbaier, M.D.; Polikar, R. Learning concept drift in nonstationary environments using an ensemble of classifiers based approach. In Proceedings of the 2008 IEEE International Joint Conference on Neural Networks (IEEE World Congress on Computational Intelligence), Hong Kong, China, 1–8 June 2008; pp. 3455–3462.

52. Barddal, J.P.; Gomes, H.M.; Enembreck, F.; Pfahringer, B. A survey on feature drift adaptation: Definition, benchmark, challenges and future directions. *J. Syst. Softw.* **2017**, *127*, 278–294. [CrossRef]

53. Gama, J.; Medas, P.; Castillo, G.; Rodrigues, P. Learning with Drift Detection. In *Advances in Artificial 605 Intelligence, Proceedings of the 17th Brazilian Symposium on Artificial Intelligence (SBIA 2004), Sao Luis, Maranhao, Brazil, 29 September–1 Ocotber 2004*; Bazzan, A.L.C., Labidi, S., Eds.; Springer: Berlin/Heidelberg, Germany, 2004; pp. 286–295.

54. Bifet, A. *Classifier concept drift Detection and the Illusion of Progress*; Springer: Cham, Switzerland, 2017; pp. 715–725.

55. Baruah, R.D.; Angelov, P. Evolving fuzzy systems for data streams: A survey. *Wiley Interdiscip. Rev. Data Min. Knowl. Discov.* **2011**, *1*, 461–476. [CrossRef]

56. Sateesh Babu, G.; Suresh, S.; Huang, G.B. Meta-cognitive Neural Network for classification problems in a sequential learning framework. *Neurocomputing* **2011**, *81*, 86–96. [CrossRef]

57. Pratama, M.; Pedrycz, W.; Lughofer, E. Evolving Ensemble Fuzzy Classifier. *IEEE Trans. Fuzzy Syst.* **2018**, *26*, 2552–2567. [CrossRef]

58. Chen, H.; Singal, V. Role of Speculative Short Sales in Price Formation: The Case of the Weekend Effect. *J. Financ.* **2003**, *58*, 685–705. [CrossRef]

59. Bifet, A.; Gavaldà, R. Learning from Time-Changing Data with Adaptive Windowing. In Proceedings of the 2007 SIAM International Conference on Data Mining, Minneapolis, MN, USA, 26–28 April 2007; pp. 443–448.

60. Bifet, A.; Gavaldà, R. Adaptive Learning from Evolving Data Streams. In *Advances in Intelligent Data Analysis VIII, Proceedings of the 8th International Symposium on Intelligent Data Analysis (IDA 2009), Lyon, France, 31 August–2 September 2009*; Adams, N.M., Robardet, C., Siebes, A., Boulicaut, J.F., Eds.; Springer: Berlin/Heidelberg, Germany, 2009; pp. 249–260.

61. Bifet, A.; Holmes, G.; Kirkby, R.; Pfahringer, B. MOA: Massive Online Analysis. *J. Mach. Learn. Res.* **2010**, *11*, 1601–1604.

62. Wilcoxon, F. Individual Comparisons by Ranking Methods. *Biom. Bull.* **1945**, *1*, 80–83. [CrossRef]

63. De Barros, R.S.M.; Hidalgo, J.I.G.; de Lima Cabral, D.R. Wilcoxon Rank Sum Test Drift Detector. *Neurocomputing* **2018**, *275*, 1954–1963. [CrossRef]

64. Rokach, L. Ensemble-based Classifiers. *Artif. Intell. Rev.* **2010**, *33*, 1–39. [CrossRef]

*Article*

# Multi-Modal Deep Hand Sign Language Recognition in Still Images Using Restricted Boltzmann Machine

**Razieh Rastgoo [1,2], Kourosh Kiani [1,*] and Sergio Escalera [2]**

[1] Electrical and Computer Engineering Department, Semnan University, Semnan 3513119111, Iran; rrastgoo@semnan.ac.ir

[2] Department of Mathematics and Informatics, University of de Barcelona and Computer Vision Center, 08007 Barcelona, Spain; sescalera@ub.edu

* Correspondence: kourosh.kiani@semnan.ac.ir; Tel.: +98-912-236-1274

Received: 12 September 2018; Accepted: 3 October 2018; Published: 23 October 2018

**Abstract:** In this paper, a deep learning approach, Restricted Boltzmann Machine (RBM), is used to perform automatic hand sign language recognition from visual data. We evaluate how RBM, as a deep generative model, is capable of generating the distribution of the input data for an enhanced recognition of unseen data. Two modalities, RGB and Depth, are considered in the model input in three forms: original image, cropped image, and noisy cropped image. Five crops of the input image are used and the hand of these cropped images are detected using Convolutional Neural Network (CNN). After that, three types of the detected hand images are generated for each modality and input to RBMs. The outputs of the RBMs for two modalities are fused in another RBM in order to recognize the output sign label of the input image. The proposed multi-modal model is trained on all and part of the American alphabet and digits of four publicly available datasets. We also evaluate the robustness of the proposal against noise. Experimental results show that the proposed multi-modal model, using crops and the RBM fusing methodology, achieves state-of-the-art results on Massey University Gesture Dataset 2012, American Sign Language (ASL). and Fingerspelling Dataset from the University of Surrey's Center for Vision, Speech and Signal Processing, NYU, and ASL Fingerspelling A datasets.

**Keywords:** hand sign language; deep learning; restricted Boltzmann machine (RBM); multi-modal; profoundly deaf; noisy image

## 1. Introduction

Profoundly deaf people have many problems in communicating with other people in society. Due to impairment in hearing and speaking, profoundly deaf people cannot have normal communication with other people. A special language is fundamental in order for profoundly deaf people to be able to communicate with others [1]. In recent years, some projects and studies have been proposed to create or improve smart systems for this population to recognize and detect the sign language from hand and face gestures in visual data. While each method provides different properties, more research is required to provide a complete and accurate model for sign language recognition. Using deep learning approaches has become common for improving the recognition accuracy of sign language models in recent years. In this work, we use a generative deep model, Restricted Boltzmann Machine (RBM), using two visual modalities, RGB and Depth, for automatic sign language recognition. A Convolutional Neural Network (CNN) model, using Faster Region-based Convolutional Neural Network (Faster-RCNN) [2], is applied for hand detection in the input image. Then, our goal is to test how a generative deep model, able to generate data from modeled data distribution probabilities, in combination with different visual modalities, can improve recognition performance of state-of-the-art alternatives for sign language recognition. The contributions of this paper are summarized as follows:

(a)   A generative model, Restricted Boltzmann Machine (RBM), is used for hand sign recognition. We benefit from the generative capabilities of the network and the need for fewer network parameters to achieve better generalization capabilities with fewer input data. Additionally, we show enhanced performance by the fusion of different RBM blocks, each one considering a different visual modality.

(b)   To improve the recognition performance against noise and missing data, our model is enriched with additional data in the form of augmentation based on cropped image regions and noisy regions.

(c)   We evaluate the robustness of the proposed model against different kinds of noise; as well as the effect of the different model hyper-parameters.

(d)   We provide state-of-the-art results on five public sign recognition datasets.

The rest of this paper is organized as follows: Section 2 reviews the related materials and methods as well as the details of the proposed model. Experimental results on four publicly available datasets are presented in Section 3. Finally, Section 4 concludes the work.

## 2. Materials and Methods

### 2.1. Related Work

Sign language recognition has seen a major breakthrough in the field of Computer Vision in recent years [3]. A detailed review of sign language recognition models can be found in [4]. The challenges of developing sign language recognition models range from the image acquisition to the classification process [3]. We present a brief review of some related models of sign language recognition in two categories:

- Deep-based models: In this category, the proposed models use deep learning approaches for accuracy improvement. A profoundly deaf sign language recognition model using the Convolutional Neural Network (CNN) was developed by Garcia and Viesca [5]. Their model classifies correctly some letters of the American alphabet when tested for the first time, and some other letters most of the time. They fine-tuned the GoogLeNet model and trained their model on American Sign Language (ASL) and the Finger Spelling Dataset from the University of Surrey's Center for Vision, Speech, and Signal Processing and Massey University Gesture Dataset 2012 [5]. Koller et al. used Deep Convolutional Neural Network (DCNN) and Hidden-Markov-Model (HMM) to model mouth shapes to recognize sign language. The classification accuracy of their model outperformed state-of-the-art mouth model recognition systems [6]. An RGB ASL Image Dataset (ASLID) and a deep learning-based model were introduced by Gattupalli et al. to improve the pose estimation of the sign language models. They measured the recognition accuracy of two deep learning-based state-of-the-art methods on the provided dataset [7]. Koller et al. proposed a hybrid model, including CNN and Hidden Markov Model (HMM), to handle the sequence data in sign language recognition. They interpreted the output of their model in a Bayesian fashion [8]. Guo et al. suggested a tree-structured Region Ensemble Network (REN) for 3D hand pose estimation by dividing the last convolution outputs of CNN into some grid regions. They achieved state-of-the-art estimation accuracy on three public datasets [9]. Deng et al. designed a 3D CNN for hand pose estimation from a single depth image. This model directly produces the 3D hand pose and does not need further processing. They achieved state-of-the-art estimation accuracy on two public datasets [10]. A model-based deep learning approach has been suggested by Zhou et al. [11]. They used a 3D CNN with a kinematics-based layer to estimate the hand geometric parameters. The report of experimental results of their model shows that they attained state-of-the-art estimation accuracy on some publicly available datasets. A Deep Neural Network (DNN) has been proposed by the LIRIS team of ChaLearn challenge 2014 for hand gesture recognition from two input modalities, RGB and Depth. They achieved the highest accuracy results of the challenge, using early fusion of joint motion features from two input modalities [12]. Koller et al. presented a new approach to classify the input frames using an embedded CNN within an iterative Expectation Maximum (EM) algorithm. The proposed model has been evaluated on over 3000 manually labelled hand shape images of 60 different classes and led to 62.8 top-1 accuracy on the input data [13]. While their model is applied not only for image input but also for frame sequences of a video, there are many rooms to improve the model performance in the case of time and complexity due to using HMMs and the EM algorithm. Guo et al. [14] proposed a simple tree-structured REN for 3D coordinate regression of depth image input. They partitioned the last convolution outputs of ConvNet into several grid regions and integrated the output of fully connected (FC) regressors from regions into another FC layer.

- Non-deep models: In this category, the proposed model does not use deep learning approaches. Philomena and Jasmin suggested a smart system composed of a group of Flex sensors, machine learning and artificial intelligence concepts to recognize hand gestures and show the suitable form of outputs. Unfortunately, this system has been defined as a research project and the experimental results have not been reported [15]. Narayan Sawant designed and implemented an Indian Sign Language recognition system to recognize the 26-character alphabet by using the HSV color model and Principal Component Analysis (PCA) algorithm. In this work, the experimental results have not been reported [16]. Ullah designed a hand gesture recognition system using the Cartesian Genetic Programming (CGP) technique for American Sign Language (ASL). Unfortunately, the designed system is still restricted and slow. Improving the recognition accuracy and learning ability of the suggested system are necessary [17]. Kalsh and Garewal proposed a real-time system for hand sign recognition using different hand shapes. They used the Canny edge detection algorithm and Gray-level images. They selected only six alphabets of ASL and achieved a recognition accuracy of 100 [18]. An Adaptive Neuro-Fuzzy Inference System (ANFIS) was designed to recognize sign language by Wankhade and Zade. They compared the performance of Neural Network, HMM, and Adaptive Neuro-Fuzzy Inference System (ANFIS) for sign language recognition. Based on their experimental results for 35 samples, ANFIS had a higher accuracy than the other methods [19]. Plawiak et al. [20] designed a system for efficient recognition of hand body language based on specialized glove sensors. Their model used Probabilistic Neural Network, Support Vector Machine, and K-Nearest Neighbor algorithms for gesture recognition. The proposed model has been evaluated on data collected from ten people performing 22 hand body languages. While the experimental results show high recognition performance, gestures with low inter-class variability use are miss-classified.

In this work, we propose a deep-based model using RBM to improve sign language recognition accuracy from two input modalities, RGB and Depth. Using three forms of the input images, original, cropped, and noisy cropped, the hands of these images are detected using CNN. While each of these forms for each modality is passed to an RBM, the output of these RBMs are fused in another RBM to recognize the output hand sign language label. Furthermore, we evaluate the noise robustness of the model by generating different test cases, including different types of noise applied to input images. Based on the labels of the input images, some states, including all or parts of the output class labels, are generated. Some of the letters, such as Z and Y, are hardly detected because of the complexities in their signs. In this regard, we generate different states in order to have the freedom to ignore these hardly detected letters in some of the states. We expect that the states that do not include the hardly detected letters or digits have good recognition accuracy. The proposed model is trained on the Massey, ASL dataset at Surrey, NYU, and ASL Fingerspelling A dataset and achieves state-of-the-art results.

### 2.2. Proposed Model

The proposed model includes the following steps:

- Inputs: The original input images are entered into the model in order to extract their features. As Figure 1 shows, we use two modalities, RGB and Depth, in the input images. In the case of one modality in the input images, we use the model illustrated in Figures 2 and 3 for depth and RGB input images.

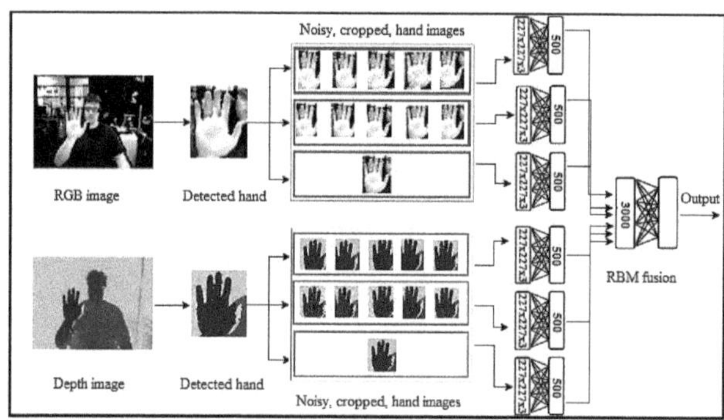

**Figure 1.** The proposed model.

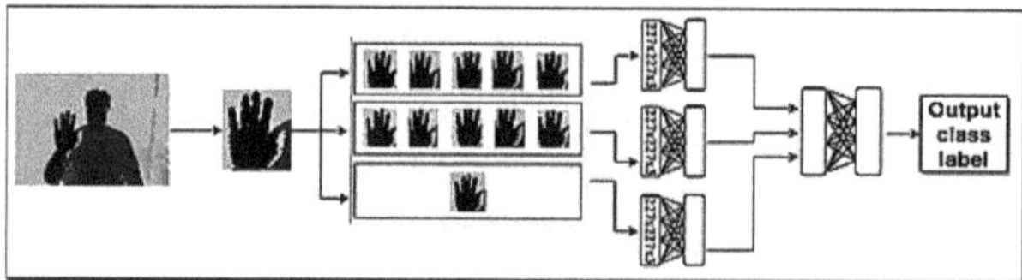

**Figure 2.** The proposed model in the case of using just depth modality in the input.

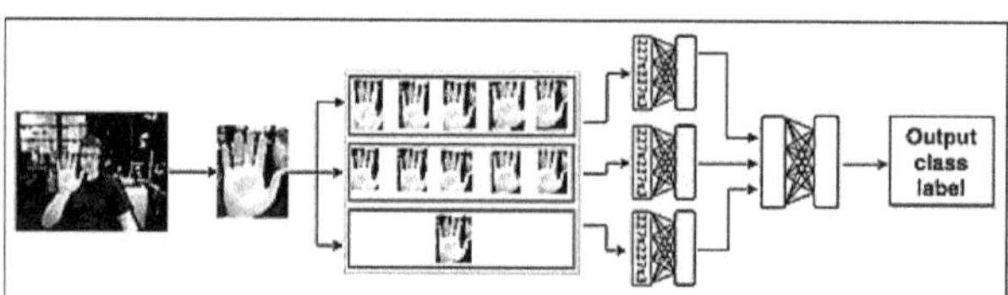

**Figure 3.** The proposed model in the case of using just RGB modality in the input.

- Hand detection: To improve the recognition accuracy of the proposed model, a fine-tuned CNN model, based on the Faster-RCNN model [2], is used to detect hands in the input images.
- Crop the images: The input images are cropped from five regions of the image using a CNN.
- Add noise: To increase the robustness of the proposed model, two types of noise, Gaussian and Salt-and-Pepper, are added to input images of the model.
- Enter into the RBM: In the proposed model, we use not only two modalities, RGB and Depth, but also three forms of input image: an original input image, a five cropped input image, and a five noisy cropped input image. For each model, we use these three forms of input image and send them to the RBM. Six RBMs are used in this step as follows:

  First RBM: The inputs of the first RBM are five RGB noisy cropped images. Each of these five noisy crops is separately input to the RBM.

  Second RBM: Five crops of RGB input image are the inputs of the second RBM.

  Third RBM: Only the original detected hand of the RGB input image is considered as the input of third RBM.

Fourth RBM: Five depth noisy cropped images are separately sent to the fourth RBM.

Fifth RBM: The inputs of the fifth RBM are five depth cropped images.

Sixth RBM: The original depth detected hand is considered as the input of the sixth RBM.

- RBM outputs fusion: We use another RBM for fusing the outputs of six RBMs used in the previous step. The outputs of six RBMs are fused and input into the seventh RBM in order not only to decrease the dimension but also to generate the distribution of data to recognize the final hand sign label. In Figure 1, we show how to use these RBMs in our model.

Details of the mentioned parts of the proposed method are explained in the following sub-sections.

### 2.2.1. Input Image

We use two modalities, RGB and depth, in the input images. In the case that we have only one modality in the input images, we use a part of the model for that input modality. In the proposed multi-modal model, Figure 4, the top part of the model, as seen in Figure 2, is the model for depth inputs and the bottom part, as see in Figure 3, is the model for RGB inputs.

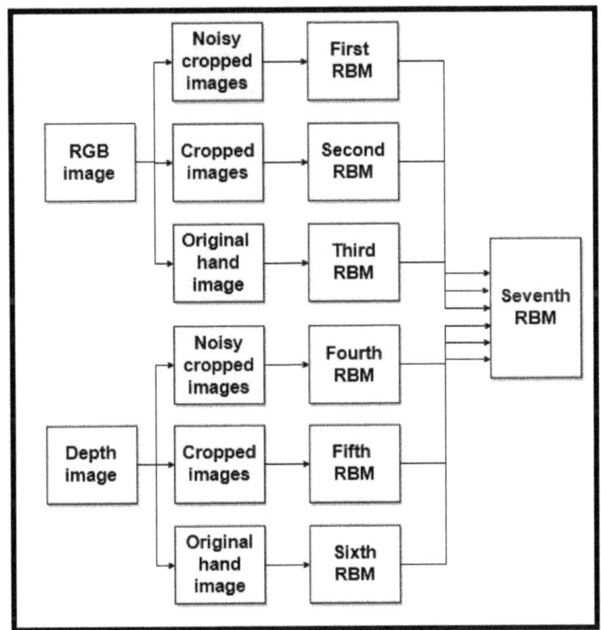


**Figure 4.** Flowchart of the proposed model.

### 2.2.2. Hand Detecting

The hands in the input image are detected using the fine-tuned Faster-RCNN [2]. Faster-RCNN is a fast framework for object detection using CNN. Faster-RCNN network takes an input image and a set of object proposals. The outputs of this network are the real-valued number-encoded refined bounding-box positions for each of the output classes in the network. Faster-RCNN uses a Region Proposal Network (RPN) to share full-image convolutional features with the detection network, which leads to providing approximately cost-free region proposals. RPN is a fully convolutional network that is used to predict the object bounds. Faster-RCNN achieved state-of-the-art object detection accuracy on some public datasets. In addition, Faster-RCNN has a high frame rate detection on very deep networks such as VGG-16. Sharing the convolutional features has led to decreasing the parameters as well as increasing the detection speed in the network. Due to a high speed and low cost in the object detection, we used the Faster-RCNN to detect the hands in the input images.

### 2.2.3. Image Cropping

To increase the accuracy of the proposed method in recognizing the hand sign language under different situations, different crops of input images are used, as Figure 5 shows. Using different crops is helpful for increasing the accuracy of the model in recognizing input images in situations where some parts of the images do not exist or have been destroyed. In addition, by using these crops, the size of the dataset is increased, being beneficial for deep learning approaches. The proposed method is evaluated by using different numbers of crops to select the suitable number of crops. Furthermore, the proposed method is trained not only on the input images without any crops but also on the cropped images. A sample generating different crops of an image is shown in Figure 6.

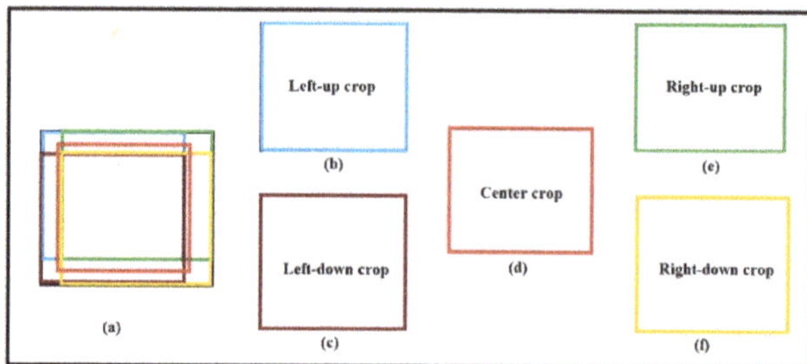


**Figure 5.** Generating different crops of the input image.

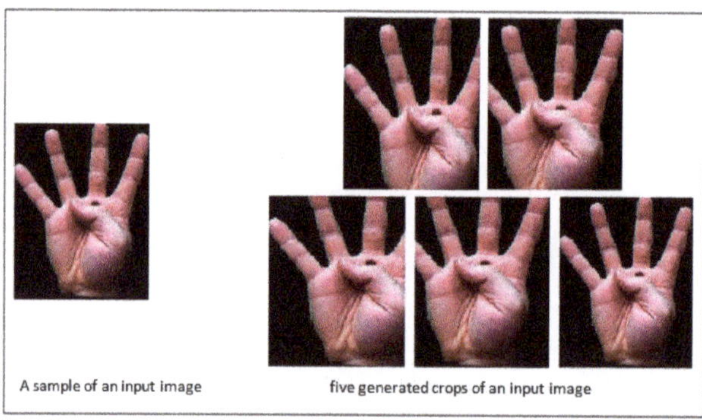


**Figure 6.** A sample image and generated crops.

### 2.2.4. Add Noise

To increase the noise robustness of the proposed method, three types of noise are added to the input images. Figure 7 shows a sample image as well as the applied noises. Gaussian, Gaussian Blur, and Salt-and-Pepper noises are selected due to some beneficial features such as being additive, independent at each pixel, and independent of signal intensity. Four test sets are generated to evaluate the noise robustness of the proposed method as follows:

1.   TSet1: In this test set, Gaussian noise is added to the data.
2.   TSet2: In this test set, Salt-and-Pepper noise is added to the data.
3.   TSet3: In this test set, Gaussian noise is added to one part of data and Salt-and-Pepper noise is added to another part of data.

4.  TSet4: In this test set, Gaussian Blur noise is added to the data.

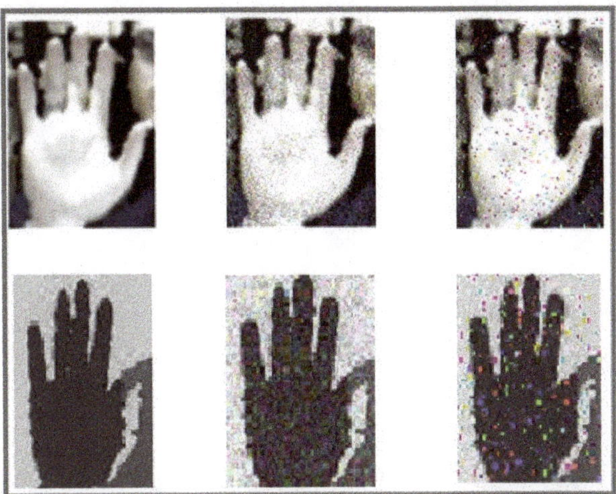

**Figure 7.** A sample image applying different kinds of noise. (**Left column**): original images, (**Internal column**): Gaussian noise, (**Right column**): Salt-and-pepper noise.

### 2.2.5. Entry into the RBM

RBM is an energy-based model that is shown via an undirected graph, as illustrated in Figure 8. RBM is used as a generative model in different types of data and applications to approximate data distribution. The RBM graph contains two layers, namely visible and hidden units. While the units of each layer are independent of each other, they are conditioned on the units of the other layer. RBM can be trained by using the Contrastive Divergence (CD) learning algorithm. To acquire a suitable estimator of the log-likelihood gradient in RBM, Gibbs sampling is used. Suitable adjustment of the parameters of RBM, such as the learning rate, the momentum, the initial values of the weights, and the number of hidden units, plays a very important role in the convergence of the model [21,22].

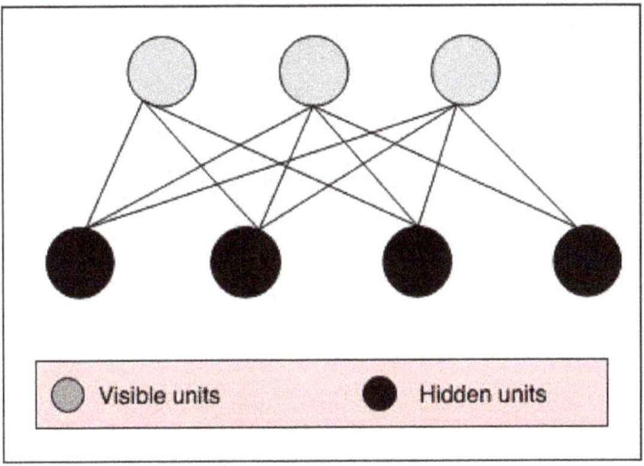


**Figure 8.** RBM network graph.

We are using a reduced set of data where CNN approaches are not able to generalize well. In this case, RBM, a deep learning model with fewer parameters on the generated dataset, can be a good alternative. In the proposed method, we use RBM for hand sign recognition. The achieved results comparing the proposed method with the CNN models shows the outperforming of the RBM model for hand sign recognition on the tested datasets. We use some RBMs in the proposed method for generating the distribution of the input data as well as the recognizing the hand sign label. For each input image modality, we use three RBMs for three forms of input images, which are: original detected hand image, five cropped detected hand images, and five noisy cropped detected hand images. While the input layer of these RBMs includes the size of the $227 \times 227 \times 3$ visible neurons, the hidden layer has 500 neurons. Figure 9 shows the RGB cropped detected hand inputs of one of the RBMs used in the proposed model.

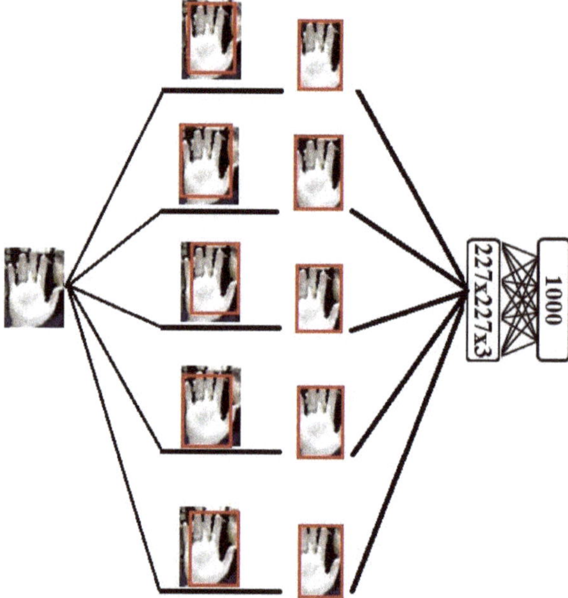

**Figure 9.** The RGB cropped detected hand inputs of one of the RBMs used in the proposed model.

## 2.2.6. Outputs Fusing

The outputs of the RBMs, used for each form of the input image for each input modality, are fused in another RBM for hand sign label recognition, while in the case of having just one modality, RGB or depth, we fused three RBM outputs of three input image forms, and fused six RBM outputs in two-modality inputs. Figure 10 shows the RBM outputs fusing for two-modality inputs of our model.

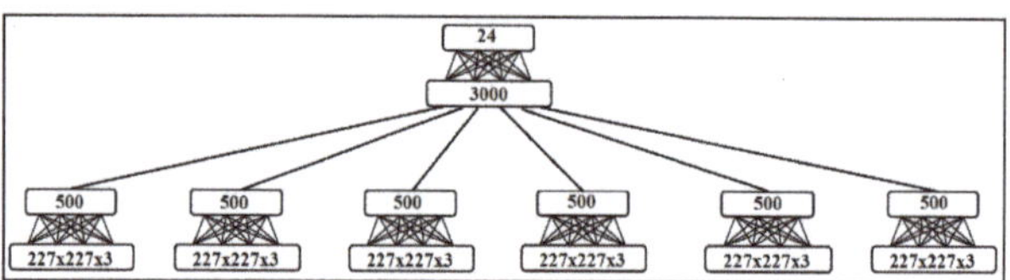

**Figure 10.** RBM outputs fusing in two-modality inputs of our model.

## 3. Results and Discussion

Details of the achieved results of the proposed method on four public datasets are discussed in this section. Results are also compared to state-of-the-art alternatives. Furthermore, we self-compared the proposed model on four used datasets.

### 3.1. Implementation Details

We implemented our model on Intel(R) Xeon(R) CPU E5-2699 (2 processors) with 30 GB RAM on Microsoft Windows 10 operating system and Matlab 2017 software on NVIDIA GPU. Training and test sets are set as defined in the public dataset description for all methods. Five crops of input images are generated and used. We use Stochastic Gradient Descent (SGD) with a mini-batch size of 128. The learning rate starts from 0.005 and is divided by 10 every 1000 epochs. The proposed model is trained for a total of 10,000 epochs. In addition, we use a weight decay of $1 \times 10^{-4}$ and a momentum of 0.92. Our model is trained from scratch with random initialization. To evaluate the noise robustness of our model, we use the Gaussian and Gaussian Blur noise with zero mean and variance equal to 0.16. The noise density parameter of the Salt-and-Pepper noise is 0.13. Details of the used parameters in the proposed method are shown in Table 1.

**Table 1.** Details of the parameters in the proposed method.

| Parameter | Value | Parameter | Value |
|---|---|---|---|
| Theta for Learning | 0.005 | Crop numbers | 5 |
| Weight Decay | $1 \times 10^{-4}$ | Batch-size | 128 |
| Iteration | 100, 1000, 5000, 10,000 | Size of the input image | $227 \times 227 \times 3$ |
| Gaussian Noise Parameters | Mean: 0, Variance: 0.16 | Salt-and-pepper noise parameter | noise density: 0.13 |

### 3.2. Datasets

The ASL Fingerspelling Dataset from the University of Surrey's Center for Vision, Speech and Signal Processing [23], Massey University Gesture Dataset 2012 [24], ASL Fingerspelling A [25], and NYU [26] datasets have been used to evaluate the proposed model. Details of these datasets are shown in Table 2. To show the effect of the background in the achieved results, we used not only the datasets without background but also the datasets including background. Figure 11 shows some samples of the ASL Fingerspelling A dataset.

**Table 2.** Details of four datasets used for the proposed model evaluation.

| Dataset | Language | Class Numbers | Samples | Type |
|---|---|---|---|---|
| Massey | American | 36 | 2524 | Image (RGB) |
| ASL Fingerspelling A | American | 24 | 131,000 | Image (RGB , Depth) |
| NYU | American | 36 | 81,009 | Image (RGB, Depth) |
| ASL Fingerspelling Dataset of the Surrey University | American | 24 | 130,000 | Image (RGB ,Depth) |

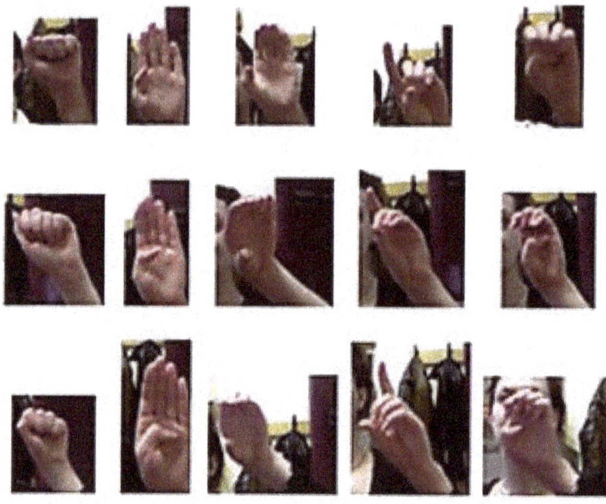

**Figure 11.** Samples of the American Sign Language (ASL) Fingerspelling A dataset.

### 3.3. Parameter Evaluation

Changing some parameters in the proposed method led to different accuracies in the method. Suitable values for the parameters are selected after testing different values for these parameters. Figure 12 shows the effect of changing the learning rate and weight decay parameters in the proposed method. After selecting the best values of the parameters, we fixed and tested the model.

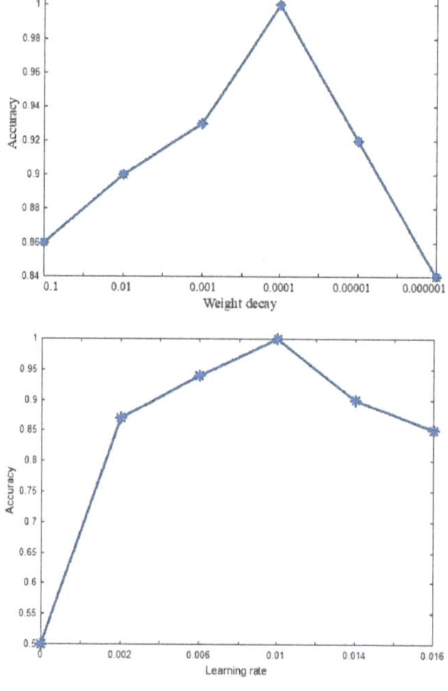

**Figure 12.** Accuracy versus Weight decay and Learning rate parameters.

Using the five crops in the training of the proposed method increases not only the size of the dataset but also the robustness of the method in coping with the missed or destroyed parts of the input images. Selecting the suitable number of the crops was done by testing the different values and analyzing the accuracy of the proposed method on the training data. After testing different numbers of crops, the number five was used. Figure 13 shows the best-achieved accuracy of the proposed method in different crops of input images. As Figure 13 shows, while the accuracy of the proposed method monotonically increases in the crop numbers ranging from 1 to 5, the accuracy is approximately fixed in the higher values of the crop number. Due to decreasing of time and cost complexity, five crop numbers were selected.

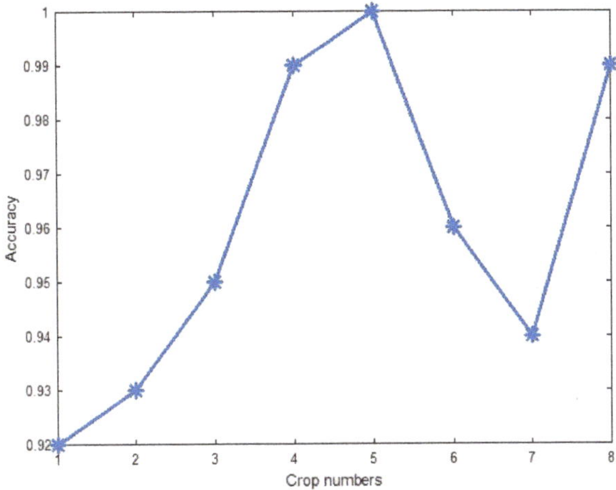

**Figure 13.** Accuracy versus number of crops of the proposed method on the Massey University Gesture Dataset 2012.

## 3.4. Self-Comparison

The proposed model is trained on four public datasets for hand sign recognition. We use two modalities in the input images, RGB and Depth. We used accuracy for model evaluation and comparison, defined as follows:

$$Acc = NT/NT + NF, \tag{1}$$

with $NT$ being the number of the input samples correctly classified and $NF$ the number of input samples miss-classified. Model has a better accuracy on Massey University Gesture Dataset 2012 than the other datasets used for evaluation. This was predictable because this dataset includes only the RGB images without background in the images. The other datasets, ASL Fingerspelling Dataset from the University of Surrey's Center for Vision, Speech and Signal Processing, NYU, and ASL Fingerspelling A, have background in their images. Table 3 shows the results of this comparison. Comparison of the results of the proposed model shows that the recognition accuracy of the proposed model on Massey University Gesture Dataset 2012, with RGB input images, were higher than the other used datasets.

**Table 3.** Recognition accuracy of the proposed model on four datasets.

| Dataset | Recognition Accuracy |
| --- | --- |
| Massey University Gesture Dataset 2012 | **99.31** |
| ASL Fingerspelling Dataset of the Surrey University | 97.56 |
| NYU | 90.01 |
| ASL Fingerspelling A | 98.13 |

*3.5. Evaluating the Robustness to Noise of the Proposed Method*

Four test sets, TSet1, TSet2, TSet3, and TSet4, are generated to evaluate the robustness to noise of the proposed method. Table 4 compares the accuracy of the proposed method in four different states, with the details of the generated test sets being as follows:

1.  TSet1: In this test set, the Gaussian noise with zero mean and variance equal to 0.16 is added.
2.  TSet2: In this test set, the Salt-and-Pepper noise with noise density equal to 0.13 is added.
3.  TSet3: In this test set, the Gaussian noise with zero mean, variance equal to 0.16 is added to one part of data, and Salt-and-Pepper noise with noise density equal to 0.13 is added to another part of data.
4.  TSet4: In this test set, the Gaussian Blur noise with zero mean and variance equal to 0.16 is added.

**Table 4.** Accuracy of the proposed method on four test sets.

| Accuracy of Proposed Method | TS1 | TS2 | TS3 | TS4 |
| --- | --- | --- | --- | --- |
| Massey University Gesture Dataset 2012 | 95.01 | 94.94 | 94.86 | **95.36** |
| ASL Fingerspelling Dataset of the Surrey University | 91.09 | 90.74 | 90.03 | 91.18 |
| NYU | 85.01 | 83.84 | 83.00 | 85.23 |
| ASL Fingerspelling A | 93.84 | 93.33 | 92.93 | 94.04 |

As Table 4 shows, the proposed model achieves higher accuracy on Massey University Gesture Dataset 2012 dataset than with the other used datasets. Due to not having background and occlusion as well as high transparency of the RGB images of this dataset, higher accuracy than the other used datasets with complex background and occlusion in the input images is expected.

*3.6. State-of-the-Art Comparison*

The proposed method is compared with state-of-the-art alternatives in hand sign recognition on four publicly available datasets. Comparison is done under the same conditions of training and testing data partitioning as in previous work, for a fair comparison. As one can observe in Table 5, the proposed model achieves the highest performance in all four datasets.

To evaluate the recognition accuracy of the proposed model for hardly detected characters such as $Z$ and $Y$, we generate three categories from the Massey University Gesture Dataset 2012 in order to compare the proposed method with the model suggested by Garcia et al. [5]. The first category includes all 26 characters. The second category includes only 11 characters and ignores the $Z$ and $Y$. Finally, the third category includes only 11 characters and ignores the $Z$ and $Y$. Details of three categories are as follows:

*   Category1: In this category, two models are trained on alphabets to include *a–y*.
*   Category2: In this category, two models are trained on alphabets to include *a–k*.
*   Category3: In this category, two models are trained on alphabets to include *a–e*.

**Table 5.** State-of-the-art comparison.

| Reference | Result | Dataset |
|---|---|---|
| [9] | 87.00 | |
| [10] | 74.00 | |
| [14] | 84.40 | NYU |
| [27] | 79.40 | |
| **Ours** | **90.01** | |
| [5] | 72.00 | Massey University |
| **Ours** | **99.31** | |
| [25] | 87.00 | ASL Fingerspelling A |
| **Ours** | **98.13** | |
| [9] | 69.00 | ASL Surrey |
| **Ours** | **97.56** | |

The results of the comparison of Top-1 and Top-5 accuracies are shown in Tables 6 and 7. The proposed method significantly outperforms the Garcia and Viesca [5] model in recognition accuracy.

**Table 6.** Comparison of Top-1 accuracy of the proposed method and Garcia [5] model in three considered categories on Massey University Gesture Dataset 2012.

| Top-1 Val Accuracy | Proposed Method | Garcia [5] |
|---|---|---|
| Alphabets [$a$–$y$] | **98.91** | 69.65 |
| Alphabets [$a$–$k$] | **99.03** | 74.30 |
| Alphabets [$a$–$e$] | **99.15** | 97.82 |

**Table 7.** Comparison of Top-5 accuracy of the proposed method and Garcia [5] model in three considered categories on Massey University Gesture Dataset 2012.

| Top-5 Val Accuracy | Proposed Method | Garcia [5] |
|---|---|---|
| Alphabets [$a$–$y$] | **99.31** | 90.76 |
| Alphabets [$a$–$k$] | **99.59** | 89.70 |
| Alphabets [$a$–$e$] | 99.78 | **100** |

## 4. Conclusions

We proposed the use of RBM as a deep generative model for sign language recognition in multi-modal RGB-Depth data. We showed the model to provide a generalization in instances of low amounts of annotated data thanks to the low number of model parameters. We also showed the model to be robust against different kinds of noise present in the data, and benefitting from the fusion of RGB and Depth visual modalities. We achieved state-of-the-art results in five public sign recognition datasets. However, the model shows difficulty recognizing characters with low visual inter-class variability, such as in the case of the high similarity of hand poses for defining $Z$ and $Y$ characters. For future work, we plan to further reduce the complexity of the whole ensemble of RBMs by defining isolated simple RBM models that can share information in early training stages. Furthermore, we plan to extend model behavior to deal with image sequences and model spatio-temporal information of sign gestures.

**Author Contributions:** This work is part of R.R., Ph.D. K.K. and S.E. are work supervisors. Conceptualization, R.R., K.K. and S.E.; Methodology, R.R.; Supervision, K.K. and S.E.; Validation, R.R.; Visualization, R.R.; Writing, R.R.; Review and editing, R.R., K.K. and S.E.

**Funding:** This research received no external funding.

**Acknowledgments:** This work has been partially supported by the Spanish project TIN2016-74946-P (MINECO/FEDER, UE), CERCA Programme/Generalitat de Catalunya, and High Intelligent Solution (HIS) company of Iran. We gratefully acknowledge the support of NVIDIA Corporation with the donation of the Titan XP GPU used for this research.

**Conflicts of Interest:** The authors declare no conflict of interest.

## Abbreviations

The following abbreviations are used in this manuscript:

| | |
|---|---|
| RBM | Restricted Boltzmann Machine |
| CNN | Convolutional Neural Network |
| DCNN | Deep Convolutional Neural Network |
| RPN | Region Proposal Network |
| CD | Contrastive Divergence |
| SGD | Stochastic Gradient Descent |
| EM | Expectation Maximum |
| PCA | Principal Component Analysis |

## References


1. Philomina, S.; Jasmin, M. Hand Talk: Intelligent Sign Language Recognition for Deaf and Dumb. *Int. J. Innov. Res. Sci. Eng. Technol.* **2015**, *4*, doi:10.15680/IJIRSET.2015.0401133. [CrossRef]
2. Ren, S.; He, K.; Girshick, R.; Sun, J. Faster R-CNN: Towards Real-Time Object Detection with Region Proposal Networks. *IEEE Trans. Pattern Anal. Mach. Intell.* **2017**, *39*, 1137–1149. [CrossRef] [PubMed]
3. Suharjito, S.; Anderson, R.; Wiryana, F.; Ariesta, M.C.; Kusuma, G.P. Sign Language Recognition Application Systems for Deaf-Mute People: A Review Based on Input-Process-Output. *Procedia Comput. Sci.* **2017**, *116*, 441–448. [CrossRef]
4. Cheok, M.J.; Omar, Z.; Jaward, M.H. A review of hand gesture and sign language recognition techniques. *Int. J. Mach. Learn. Cybern.* **2017**, 1–23, doi:10.1007/s13042-017-0705-5. [CrossRef]
5. Garcia, B.; Viesca, S. Real-time American Sign Language Recognition with Convolutional Neural Networks. Reports, Stanford University, Stanford, CA, USA, 2016. Available online: http://cs231n.stanford.edu/reports/2016/pdfs/214_Report.pdf (accessed on 5 October 2018).
6. Koller, O.; Ney, H.; Bowden, R. Deep Learning of Mouth Shapes for Sign Language. In Proceedings of the 2015 IEEE International Conference on Computer Vision Workshop (ICCVW), Santiago, Chile, 7–13 December 2015.
7. Gattupalli, S.; Ghaderi, A.; Athitsos, V. Evaluation of Deep Learning based Pose Estimation for Sign Language Recognition. In Proceedings of the 9th ACM International Conference on Technologies Related to Assistive Environments, Corfu, Greece, 29 June–1 July 2016.
8. Koller, O.; Zargaran, O.; Ney, H.; Bowden, R. Deep Sign: Hybrid CNN-HMM for Continuous Sign Language Recognition. In Proceedings of the 2016 British Machine Vision Conference (BMVC), York, UK, 19–22 September 2016; pp. 3793–3802.
9. Guo, H.; Wang, G.; Chen, X.; Zhang, C.; Qiao, F.; Yang, H. Region ensemble network: Improving convolutional network for hand pose estimation. In Proceedings of the 2017 IEEE International Conference on Image Processing (ICIP), Beijing, China, 17–20 September 2017; pp. 4512–4516.
10. Deng, X.; Yang, S.; Zhang, Y.; Tan, P.; Chang, L.; Wang, H. Hand3D: Hand Pose Estimation using 3D Neural Network. *arXiv* **2017**, arXiv:1704.02224.
11. Zhou, X.; Wan, Q.; Zhang, W.; Xue, X.; Wei, Y. Model-based Deep Hand Pose Estimation. *arXiv* **2016**, arXiv:1606.06854v1.
12. Escalera, S.; Baró, X.; Gonzalez, J.; Bautista, M.A.; Madadi, M.; Reyes, M.; Ponce-López, V.; Escalante, H.J.; Shotton, J.; Guyon, I. ChaLearn Looking at People Challenge 2014: Dataset and Results. In *ECCV 2014: Computer Vision—ECCV 2014 Workshops*; Springer: Cham, Switzerland, 2014.
13. Koller, O.; Ney, H.; Bowden, R. Deep Hand: How to Train a CNN on 1 Million Hand Images When Your Data Is Continuous and Weakly Labelled. In Proceedings of the 29th IEEE Conference on Computer Vision and Pattern Recognition (CVPR), Las Vegas, NV, USA, 26 June–1 July 2016.

14. Guo, H.; Wang, G.; Chen, X.; Zhang, C. Towards Good Practices for Deep 3D Hand Pose Estimation. *arXiv* **2017**, arXiv:1707.07248v1.

15. Kumar, P.; Gauba, H.; Roy, P.P.; Dogra, D.P. Coupled HMM-based multi-sensor data fusion for sign language recognition. *Pattern Recognit. Lett.* **2017**, *86*, 1–8. [CrossRef]

16. Sawant, S.N. Sign Language Recognition System to aid Deaf-dumb People Using PCA. *Int. J. Comput. Sci. Eng. Technol. (IJCSET)* **2014**, *5*, 570–574.

17. Ullah, F. American Sign Language Recognition System for Hearing Impaired People Using Cartesian Genetic Programming. In Proceedings of the 5th International Conference on Automation, Robotics and Applications, Wellington, New Zealand, 6–8 December 2011.

18. Kalsh, E.A.; Garewal, N.S. Sign Language Recognition System. *Int. J. Comput. Eng. Res.* **2013**, *3*, 15–21.

19. Wankhade, M.K.A; Zade, G.N. Sign Language Recognition For Deaf And Dumb People Using ANFIS. *Int. J. Sci. Eng. Technol. Res. (IJSETR)* **2014**, *3*, 1206–1210.

20. Pławiak, P.; Sośnicki, T.; Niedźwiecki, M.; Tabor, Z.; Rzecki, K. Hand Body Language Gesture Recognition Based on Signals From Specialized Glove and Machine Learning Algorithms. *IEEE Trans. Ind. Inform.* **2016**, *12*, 1104–1113. [CrossRef]

21. Hinton, G.E. A Practical Guide to Training Restricted Boltzmann Machines. In *Neural Networks: Tricks of the Trade*; Montavon, G., Orr, G.B., Müller, K.R., Eds.; (Lecture Notes in Computer Science); Springer: Berlin/Heidelberg, Germany, 2012; Volume 7700, pp. 599–619.

22. Jampani, V. Learning Inference Models for Computer Vision. *arXiv* **2017**, arXiv:1709.00069v1.

23. Ong, E.J.; Cooper, H.; Pugeault, N.; Bowden, R. Sign Language Recognition using Sequential Pattern Trees. In Proceedings of the IEEE Computer Society Conference on Computer Vision and Pattern Recognition, Providence, RI, USA, 16–21 June 2012.

24. Barczak, A.L.C.; Reyes, N.H.; Abastillas, M.; Piccio, A.; Susnjak, T. A New 2D Static Hand Gesture Colour Image Dataset for ASL Gestures. *Res. Lett. Inf. Math. Sci.* **2011**, *15*, 12–20.

25. Pugeault, N.; Bowden, R. Spelling It Out: Real-Time ASL Fingerspelling Recognition. In Proceedings of the 1st IEEE Workshop on Consumer Depth Cameras for Computer Vision, jointly with ICCV, Barcelona, Spain, 6–13 November 2011.

26. Tompson, J.; Stein, M.; Lecun, Y.; Perlin, K. Real-time continuous pose recovery of human hands using convolutional networks. *ACM Trans. Graph. (ToG)* **2014**, *33*. [CrossRef]

27. Yuan, S.; Ye, Q.; Stenger, B.; Jain, S.; Kim, T.K. BigHand2.2M Benchmark: Hand Pose Dataset and State of the Art Analysis. *arXiv* **2014**, arXiv:1704.02612.

*Article*

# Multi-Objective Evolutionary Rule-Based Classification with Categorical Data

**Fernando Jiménez** [1,*], **Carlos Martínez** [1], **Luis Miralles-Pechuán** [2] and **Gracia Sánchez** [1] and **Guido Sciavicco** [3]

[1]  Department of Information and Communication Engineering, University of Murcia, 30071 Murcia, Spain; carlos.martinez6@um.es (C.M.); gracia@um.es (G.S.)

[2]  Centre for Applied Data Analytics Research (CeADAR), University College Dublin, D04 Dublin 4, Ireland; luis.miralles@ucd.ie

[3]  Department of Mathematics and Computer Science, University of Ferrara, 44121 Ferrara, Italy; scvgdu@unife.it

*  Correspondence: fernan@um.es; Tel.: +34-868-884630

Received: 30 July 2018; Accepted: 6 September 2018; Published: 7 September 2018

**Abstract:** The ease of interpretation of a classification model is essential for the task of validating it. Sometimes it is required to clearly explain the classification process of a model's predictions. Models which are inherently easier to interpret can be effortlessly related to the context of the problem, and their predictions can be, if necessary, ethically and legally evaluated. In this paper, we propose a novel method to generate rule-based classifiers from categorical data that can be readily interpreted. Classifiers are generated using a multi-objective optimization approach focusing on two main objectives: maximizing the performance of the learned classifier and minimizing its number of rules. The multi-objective evolutionary algorithms *ENORA* and *NSGA-II* have been adapted to optimize the performance of the classifier based on three different machine learning metrics: accuracy, area under the *ROC* curve, and root mean square error. We have extensively compared the generated classifiers using our proposed method with classifiers generated using classical methods such as *PART*, *JRip*, *OneR* and *ZeroR*. The experiments have been conducted in full training mode, in 10-fold cross-validation mode, and in train/test splitting mode. To make results reproducible, we have used the well-known and publicly available datasets *Breast Cancer*, *Monk's Problem 2*, *Tic-Tac-Toe-Endgame*, *Car*, *kr-vs-kp* and *Nursery*. After performing an exhaustive statistical test on our results, we conclude that the proposed method is able to generate highly accurate and easy to interpret classification models.

**Keywords:** multi-objective evolutionary algorithms; rule-based classifiers; interpretable machine learning; categorical data

## 1. Introduction

*Supervised Learning* is the branch of *Machine Learning* (*ML*) [1] focused on modeling the behavior of systems that can be found in the environment. Supervised models are created from a set of past records, each one of which, usually, consists of an input vector labeled with an output. A supervised model is an algorithm that simulates the function that maps inputs with outputs [2]. The best models are those that predict the output of new inputs in the most accurate way. Thanks to modern computing capabilities, and to the digitization of ever-increasing quantities of data, nowadays, supervised learning techniques play a leading role in many applications. The first classification systems date back to the 1990s; in those days, researchers were focused on both precision and interpretability, and the systems to be modeled were relatively simple. Years later, when it became necessary to model more difficult behaviors, the researchers focused on developing more and more precise models, leaving aside the interpretability. *Artificial Neural Networks (ANN)* [3], and, more recently, *Deep Learning Neural Networks (DLNN)* [4], as well as *Support Vector Machines (SVM)* [5], and *Instance-based Learning (IBL)* [6] are archetypical examples of this approach. A *DLNN*, for example, is a large mesh of ordered nodes arranged in a hierarchical manner and composed of a huge number of variables. *DLNNs* are capable of modeling very complex behaviors, but it is extremely difficult to understand the logic behind their predictions, and similar considerations can be drawn for *SVNs* and *IBLs*, although the underlying principles are different. These models are known as *black-box* methods. While there are applications in which knowing the ratio behind a prediction is not necessarily relevant, (e.g., predicting a currency's future value, whether or not a user clicks on an advert or the amount of rain in a certain area), there are other situations where the interpretability of a model plays a key role.

The *interpretability* of classification systems refers to the ability they have to explain their behavior in a way that is easily understandable by a user [7]. In other words, a model is considered interpretable when a human is able to understand the logic behind its prediction. In this way, Interpretable classification models allow external validation by an expert. Additionally, there are certain disciplines such as medicine, where it is essential to provide information about decision making for ethical and human reasons. Likewise, when a public institution asks an authority for permission to investigate an alleged offender, or when the CEO of a certain company wants to take a difficult decision which can seriously change the direction of the company, some kind of explanations to justify these decisions may be required. In these situations, using transparent (also called grey-box) models is recommended. While there is a general consensus on how the performance of a classification system is measured (popular metrics include *accuracy, area under the ROC curve,* and *root mean square error*), there is no universally accepted metric to measure the interpretability of the models. Nor is there an ideal balance between the interpretability and performance of classification systems but this depends on the specific application domain. However, the rule of thumb says that the simpler a classification system is, the easier it is to interpret. *Rule-based Classifiers (RBC)* [8,9] are among the most popular interpretable models, and some authors define the degree of interpretability of an *RBC* as the number of its rules or as the number of axioms that the rules have. These metrics tend to reward models with fewer rules as simple as possible [10,11]. In general, *RBCs* are classification learning systems that achieve a high level of interpretability because they are based on a human-like logic. Rules follow a very simple schema:

*IF (Condition 1) and (Condition 2) and ... (Condition N) THEN (Statement)*

and the fewer rules the models have and the fewer conditions and attributes the rules have, the easier it will be for a human to understand the logic behind each classification. In fact, *RBCs* are so natural in some applications that they are used to interpret other classification models such as *Decision Trees (DT)* [12]. *RBCs* constitute the basis of more complex classification systems based on fuzzy logic [13] such as *LogitBoost* or *AdaBoost* [14].

Our approach investigates the conflict between accuracy and interpretability as a *multi-objective optimization problem*. We define a solution as a set of rules (that is, a classifier), and establish two objectives to be maximized: interpretability and accuracy. We decided to solve this problem by applying *multi-objective evolutionary algorithms (MOEA)* [15,16] as meta-heuristics, and, in particular, two known algorithms: *NSGA-II* [15] and *ENORA* [17]. They are both state-of-the-art evolutionary algorithms which have been applied, and compared, on several occasions [18–20]. *NSGA-II* is very well-known and has the advantage of being available in many implementations, while *ENORA* generally has a higher performance. In the current literature, *MOEAs* are mainly used for learning *RBCs* based on fuzzy logic [18,21–26]. However, *Fuzzy RBCs* are designed for numerical data, from which fuzzy sets are constructed and represented by linguistic labels. In this paper, on the contrary, we are interested in *RBCs* for categorical data, for which a novel approach is necessary.

This paper is organized as follows. In Section 2, we introduce multi-objective constrained optimization, the evolutionary algorithms *ENORA* and *NSGA-II*, and the well-known rule-based classifier learning systems *PART, JRip, OneR* and *ZeroR*. In Section 3, we describe the structure of an *RBC* for categorical data, and we propose the use of multi-objective optimization for the task of learning a classifier. In Section 4, we show the result of our experiments, performed on the well-known publicly accessible datasets *Breast Cancer, Monk's Problem 2, Tic-Tac-Toe-Endgame, Car, kr-vs-kp* and *Nursery*. The experiments allow a comparison among the performance of the classifiers learned by our technique against those of classifiers learned by *PART, JRip, OneR* and *ZeroR*, as well as a comparison between *ENORA* and *NSGA-II* for the purposes of this task. In Section 5, the results are analyzed and discussed, before concluding in Section 6. Appendices A and B show the tables of the statistical tests results. Appendix C shows the symbols and the nomenclature used in the paper.

## 2. Background

### 2.1. Multi-Objective Constrained Optimization

The term *optimization* [27] refers to the selection of the best element, with regard to some criteria, from a set of alternative elements. *Mathematical programming* [28] deals with the theory, algorithms, methods and techniques to represent and solve optimization problems. In this paper, we are interested in a class of mathematical programming problems called *multi-objective constrained optimization problems* [29], which can be formally defined, for $l$ objectives and $m$ constraints, as follows:

$$
\begin{aligned}
Min./Max. \quad & f_i(\mathbf{x}), \qquad i = 1, \ldots, l \\
subject\ to \quad & g_j(\mathbf{x}) \leq 0, \quad j = 1, \ldots, m
\end{aligned}
\tag{1}
$$

where $f_i(\mathbf{x})$ (usually called *objectives*) and $g_j(\mathbf{x})$ are arbitrary functions. Optimization problems can be naturally separated into two categories: those with discrete variables, which we call *combinatorial*, and those with continuous variables. In combinatorial problems, we are looking for objects from a finite, or countably infinite, set $\mathcal{X}$, where objects are typically integers, sets, permutations, or graphs. In problems with continuous variables, instead, we look for real parameters belonging to some continuous domain. In Equation (1), $\mathbf{x} = \{x_1, x_2, \ldots, x_w\} \in \mathcal{X}^w$ represents the set of decision variables, where $\mathcal{X}$ is the domain for each variable $x_k, k = 1, \ldots, w$.

Now, let $\mathcal{F} = \{\mathbf{x} \in \mathcal{X}^w \mid g_j(\mathbf{x}) \leq 0, j = 1, \ldots, m\}$ be the set of all feasible solutions to Equation (1). We want to find a subset of solutions $\mathcal{S} \subseteq \mathcal{F}$ called *non-dominated set* (or *Pareto optimal set*). A solution $\mathbf{x} \in \mathcal{F}$ is *non-dominated* if there is no other solution $\mathbf{x}' \in \mathcal{F}$ that dominates $\mathbf{x}$, and a solution $\mathbf{x}'$ *dominates* $\mathbf{x}$ if and only if there exists $i$ $(1 \leq i \leq l)$ such that $f_i(\mathbf{x}')$ improves $f_i(\mathbf{x})$, and for every $i$ $(1 \leq i \leq l)$, $f_i(\mathbf{x})$ does not improve $f_i(\mathbf{x}')$. In other words, $\mathbf{x}'$ *dominates* $\mathbf{x}$ if and only if $\mathbf{x}'$ is better than $\mathbf{x}$ for at least one objective, and not worse than $\mathbf{x}$ for any other objective. The set $\mathcal{S}$ of non dominated solutions of Equation (1) can be formally defined as:

$$
\mathcal{S} = \{\mathbf{x} \in \mathcal{F} \mid \nexists \mathbf{x}'(\mathbf{x}' \in \mathcal{F} \wedge \mathcal{D}(\mathbf{x}', \mathbf{x}))\}
$$

where:

$$
\mathcal{D}(\mathbf{x}', \mathbf{x}) = \exists i(1 \leq i \leq l, f_i(\mathbf{x}') < f_i(\mathbf{x})) \wedge \forall i(1 \leq i \leq l, f_i(\mathbf{x}') \leq f_i(\mathbf{x})).
$$

Once the set of optimal solutions is available, the most satisfactory one can be chosen by applying a preference criterion. When all the functions $f_i$ are linear, then the problem is a *linear programming problem* [30], which is the classical mathematical programming problem and for which extremely efficient algorithms to obtain the optimal solution exist (e.g., the *simplex method* [31]). When any of the functions $f_i$ is non-linear then we have a *non-linear programming problem* [32]. A non-linear programming problem in which the objectives are arbitrary functions is, in general, intractable. In principle, any search algorithm can be used to solve combinatorial optimization problems, although it is not guaranteed that they will find an optimal solution. *Metaheuristics* methods such as *evolutionary algorithms* [33] are typically used to find approximate solutions for complex multi-objective optimization problems, including feature selection and fuzzy classification.

### 2.2. The Multi-Objective Evolutionary Algorithms ENORA and NSGA-II

The *MOEA ENORA* [17] and *NSGA-II* [15] use a $(\mu + \lambda)$ strategy (Algorithm 1) with $\mu = \lambda = popsize$, where $\mu$ corresponds to the number of parents and $\lambda$ refers to the number of children (*popsize* is the population size), with *binary tournament selection* (Algorithm 2) and a rank function based on Pareto fronts and *crowding* (Algorithms 3 and 4). The difference between *NSGA-II* and *ENORA* is how the calculation of the ranking of the individuals in the population is performed. In *ENORA*, each individual belongs to a slot (as established in [34]) of the objective search space, and the rank of an individual in a population is the non-domination level of the individual in its slot. On the other hand, in *NSGA-II*, the rank of an individual in a population is the non-domination level of the individual in the whole population. Both *ENORA* and *NSGA-II* MOEAs use the same non-dominated sorting algorithm, the *fast non-dominated sorting* [35]. It compares each solution with the rest of the solutions and stores the results so as to avoid duplicate comparisons between every pair of solutions. For a problem with $l$ objectives and a population with $N$ solutions, this method needs to conduct $l \cdot N \cdot (N - 1)$ objective comparisons, which means that it has a time complexity of $O(l \cdot N^2)$ [36]. However, *ENORA* distributes the population in $N$ slots (in the best case), therefore, the time complexity of *ENORA* is $O(l \cdot N^2)$ in the worst case and $O(l \cdot N)$ in the best case.

---

**Algorithm 1** $(\mu + \lambda)$ strategy for multi-objective optimization.

---

**Require:** $T > 1$ {Number of generations}
**Require:** $N > 1$ {Number of individuals in the population}
 1: Initialize $P$ with $N$ individuals
 2: Evaluate all individuals of $P$
 3: $t \leftarrow 0$
 4: **while** $t < T$ **do**
 5:    $Q \leftarrow \varnothing$
 6:    $i \leftarrow 0$
 7:    **while** $i < N$ **do**
 8:       $Parent1 \leftarrow$ Binary tournament selection from $P$
 9:       $Parent2 \leftarrow$ Binary tournament selection from $P$
 10:       $Child1, Child2 \leftarrow Crossover(Parent1, Parent2)$
 11:       $Offspring1 \leftarrow Mutation(Child1)$
 12:       $Offspring2 \leftarrow Mutation(Child2)$
 13:       Evaluate $Offspring1$
 14:       Evaluate $Offspring2$
 15:       $Q \leftarrow Q \cup \{Offspring1, Offspring2\}$
 16:       $i \leftarrow i + 2$
 17:    **end while**
 18:    $R \leftarrow P \cup Q$
 19:    $P \leftarrow N$ best individuals from $R$ according to the *rank-crowding* function in population $R$
 20:    $t \leftarrow t + 1$
 21: **end while**
 22: **return** Non-dominated individuals from $P$

---

---

**Algorithm 2** Binary tournament selection.

---

**Require:** $P$ {Population}

1: $I \leftarrow$ Random selection from $P$
2: $J \leftarrow$ Random selection from $P$
3: **if** $I$ is better than $J$ according to the *rank-crowding* function in population $P$ **then**
4:    **return** $I$
5: **else**
6:    **return** $J$
7: **end if**

---


---

**Algorithm 3** Rank-crowding function.

---

**Require:** $P$ {Population}
**Require:** $I, J$ {Individuals to compare}

1: **if** $rank\,(P, I) < rank\,(P, J)$ **then**
2:    **return** *True*
3: **end if**
4: **if** $rank(P, J) < rank(P, I)$ **then**
5:    **return** *False*
6: **end if**
7: **return** $Crowding\_distance\,(P, I) > Crowding\_distance\,(P, J)$

---

The main reason *ENORA* and *NSGA-II* behave differently is as follows. *NSGA-II* never selects the individual dominated by the other in the binary tournament, while, in *ENORA*, the individual dominated by the other may be the winner of the tournament. Figure 1 shows this behavior graphically. For example, if individuals $B$ and $C$ are selected for a binary tournament with *NSGA-II*, individual $B$ beats $C$ because $B$ dominates $C$. Conversely, individual $C$ beats $B$ with *ENORA* because individual $C$ has a better rank in his slot than individual $B$. In this way, *ENORA* allows the individuals in each slot to evolve towards the Pareto front encouraging diversity. Even though in *ENORA* the individuals of each slot may not be the best of the total individuals, this approach generates a better hypervolume than that of *NSGA-II* throughout the evolution process.

*ENORA* is our *MOEA*, on which we are intensively working over the last decade. We have applied *ENORA* to constrained real-parameter optimization [17], fuzzy optimization [37], fuzzy classification [18], feature selection for classification [19] and feature selection for regression [34]. In this paper, we apply it to rule-based classification. *NSGA-II* algorithm was designed by Deb et al. and has been proved to be a very powerful and fast algorithm in multi-objective optimization contexts of all kinds. Most researchers in multi-objective evolutionary computation use *NSGA-II* as a baseline to compare the performance of their own algorithms. Although *NSGA-II* was developed in 2002 and remains a state-of-the-art algorithm, it is still a challenge to improve on it. There is a recently updated improved version for *many-objective optimization* problems called *NSGA-III* [38].

---

**Algorithm 4** Crowding_distance function.

---

**Require:** $P$ {Population}

**Require:** $I$ {Individual}

**Require:** $l$ {Number of objectives}

1: **for** $j = 1$ to $l$ **do**

2:     $f_j^{max} \leftarrow \max_{I \in P}\{f_j^I\}$

3:     $f_j^{min} \leftarrow \min_{I \in P}\{f_j^I\}$

4:     $f_j^{sup_j^I} \leftarrow$ value of the $j$th objective for the individual higher adjacent in the $j$th objective to the individual $I$

5:     $f_j^{inf_j^I} \leftarrow$ value of the $j$th objective for the individual lower adjacent in the $j$th objective to the individual $I$

6: **end for**

7: **for** $j = 1$ to $l$ **do**

8:     **if** $f_j^I = f_j^{max}$ or $f_j^I = f_j^{min}$ **then**

9:       **return** $\infty$

10:     **end if**

11: **end for**

12: $CD \leftarrow 0.0$

13: **for** $j = 1$ to $l$ **do**

14:     $CD \leftarrow CD + \dfrac{f_j^{sup_j^I} - f_j^{inf_j^I}}{f_j^{max} - f_j^{min}}$

15: **end for**

16: **return** $CD$

---

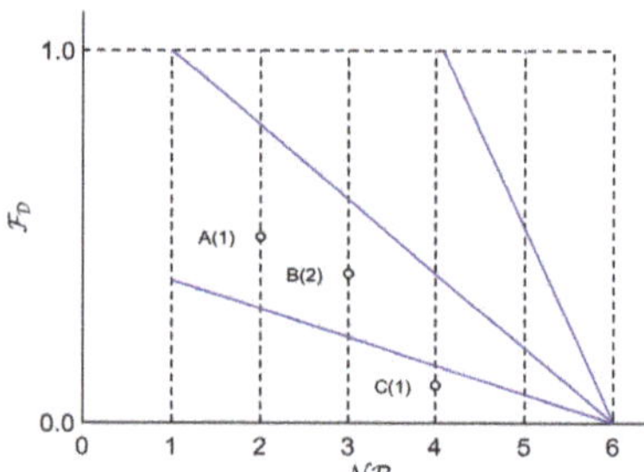
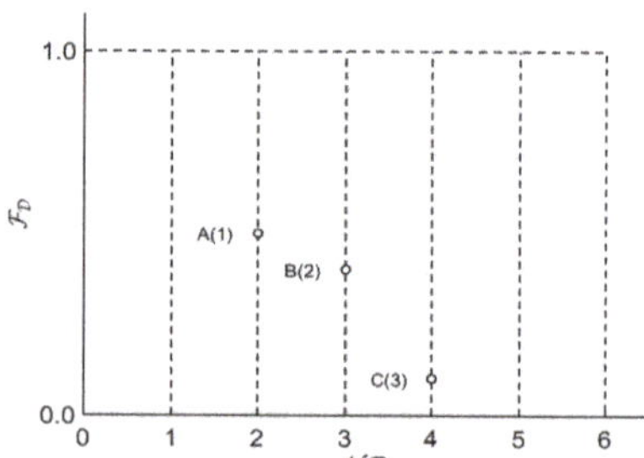


**Figure 1.** Rank assignment of individuals with *ENORA* vs. *NSGA-II*.

### 2.3. PART

*PART* (*Partial DT Method* [39]) is a widely used rule learning algorithm that was developed at the University of Waikato in New Zealand [40]. Experiments show that it is a very efficient algorithm in terms of both computational performance and results. *PART* combines the divide-and-conquer strategy typical of decision tree learning with the separate-and-conquer strategy [41] typical of rule learning, as follows. A decision tree is first constructed (using *C4.5* algorithm [42]), and the leaf with the highest coverage is converted into a rule. Then, the set of instances that are covered by that rule are discarded, and the process starts over. The result is an ordered set of rules, completed by a *default* rule that applies to instances that do not meet any previous rule.

### 2.4. JRip

*JRip* is a fast and optimized implementation in *Weka* of the famous *RIPPER* (*Repeated Incremental Pruning to Produce Error Reduction*) algorithm [43]. *RIPPER* was proposed in [44] as a more efficient version of the incrementally reduced error pruning (*IREP*) rule learner developed in [45]. *IREP* and *RIPPER* work in a similar manner. They begin with a default rule and, using a training dataset, attempt to learn rules that predict exceptions to the default. Each rule learned is a conjunction of propositional literals. Each literal corresponds to a split of the data based on the value of a single feature. This family of algorithms, similar to decision trees, has the advantage of being easy to interpret, and experiments show that *JRip* is particularly efficient in large datasets. *RIPPER* and *IREP* use a strategy based on the separate-and-conquer method to generate an ordered set of rules that are extracted directly from the dataset. The classes are examined one by one, prioritizing those that have more elements. These algorithms are based on four basic steps (growing, pruning, optimizing and selecting) applied repetitively to each class until a stopping condition is met [44]. These steps can be summarized as follows. In the growing phase, rules are created taking into account an increasing number of predictors until the stopping criterion is satisfied (in the *Weka* implementation, the procedure selects the condition with the highest information gain). In the pruning phase redundancy is eliminated and long rules are reduced. In the optimization phase, the rules generated in the previous steps are improved (if possible) by adding new attributes or by adding new rules. Finally, in the selection phase, the best rules are selected and the others discarded.

### 2.5. OneR

*OneR* (*One Rule*) is a very simple, while reasonably accurate, classifier based on a frequency table. First, *OneR* generates a set of rules for each attribute of the dataset, and, then, it selects only one rule from that set—the one with the lowest error rate [46]. The set of rules is created using a frequency table constructed for each predictor of the class, and numerical classes are converted into categorical values.

### 2.6. ZeroR

Finally, *ZeroR* (*Zero Rules* [40]) is a classifier learner that does not create any rules and uses no attributes. *ZeroR* simply creates the class classification table by selecting the most frequent value. Such a classifier is obviously the simplest possible one, and its capabilities are limited to the prediction of the majority class. In the literature, it is not used for practical classifications tasks, but as a generic reference to measure the performance of other classifiers.

## 3. Multi-Objective Optimization for Categorical Rule-Based Classification

In this section, we propose a general schema for an *RBC* specifically designed for categorical data. Then, we propose and describe a multi-objective optimization solution to obtain optimal categorical *RBCs*.

### 3.1. Rule-Based Classification for Categorical Data

Let $\Gamma$ be a classifier composed by $M$ rules, where each rule $R_i^\Gamma$, $i = 1, \ldots, M$, has the following structure:

$$R_i^\Gamma: \quad IF \quad x_1 = b_{i1}^\Gamma \quad AND \quad , \ldots, \quad AND \quad x_p = b_{ip}^\Gamma \quad THEN \quad y = c_i^\Gamma \tag{2}$$

where for $j = 1, \ldots, p$ the attribute $b_{ij}^\Gamma$ (called *antecedent*) takes values in a set $\{1, \ldots, v_j\}$ $(v_j > 1)$, and $c_i^\Gamma$ (called *consequent*) takes values in $\{1, \ldots, w\}$ $(w > 1)$. Now, let $\mathbf{x} = \{x_1, \ldots, x_p\}$ be an observed example, with $x_j \in \{1, \ldots, v_j\}$, for each $j = 1, \ldots, p$. We propose *maximum matching* as *reasoning method*, where the *compatibility degree* of the rule $R_i^\Gamma$ for the example **x** (denoted by $\varphi_i^\Gamma(\mathbf{x})$) is calculated as the number of attributes whose value coincides with that of the corresponding antecedent in $R_i^\Gamma$, that is

$$\varphi_i^\Gamma(\mathbf{x}) = \sum_{j=1}^{p} \mu_{ij}^\Gamma(\mathbf{x})$$

where:

$$\mu_{ij}^\Gamma(\mathbf{x}) = \begin{cases} 1 & if\ x_j = b_{ij}^\Gamma \\ 0 & if\ x_j \neq b_{ij}^\Gamma \end{cases}$$

The *association degree* for the example **x** with a class $c \in \{1, \ldots, w\}$ is computed by adding the compatibility degrees for the example **x** of each rule $R_i^\Gamma$ whose consequent $c_i^\Gamma$ is equal to class $c$, that is:

$$\lambda_c^\Gamma(\mathbf{x}) = \sum_{i=1}^{M} \eta_{ic}^\Gamma(\mathbf{x})$$

where:

$$\eta_{ic}^\Gamma(\mathbf{x}) = \begin{cases} \varphi_i^\Gamma(\mathbf{x}) & if\ c = c_i^\Gamma \\ 0 & if\ c \neq c_i^\Gamma \end{cases}$$

Therefore, the *classification* (or output) of the classifier $\Gamma$ for the example **x** corresponds to the class whose association degree is maximum, that is:

$$f^\Gamma(\mathbf{x}) = \arg_c \max_{c=1}^{w} \lambda_c^\Gamma(\mathbf{x})$$

### 3.2. A Multi-Objective Optimization Solution

Let $\mathcal{D}$ be a dataset of $K$ instances with $p$ categorical input attributes, $p > 0$, and a categorical output attribute. Each input attribute $j$ can take a category $x_j \in \{1, \ldots, v_j\}$, $v_j > 1$, $j = 1, \ldots, p$, and the output attribute can take a class $c \in \{1, \ldots, w\}$, $w > 1$. The problem of finding an optimal classifier $\Gamma$, as described in the previous section, can be formulated as an instance of the multi-objective constrained problem in Equation (1) with two objectives and two constraints:

$$\begin{aligned} Max./Min. & \quad \mathcal{F}_\mathcal{D}(\Gamma) \\ Min. & \quad \mathcal{NR}(\Gamma) \\ subject\ to: & \quad \mathcal{NR}(\Gamma) \geq w \\ & \quad \mathcal{NR}(\Gamma) \leq M_{max} \end{aligned} \tag{3}$$

In the problem (Equation (3)), the function $\mathcal{F}_\mathcal{D}(\Gamma)$ is a performance measure of the classifier $\Gamma$ over the dataset $\mathcal{D}$, the function $\mathcal{NR}(\Gamma)$ is the number of rules of the classifier $\Gamma$, and the constraints $\mathcal{NR}(\Gamma) \geq w$ and $\mathcal{NR}(\Gamma) \leq M_{max}$ limit the number of rules of the classifier $\Gamma$ to the interval $[w, M_{max}]$, where $w$ is the number of classes of the output attribute and $M_{max}$ is given by a user. Objectives $\mathcal{F}_\mathcal{D}(\Gamma)$ and $\mathcal{NR}(\Gamma)$ are in conflict. The fewer rules the classifier has, the fewer instances it can cover, that is, if the classifier is simpler it will have less capacity for prediction. There is, therefore, an intrinsic conflict between problem objectives (e.g., maximize accuracy and minimize model complexity) which cannot be easily aggregated to a single objective. Both objectives are typically optimized simultaneously in many other classification systems, such as neural networks or decision trees [47,48]. Figure 2 shows the Pareto front of a dummy binary classification problem described as in Equation (3), with $M_{max} = 6$ rules, where $\mathcal{F}_\mathcal{D}(\Gamma)$ is maximized. This front is composed of three non-dominated solutions (three possible classifiers) with two, three and four rules, respectively. The solutions with five and six rules are dominated (both by the solution with four rules).

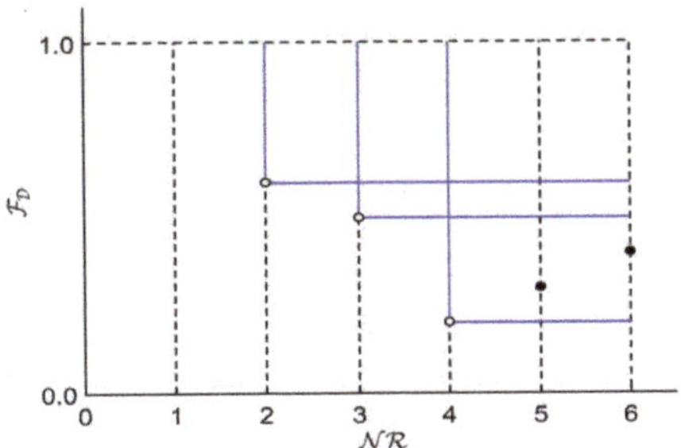

**Figure 2.** A Pareto front of a binary classification problem as formulated in Equation (3) where $\mathcal{F}_D(\Gamma)$ is minimized and $\mathcal{NR}(\Gamma)$ is minimized.

Both *ENORA* and *NSGA-II* have been adapted to solve the problem described in Equation (3) with *variable-length representation* based on a *Pittsburgh approach, uniform random initialization, binary tournament selection, handling constraints,* ranking based on *non-domination level* with *crowding distance,* and *self-adaptive variation operators. Self-adaptive variation operators* work on different levels of the classifier: *rule crossover, rule incremental crossover, rule incremental mutation,* and *integer mutation.*

### 3.2.1. Representation

We use a variable-length representation based on a Pittsburgh approach [49], where each individual $I$ of a population contains a variable number of rules $M_I$, and each rule $R_i^I$, $i = 1, \ldots$, is codified in the following components:

- Integer values associated to the antecedents $b_{ij}^I \in \{1, \ldots, v_j\}$, for $i = 1, \ldots, M_I$ and $j = 1, \ldots, p$.
- Integer values associated to the consequent $c_i^I \in \{1, \ldots, w\}$, for $i = 1, \ldots, M_I$.

Additionally, to carry out self-adaptive crossing and mutation, each individual has two discrete parameters $d_I \in \{0, \ldots, \delta\}$ and $e_I \in \{0, \ldots, \epsilon\}$ associated with crossing and mutation, where $\delta \geq 0$ is the number of crossing operators and $\varepsilon \geq 0$ is the number of mutation operators. Values $d_I$ and $e_I$ for self-adaptive variation are randomly generated from $\{0, \delta\}$ and $\{0, \epsilon\}$, respectively. Table 1 summarizes the representation of an individual.

**Table 1.** Chromosome coding for an individual $I$.

| Codification for Rule Set | | | | Codification for Adaptive Crossing and Mutation | |
|---|---|---|---|---|---|
| Antecedents | | | Consequent | Associated Crossing | Associated Mutation |
| $b_{11}^I$ | $b_{21}^I$ $\cdots$ | $b_{q1}^I$ | $c_1^I$ | | |
| $\vdots$ | $\vdots$ $\quad\vdots$ | $\vdots$ | $\vdots$ | $d_I$ | $e_I$ |
| $b_{1M_I}^I$ | $b_{2M_I}^I$ $\cdots$ | $b_{qM_I}^I$ | $c_{M_I}^I$ | | |

### 3.2.2. Constraint Handling

The constraints $\mathcal{NR}(\Gamma) \geq w$ and $\mathcal{NR}(\Gamma) \leq M_{max}$ are satisfied by means of specialized initialization and variation operators, which always generate individuals with a number of rules between $w$ and $M_{max}$.

### 3.2.3. Initial Population

The initial population (Algorithm 5) is randomly generated with the following conditions:

- Individuals are uniformly distributed with respect to the number of rules with values between $w$ and $M_{max}$, and with an additional constraint that specifies that there must be at least one individual for each number of rules (Steps 4–8). This ensures an adequate initial diversity in the search space in terms of the second objective of the optimization model.
- All individuals contain at least one rule for any output class between 1 and $w$ (Steps 16–20).

---

**Algorithm 5** Initialize population.

---

**Require:** $p > 0$ {Number of categorical input attributes}
**Require:** $v_1, \ldots, v_p, v_j > 1, j = 1, \ldots, p$ {Number of categories for the input attributes}
**Require:** $w > 1$, {Number of classes for the output attribute}
**Require:** $\delta > 0$ {Number of crossing operators}
**Require:** $\epsilon > 0$ {Number of mutation operators}
**Require:** $M_{max} \geq w$ {Maximum number of rules}
**Require:** $N > 1$ {Number of individuals in the population}

  1:   $P \leftarrow \emptyset$
  2:   **for** $k = 1$ to $N$ **do**
  3:      $I \leftarrow$ new Individual
  4:      **if** $k \leq M_{max} - w + 1$ **then**
  5:         $M_I \leftarrow k + w - 1$
  6:      **else**
  7:         $M_I \leftarrow$ Int $Random(w, M_{max})$
  8:      **end if**
  9:      {Random rule $R_i^I$}
10:      **for** $i = 1$ to $M_I$ **do**
11:         {Random integer values associated with the antecedents}
12:         **for** $j = 1$ to $p$ **do**
13:            $b_{ij}^I \leftarrow Random(1, v_j)$
14:         **end for**
15:         {Random integer value associated with the consequent}
16:         **if** $i < w$ **then**
17:            $c_i^I = j$
18:         **else**
19:            $c_i^I \leftarrow Random(1, w)$
20:         **end if**
21:      **end for**
22:      {Random integer values for adaptive variation}
23:      $d_I \leftarrow Random(0, \delta)$
24:      $e_I \leftarrow Random(0, \epsilon)$
25:      $P \leftarrow P \cup I$
26: **end for**
27: **return** $P$

---

### 3.2.4. Fitness Functions

Since the optimization model encompasses two objectives, each individual must be evaluated with two fitness functions, which correspond to the objective functions $\mathcal{F}_D(\Gamma)$ and $\mathcal{NR}(\Gamma)$ of the problem (Equation (3)). The selection of the best individuals is done using the Pareto concept in a binary tournament.

### 3.2.5. Variation Operators

We use *self-adaptive crossover and mutation*, which means that the selection of the operators is made by means of an adaptive technique. As we have explained (cf. Section 3.2.1), each individual $I$ has two integer parameters $d_I \in \{0, \ldots, \delta\}$ and $e_I \in \{0, \ldots, \epsilon\}$ to indicate which crossover or mutation is carried out. In our case, $\delta = 2$ and $\epsilon = 2$ are two crossover operators and two mutation operators, so that $d_I, e_I \in \{0, 1, 2\}$. Note that value 0 indicates that no crossover or no mutation is performed. Self-adaptive variation (Algorithm 6) generates two children from two parents by self-adaptive crossover (Algorithm 7) and self-adaptive mutation (Algorithm 8). Self-adaptive crossover of individuals $I$, $J$ and self-adaptive mutation of individual $I$ are similar to each other. First, with a probability $p_v$, the values $d_I$ and $e_I$ are replaced by a random value. Additionally, in the case of crossover, the value $d_J$ is replaced by $d_I$. Then, the crossover indicated by $d_I$ or the mutation indicated by $e_I$ is performed. In summary, if an individual comes from a given crossover or a given mutation, that specific crossover and mutation are preserved to their offspring with probability $p_v$, so the value of $p_v$ must be small enough to ensure a controlled evolution (in our case, we use $p_v = 0.1$). Although the probability of the crossover and mutation is not explicitly represented, it can be computed as the ratio of the individuals for which crossover and mutation values are set to 1. As the population evolves, individuals with more successful types of crossover and mutation will be more common, so that the probability of selecting the more successful crossover and mutation types will increase. Using self-adaptive crossover and mutation operators helps to realize the goals of maintaining diversity in the population and sustaining the convergence capacity of the evolutionary algorithm, also eliminating the need of setting an a priori operator probability to each operator. In other approaches (e.g., [50]), the probabilities of crossover and mutation vary depending on the fitness value of the solutions.

Both *ENORA* and *NSGA-II* have been implemented with two crossover operators, *rule crossover* (Algorithm 9) and *rule incremental crossover* (Algorithm 10), and two mutation operators: *rule incremental mutation* (Algorithm 11) and *integer mutation* (Algorithm 12). *Rule crossover* randomly exchanges two rules selected from the parents, and *rule incremental crossover* adds to each parent a rule randomly selected from the other parent if its number of rules is less than the maximum number of rules. On the other hand, *rule incremental mutation* adds a new rule to the individual if the number of rules of the individual is less than the maximum number of rules, while *integer mutation* carries out a uniform mutation of a random antecedent belonging to a randomly selected rule.

---

**Algorithm 6** Variation.

---

**Require:** *Parent1, Parent2* {Individuals for variation}

  1: *Child1* ← *Parent1*

  2: *Child1* ← *Parent2*

  3: Self-adaptive crossover *Child1, Child2*

  4: Self-adaptive mutation *Child1*

  5: Self-adaptive mutation *Child2*

  6: **return** *Child1, Child2*

---

---

**Algorithm 7** Self-adaptive crossover.

---

**Require:** $I, J$ {Individuals for crossing}
**Require:** $p_v$ $(0 < p_v < 1)$ {Probability of variation}
**Require:** $\delta > 0$ {Number of different crossover operators}
  1: **if** a random Bernoulli variable with probability $p_v$ takes the value 1 **then**
  2:    $d_I \leftarrow Random(0,\delta)$
  3: **end if**
  4: $d_J \leftarrow d_I$
  5: Carry out the type of crossover specified by $d_I$:
    {0: No cross}
    {1: Rule crossover}
    {2: Rule incremental crossover}

---

**Algorithm 8** Self-adaptive mutation.

---

**Require:** $I$ {Individual for mutation}
**Require:** $p_v$ $(0 < p_v < 1)$ {Probability of variation}
**Require:** $\epsilon > 0$ {Number of different mutation operators}
  1: **if** a random Bernoulli variable with probability $p_v$ takes the value 1 **then**
  2:    $e_I \leftarrow Random(0,\epsilon)$
  3: **end if**
  4: Carry out the type of mutation specified by $e_I$:
    {0: No mutation}
    {1: Rule incremental mutation}
    {2: Integer mutation}

---

**Algorithm 9** Rule crossover.

---

**Require:** $I, J$ {Individuals for crossing}
  1: $i \leftarrow Random(1,M_I)$
  2: $j \leftarrow Random(1,M_J)$
  3: Exchange rules $R_i^I$ and $R_j^J$

---

**Algorithm 10** Rule incremental crossover.

---

**Require:** $I, J$ {Individuals for crossing}
**Require:** $M_{max}$ {Maximum number of rules}
  1: **if** $M_I < M_{max}$ **then**
  2:    $j \leftarrow Random(1,M_J)$
  3:    Add $R_j^J$ to individual $I$
  4: **end if**
  5: **if** $M_J < M_{max}$ **then**
  6:    $i \leftarrow Random(1,M_I)$
  7:    Add $R_i^I$ to individual $J$
  8: **end if**

---

---

**Algorithm 11** Rule incremental mutation.

---

**Require:** $I$ {Individual for mutation}

**Require:** $M_{max}$ {Maximum number of rules}

1: **if** $M_I < M_{max}$ **then**

2:     Add a new random rule to $I$

3: **end if**

---

---

**Algorithm 12** Integer mutation.

---

**Require:** $I$ {Individual for mutation}

**Require:** $p > 0$ {Number of categorical input attributes}

**Require:** $v_1, \ldots, v_p, v_j > 1, j = 1, \ldots, p$ {Number of categories for the input attributes}

1: $i \leftarrow Random(1, M_I)$

2: $j \leftarrow Random(1, p)$

3: $b_{ij}^I \leftarrow Random(1, v_j)$

---

## 4. Experiment and Results

To ensure the reproducibility of the experiments, we have used publicly available datasets. In particular, we have designed two sets of experiments, one using the *Breast Cancer* [51] dataset, and the other using the *Monk's Problem 2* [52] dataset.

### 4.1. The Breast Cancer Dataset

*Breast Cancer* encompasses 286 instances. Each instance corresponds to a patient who suffered from breast cancer and uses nine attributes to describe each patient. The class to be predicted is binary and represents whether the patient has suffered a recurring cancer event. In this dataset, 85 instances are positive and 201 are negative. Table 2 summarizes the attributes of the dataset. Among all instances, nine present some missing values; in the pre-processing phase, these have been replaced by the mode of the corresponding attribute.

**Table 2.** Attribute description of the *Breast Cancer* dataset.

| # | Attribute Name | Type | Possible Values |
|---|---|---|---|
| 1 | age | categorical | 10–19, 20–29, 30–39, 40–49, 50–59, 60–69, 70–79, 80–89, 90–99. |
| 2 | menopause | categorical | lt40, ge40, premeno |
| 3 | tumour-size | categorical | 0–4, 5–9, 10–14, 15–19, 20–24, 25–29, 30–34, 35–39, 40–44, 45–49, 50–54, 55–59 |
| 4 | inv-nodes | categorical | 0–2, 3–5, 6–8, 9–11, 12–14, 15–17, 18–20, 21–23, 24–26, 27–29, 30–32, 33–35, 36–39 |
| 5 | node-caps | categorical | yes, no |
| 6 | deg-malign | categorical | 1, 2, 3 |
| 7 | breast | categorical | left, right |
| 8 | breast-quad | categorical | left-up, left-low, right-up, right-low, central |
| 9 | irradiat | categorical | yes, no |
| 10 | class | categorical | no-recurrence-events, recurrence-events |

### 4.2. The Monk's Problem 2 Dataset

In July 1991, the monks of *Corsendonk Priory* attended a summer course that was being held in their priory, namely the 2nd European Summer School on Machine Learning. After a week, the monks could not yet clearly identify the best *ML* algorithms, or which algorithms to avoid in which cases. For this reason, they decided to create the three so-called *Monk's problems*, and used them to determine which *ML* algorithms were the best. These problems, rather simple and completely artificial, became later famous (because of their peculiar origin), and have been used as a comparison for many algorithms on several occasions. In particular, in [53], they have been used to test the performance of state-of-the-art (at that time) learning algorithms such as *AQ17-DCI, AQ17-HCI, AQ17-FCLS, AQ14-NT, AQ15-GA, Assistant Professional, mFOIL, ID5R, IDL, ID5R-hat, TDIDT, ID3, AQR, CN2, WEB CLASS, ECOBWEB, PRISM, Backpropagation,* and *Cascade Correlation.* For our research, we have used the *Monk's Problem 2,* which contains six categorical input attributes and a binary output attribute, summarized in Table 3. The target concept associated with the *Monk's Problem 2* is the binary outcome of the logical formula:

*Exactly two of:*
{heap_shape= round, body_shape=round, is_smiling=yes, holding=sword, jacket_color=red, has_tie=yes}

In this dataset, the original training and testing sets were merged to allow other sampling procedures. The set contains a total of 601 instances, and no missing values.

**Table 3.** Attribute description of the *MONK's Problem 2* dataset.

| # | Atttribute Name | Type | Possible Values |
|---|---|---|---|
| 1 | head_shape | categorical | round, square, octagon |
| 2 | body_shape | categorical | round, square, octagon |
| 3 | is_smiling | categorical | yes, no |
| 4 | holding | categorical | sword, balloon, flag |
| 5 | jacket_color | categorical | red, yellow, green, blue |
| 6 | has_tie | categorical | yes, no |
| 7 | class | categorical | yes, no |

*4.3. Optimization Models*

We have conducted different experiments with different optimization models to calculate the overall performance of our proposed technique and to see the effect of optimizing different objectives for the same problem. First, we have designed a multi-objective constrained optimization model based on the *accuracy*:

$$
\begin{aligned}
&\text{Max.} &&\mathcal{ACC}_{\mathcal{D}}(\Gamma) \\
&\text{Min.} &&\mathcal{NR}(\Gamma) \\
&\text{subject to}: &&\mathcal{NR}(\Gamma) \geq w \\
& &&\mathcal{NR}(\Gamma) \leq M_{max}
\end{aligned}
\tag{4}
$$

where $\mathcal{ACC}_{\mathcal{D}}(\Gamma)$ is the proportion of correctly classified instances (both true positives and true negatives) among the total number of instances [54] obtained with the classifier $\Gamma$ for the dataset $\mathcal{D}$. $\mathcal{ACC}_{\mathcal{D}}(\Gamma)$ is defined as:

$$
\mathcal{ACC}_{\mathcal{D}}(\Gamma) = \frac{1}{K} \sum_{i=1}^{K} T_{\mathcal{D}}(\Gamma, i)
$$

where $K$ is the number of instances of the dataset $\mathcal{D}$, and $T_{\mathcal{D}}(\Gamma, i)$ is the result of the classification of the instance $i$ in $\mathcal{D}$ with the classifier $\Gamma$, that is:

$$
T_{\mathcal{D}}(\Gamma, i) = \begin{cases} 1 & if \ \hat{c}_i^{\Gamma} = c_{\mathcal{D}}^i \\ 0 & if \ \hat{c}_i^{\Gamma} \neq c_{\mathcal{D}}^i \end{cases}
$$

where $\hat{c}_i^{\Gamma}$ is the predicted value of the *i*th instance in $\Gamma$, and $c_{\mathcal{D}}^i$ is the corresponding true value in $\mathcal{D}$. Our second optimization model is based on the *area under the ROC curve*:

$$\begin{aligned}
&\textit{Max.} && \mathcal{AUC}_{\mathcal{D}}(\Gamma) \\
&\textit{Min.} && \mathcal{NR}(\Gamma) \\
&\textit{subject to}: && \mathcal{NR}(\Gamma) \geq w \\
& && \mathcal{NR}(\Gamma) \leq M_{max}
\end{aligned} \tag{5}$$

where $\mathcal{AUC}_{\mathcal{D}}(\Gamma)$ is the area under the *ROC* curve obtained with the classifier $\Gamma$ with the dataset $\mathcal{D}$. The *ROC* (*Receiver Operating Characteristic*) curve [55] is a graphical representation of the *sensitivity* versus the *specificity* index for a classifier varying the *discrimination threshold* value. Such a curve can be used to generate statistics that summarize the performance of a classifier, and it has been shown in [54] to be a simple, yet complete, empirical description of the decision threshold effect, indicating all possible combinations of the relative frequencies of the various kinds of correct and incorrect decisions. The area under the *ROC* curve can be computed as follows [56]:

$$\mathcal{AUC}_{\mathcal{D}}(\Gamma) = \int_0^1 S_{\mathcal{D}}(\Gamma, E_{\mathcal{D}}^{-1}(\Gamma, v))dv$$

where $S_{\mathcal{D}}(\Gamma, t)$ (*sensitivity*) is the proportion of positive instances classified as positive by the classifier $\Gamma$ in $\mathcal{D}$, $1 - E_{\mathcal{D}}(\Gamma, t)$ (*specificity*) is the proportion of negative instances classified as negative by $\Gamma$ in $\mathcal{D}$, and $t$ is the discrimination threshold. Finally, our third constrained optimization model is based on the *root mean square error* (*RMSE*):

$$\begin{aligned}
&\textit{Max./Min.} && \mathcal{RMSE}_{\mathcal{D}}(\Gamma) \\
&\textit{Min.} && \mathcal{NR}(\Gamma) \\
&\textit{subject to}: && \mathcal{NR}(\Gamma) \geq w \\
& && \mathcal{NR}(\Gamma) \leq M_{max}
\end{aligned} \tag{6}$$

where $\mathcal{RMSE}_{\mathcal{D}}(\Gamma)$ is defined as the square root of the *mean square error* obtained with a classifier $\Gamma$ in the dataset $\mathcal{D}$:

$$\mathcal{RMSE}_{\mathcal{D}}(\Gamma) = \frac{1}{K}\sqrt{\sum_{i=1}^{K}(\hat{c}_i^{\Gamma} - c_{\mathcal{D}}^i)^2}$$

where $\hat{c}_i^{\Gamma}$ is the predicted value of the $i$th instance for the classifier $\Gamma$, and $c_{\mathcal{D}}^i$ is the corresponding output value in the database $\mathcal{D}$. Accuracy, area under the *ROC* curve, and root mean square error are all well-accepted measures used to evaluate the performance of a classifier. Therefore, it is natural to use such measures as fitting functions. In this way, we can establish which one behaves better in the optimization phase, and we can compare the results with those in the literature.

### 4.4. Choosing the Best Pareto Front

To compare the performance of *ENORA* and *NSGA-II* as metaheuristics in this particular optimization task, we use the *hypervolume metric* [57,58]. The hypervolume measures, simultaneously, the diversity and the optimality of the non-dominated solutions. The main advantage of using hypervolume against other standard measures, such as the *error ratio*, the *generational distance*, the *maximum Pareto-optimal front error*, the *spread*, the *maximum spread*, or the *chi-square-like deviation*, is that it can be computed without an optimal population, which is not always known [15]. The hypervolume is defined as the volume of the search space dominated by a population $P$, and is formulated as:

$$HV(P) = \bigcup_{i=1}^{|Q|} v_i \tag{7}$$

where $Q \subseteq P$ is the set of non-dominated individuals of $P$, and $v_i$ is the volume of the individual $i$. Subsequently, the *hypervolume ratio (HVR)* is defined as the ratio of the volume of the non-dominated search space over the volume of the entire search space, and is formulated as follows:

$$HVR\left(P\right) = 1 - \frac{H\left(P\right)}{VS} \tag{8}$$

where $VS$ is the volume of the search space. Computing $HVR$ requires reference points that identify the maximum and minimum values for each objective. For $RBC$ optimization, as proposed in this work, the following minimum ($\mathcal{F}_D{}^{lower}, \mathcal{NR}^{lower}$) and maximum ($\mathcal{F}_D{}^{upper}, \mathcal{NR}^{upper}$) points, for each objective, are set in the multi-objective optimization models in Equations (4)–(6):

$$\mathcal{F}_D{}^{lower} = 0, \quad \mathcal{F}_D{}^{upper} = 1, \quad \mathcal{NR}^{lower} = w, \quad \mathcal{NR}^{upper} = M_{max}$$

A first single execution of all six models (three driven by $ENORA$, and three driven by $NSGA$-$II$), over both datasets, has been designed for the purpose of showing the aspect of the final Pareto front, and compare the hypervolume ratio of the models. The results of this single execution, with population size equal to 50 and 20,000 generations (1,000,000 evaluations in total), are shown in Figures 3 and 4 (by default, $M_{max}$ is set to 10, to which we add 2, because both datasets have a binary class). Regarding the configuration of the number of generations and the size of the population, our criterion has been established as follows: once the number of evaluations is set to 1,000,000, we can decide to use a population size of 100 individuals and 10,000 generations, or to use a population size of 50 individuals and 20,000 generations. The first configuration (100 × 10,000) allows a greater diversity with respect to the number of rules of the classifiers, while the second one (50 × 20,000) allows a better adjustment of the classifier parameters and therefore, a greater precision. Given the fact that the maximum number of rules of the classifiers is not greater than 12, we think that 50 individuals are sufficient to represent four classifiers on average for each number of rules (4 × 12 = 48~50). Thus, we prefer the second configuration (50 × 20,000) because having more generations increases the chances of building classifiers with a higher precision.

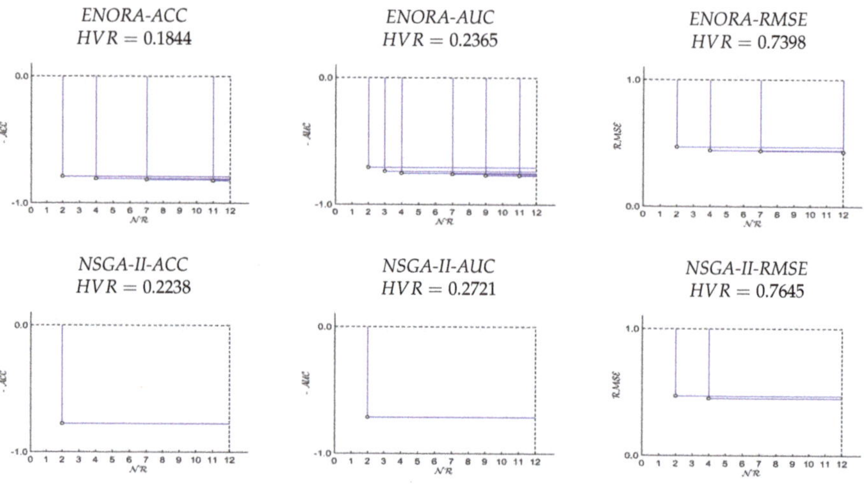

**Figure 3.** Pareto fronts of one execution of *ENORA* and *NSGA-II*, with $M_{max} = 12$, on the *Breast Cancer* dataset, and their respective *HVR*. Note that in the case of multi-objective classification where $\mathcal{F}_D$ is maximized ($\mathcal{ACC}_D$ and $\mathcal{AUC}_D$), function $\mathcal{F}_D$ has been converted to minimization for a better understanding of the Pareto front.

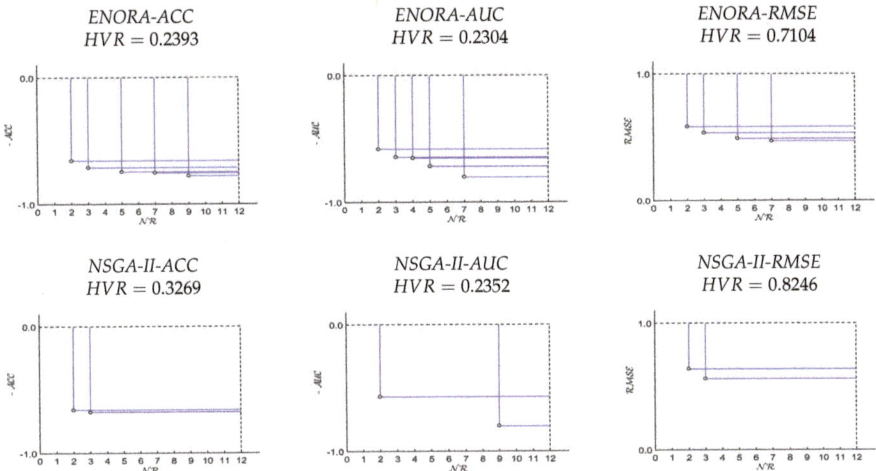

**Figure 4.** Pareto fronts of one execution of *ENORA* and *NSGA-II*, with $M_{max} = 12$, on the *Monk's Problem 2* dataset, and their respective *HVR*. Note that in the case of multi-objective classification where $\mathcal{F}_{\mathcal{D}}$ is maximized ($\mathcal{ACC}_{\mathcal{D}}$ and $\mathcal{AUC}_{\mathcal{D}}$), function $\mathcal{F}_{\mathcal{D}}$ has been converted to minimization for a better understanding of the Pareto front.

Experiments were executed in a computer x64-based PC with one processor Intel64 Family 6 Model 60 Stepping 3 GenuineIntel 3201 Mhz, RAM 8131 MB. Table 4 shows the run time for each method over both datasets. Note that, although *ENORA* has less algorithmic complexity than *NSGA-II*, it has taken longer in experiments than *NSGA-II*. This is because the evaluation time of individuals in *ENORA* is higher than that of *NSGA-II* since *ENORA* has more diversity than *NSGA-II*, and therefore *ENORA* evaluates classifiers with more rules than *NSGA-II*.

**Table 4.** Run times of *ENORA* and *NSGA-II* for *Breast Cancer* and *Monk's Problem 2* datasets.

| Method | Breast Cancer | Monk's Problem 2 |
|---|---|---|
| *ENORA-ACC* | 244.92 s. | 428.14 s. |
| *ENORA-AUC* | 294.75 s. | 553.11 s. |
| *ENORA-RMSE* | 243.30 s. | 414.42 s. |
| *NSGA-II-ACC* | 127.13 s. | 260.83 s. |
| *NSGA-II-AUC* | 197.07 s. | 424.83 s. |
| *NSGA-II-RMSE* | 134.87 s. | 278.19 s. |

From these results, we can deduce that, first, *ENORA* maintains a higher diversity of the population, and achieves a better hypervolume ratio with respect to *NSGA-II*, and, second, using accuracy as the first objective generates better fronts than using the area under the *ROC* curve, which, in turn, performs better than using the root mean square error.

### 4.5. Comparing Our Method with Other Classifier Learning Systems (Full Training Mode)

To perform an initial comparison between the performance of the classifiers obtained with the proposed method and the ones obtained with classical methods (*PART*, *JRip*, *OneR* and *ZeroR*), we have executed again the six models in full training mode.

The parameters have been configured as in the previous experiment (population size equal to 50 and 20,000 generations), excepting the $M_{max}$ parameter that was set to 2 for the *Breast Cancer* dataset (this case), while, for the *Monk's Problem 2*, it was set to 9. Observe that, since $M_{min} = 2$ in both cases, executing the optimization models using $M_{max} = 2$ leads to a single objective search for the *Breast Cancer* dataset. In fact, after the preliminary experiments were run, it turned out that the classical classifier learning systems tend to return very small, although not very precise, set of rules on *Breast Cancer*, and that justifies our choice. On the other hand, executing the classical rule learners on *Monk's Problem 2* returns more diverse sets of rules, which justifies choosing a higher $M_{max}$ in that case. To decide, a posteriori, which individual is chosen from the final front, we have used the default algorithm: the individual with the best value on the first objective is returned. In the case of *Monk's Problem 2*, that individual has seven rules. The comparison is shown in Tables 5 and 6, which show, for each classifier, the following information: *number of rules, percent correct, true positive rate, false positive rate, precision, recall, F-measure, Matthews correlation coefficient, area under the ROC curve, area under precision-recall curve, and root mean square error*. As for the *Breast Cancer* dataset (observe that the best result emerged from the proposed method), in the optimization model driven by *NSGA-II*, with root mean square error as the first objective (see Table 7), only *PART* was able to achieve similar results, although slightly worse, but at the price of having 15 rules, making the system clearly not interpretable. In the case of the *Monk's Problem 2* dataset, *PART* returned a model with 47 rules, which is not interpretable by any standard, although it is very accurate. The best interpretable result is the one with seven rules returned by *ENORA*, driven by the root mean square error (see Table 8). The experiments for classical learners have been conducted using the default parameters.

**Table 5.** Comparison of the performance of the learning models in full training mode—*Breast Cancer* dataset.

| Learning Model | Number of Rules | Percent Correct | TP Rate | FP Rate | Precision | Recall | F-Measure | MCC | ROC Area | PRC Area | RMSE |
|---|---|---|---|---|---|---|---|---|---|---|---|
| ENORA-ACC | 2 | 79.02 | 0.790 | 0.449 | 0.796 | 0.790 | 0.762 | 0.455 | 0.671 | 0.697 | 0.458 |
| ENORA-AUC | 2 | 75.87 | 0.759 | 0.374 | 0.751 | 0.759 | 0.754 | 0.402 | 0.693 | 0.696 | 0.491 |
| ENORA-RMSE | 2 | 77.62 | 0.776 | 0.475 | 0.778 | 0.776 | 0.744 | 0.410 | 0.651 | 0.680 | 0.473 |
| NSGA-II-ACC | 2 | 77.97 | 0.780 | 0.501 | 0.805 | 0.780 | 0.738 | 0.429 | 0.640 | 0.679 | 0.469 |
| NSGA-II-AUC | 2 | 75.52 | 0.755 | 0.368 | 0.749 | 0.755 | 0.752 | 0.399 | 0.693 | 0.696 | 0.495 |
| NSGA-II-RMSE | 2 | 79.37 | 0.794 | 0.447 | 0.803 | 0.794 | 0.765 | 0.467 | 0.673 | 0.700 | 0.454 |
| PART | 15 | 78.32 | 0.783 | 0.397 | 0.773 | 0.783 | 0.769 | 0.442 | 0.777 | 0.793 | 0.398 |
| JRip | 3 | 76.92 | 0.769 | 0.471 | 0.762 | 0.769 | 0.740 | 0.389 | 0.650 | 0.680 | 0.421 |
| OneR | 1 | 72.72 | 0.727 | 0.563 | 0.703 | 0.727 | 0.680 | 0.241 | 0.582 | 0.629 | 0.522 |
| ZeroR | - | 70.27 | 0.703 | 0.703 | 0.494 | 0.703 | 0.580 | 0.000 | 0.500 | 0.582 | 0.457 |

**Table 6.** Comparison of the performance of the learning models in full training mode—*Monk's Problem 2* dataset.

| Learning Model | Number of Rules | Percent Correct | TP Rate | FP Rate | Precision | Recall | F-Measure | MCC | ROC Area | PRC Area | RMSE |
|---|---|---|---|---|---|---|---|---|---|---|---|
| ENORA-ACC | 7 | 75.87 | 0.759 | 0.370 | 0.753 | 0.759 | 0.745 | 0.436 | 0.695 | 0.680 | 0.491 |
| ENORA-AUC | 7 | 68.71 | 0.687 | 0.163 | 0.836 | 0.687 | 0.687 | 0.523 | 0.762 | 0.729 | 0.559 |
| ENORA-RMSE | 7 | 77.70 | 0.777 | 0.360 | 0.777 | 0.777 | 0.762 | 0.481 | 0.708 | 0.695 | 0.472 |
| NSGA-II-ACC | 7 | 68.38 | 0.684 | 0.588 | 0.704 | 0.684 | 0.597 | 0.203 | 0.548 | 0.580 | 0.562 |
| NSGA-II-AUC | 7 | 66.38 | 0.664 | 0.175 | 0.830 | 0.664 | 0.661 | 0.497 | 0.744 | 0.715 | 0.580 |
| NSGA-II-RMSE | 7 | 68.71 | 0.687 | 0.591 | 0.737 | 0.687 | 0.595 | 0.226 | 0.548 | 0.583 | 0.559 |
| PART | 47 | 94.01 | 0.940 | 0.087 | 0.940 | 0.940 | 0.940 | 0.866 | 0.980 | 0.979 | 0.218 |
| JRip | 1 | 65.72 | 0.657 | 0.657 | 0.432 | 0.657 | 0.521 | 0.000 | 0.500 | 0.549 | 0.475 |
| OneR | 1 | 65.72 | 0.657 | 0.657 | 0.432 | 0.657 | 0.521 | 0.000 | 0.500 | 0.549 | 0.585 |
| ZeroR | - | 65.72 | 0.657 | 0.657 | 0.432 | 0.657 | 0.521 | 0.000 | 0.500 | 0.549 | 0.475 |

**Table 7.** Rule-based classifier obtained with *NSGA-II-RMSE* for *Breast Cancer* dataset.

| Rule | Antecedents | | | | | | | Consequent |
|---|---|---|---|---|---|---|---|---|
| $R_1$: | IF | age = 50–59 | AND | inv-nodes = 0–2 | AND | node-caps = no | | |
| | AND | deg-malig = 1 | AND | breast = right | AND | breast-quad = left-low | THEN | class = no-recurrence-events |
| $R_2$: | IF | age = 60–69 | AND | inv-nodes = 18–20 | AND | node-caps = yes | | |
| | AND | deg-malig = 3 | AND | breast = left | AND | breast-quad = right-up | THEN | class = recurrence-events |

**Table 8.** Rule-based classifier obtained with *ENORA-RMSE* for *Monk's Problem 2* dataset.

| Rule | Antecedents | | | | | | | | | | Consequent |
|------|----|-----------------------|-----|-----------------------|-----|------------------|------|---------|------|------------|
| $R_1$: | IF | head_shape = round | AND | body_shape = round | AND | is_smiling = no | | | | |
| | AND | holding = sword | AND | jacket_color = red | AND | has_tie = yes | THEN | class = yes |
| $R_2$: | IF | head_shape = octagon | AND | body_shape = round | AND | is_smiling = no | | | | |
| | AND | holding = sword | AND | jacket_color = red | AND | has_tie = no | THEN | class = yes |
| $R_3$: | IF | head_shape = round | AND | body_shape = round | AND | is_smiling = no | | | | |
| | AND | holding = sword | AND | jacket_color = yellow | AND | has_tie = yes | THEN | class = yes |
| $R_4$: | IF | head_shape = round | AND | body_shape = round | AND | is_smiling = no | | | | |
| | AND | holding = sword | AND | jacket_color = red | AND | has_tie = no | THEN | class = yes |
| $R_5$: | IF | head_shape = square | AND | body_shape = square | AND | is_smiling = yes | | | | |
| | AND | holding = flag | AND | jacket_color = yellow | AND | has_tie = no | THEN | class = no |
| $R_6$: | IF | head_shape = octagon | AND | body_shape = round | AND | is_smiling = yes | | | | |
| | AND | holding = balloon | AND | jacket_color = blue | AND | has_tie = no | THEN | class = no |
| $R_7$: | IF | head_shape = octagon | AND | body_shape = octagon | AND | is_smiling = yes | | | | |
| | AND | holding = sword | AND | jacket_color = green | AND | has_tie = no | THEN | class = no |

*4.6. Comparing Our Method with Other Classifier Learning Systems (Cross-Validation and Train/Test Percentage Split Mode)*

To test the capabilities of our methodology in a more significant way, we proceeded as follows. First, we designed a *cross-validated* experiment for the *Breast Cancer* dataset, in which we iterated three times a 10-fold cross-validation learning process [59] and considered the average value of the performance metrics *percent correct, area under the ROC curve,* and *serialized model size* of all results. Second, we designed a *train/test percentage split* experiment for the *Monk's Problem 2* dataset, in which we iterated ten times a 66% (training) versus 33% (testing) split and considered, again, the average result of the same metrics. Finally, we performed a statistical test over on results, to understand if they show any statistically significant difference. An execution of our methodology, and of standard classical learners, has been performed to obtain the models to be tested precisely under the same conditions of the experiment Section 4.5. It is worth observing that using two different types of evaluations allows us to make sure that our results are not influenced by the type of experiment. The results of the experiments are shown in Tables 9 and 10.

**Table 9.** Comparison of the performance of the learning models in 10-fold cross-validation mode (three repetitions)—*Breast Cancer* dataset.

| Learning Model | Percent Correct | ROC Area | Serialized Model Size |
|----------------|-----------------|----------|-----------------------|
| *ENORA-ACC* | 73.45 | 0.61 | 9554.80 |
| *ENORA-AUC* | 70.16 | 0.62 | 9554.63 |
| *ENORA-RMSE* | 72.39 | 0.60 | 9557.77 |
| *NSGA-II-ACC* | 72.50 | 0.60 | 9556.20 |
| *NSGA-II-AUC* | 70.03 | 0.61 | 9555.70 |
| *NSGA-II-RMSE* | 73.34 | 0.60 | 9558.60 |
| *PART* | 68.92 | 0.61 | 55,298.13 |
| *JRip* | 71.82 | 0.61 | 7664.07 |
| *OneR* | 67.15 | 0.55 | 1524.00 |
| *ZeroR* | 70.30 | 0.50 | 915.00 |

The statistical tests aim to verify if there are significant differences among the means of each metric: *percent correct, area under the ROC curve* and *serialized model size*. We proceeded as follows. First, we checked normality and sphericity of each sample by means of the *Shapiro–Wilk normality test*. Then, if normality and sphericity conditions were met, we applied *one way repeated measures ANOVA*; otherwise, we applied the *Friedman test*. In the latter case, when statistically significant differences were detected, we applied the *Nemenyi post-hoc test* to locate where these differences were. Tables A1–A12 in Appendix A show the results of the performed tests for the *Breast Cancer* dataset for each of the three metrics, and Tables A13–A24 in Appendix B show the results for the *Monk's Problem 2* dataset.

**Table 10.** Comparison of the performance of the learning models in split mode—*Monk's problem 2* dataset.

| Learning Model | Percent Correct | *ROC* Area | Serialized Model Size |
|---|---|---|---|
| ENORA-ACC | 76.69 | 0.70 | 9586.50 |
| ENORA-AUC | 72.82 | 0.79 | 9589.30 |
| ENORA-RMSE | 75.66 | 0.68 | 9585.30 |
| NSGA-II-ACC | 70.07 | 0.59 | 9590.60 |
| NSGA-II-AUC | 67.08 | 0.70 | 9619.70 |
| NSGA-II-RMSE | 67.63 | 0.54 | 9565.90 |
| PART | 73.51 | 0.79 | 73,115.90 |
| JRip | 64.05 | 0.50 | 5956.90 |
| OneR | 65.72 | 0.50 | 1313.00 |
| ZeroR | 65.72 | 0.50 | 888.00 |

*4.7. Additional Experiments*

Finally, we show the results of the evaluation with 10-fold cross-validation for *Monk's problem 2* dataset and for the following four other datasets:

1. *Tic-Tac-Toe-Endgame* dataset, with 9 input attributes, 958 instances, and binary class (Table 11).
2. *Car* dataset, with 6 input attributes, 1728 instances, and 4 output classes (Table 12).
3. *Chess (King-Rook vs. King-Pawn) (kr-vs-kp)*, with 36 input attributes, 3196 instances, and binary class (Table 13).
4. *Nursery* dataset, with 8 input attributes, 12,960 instances, and 5 output classes (Table 14).

**Table 11.** Attribute description of the *Tic-Tac-Toe-Endgame* dataset.

| # | Attribute Name | Type | Possible Values |
|---|---|---|---|
| 1 | top-left-square | categorical | x, o, b |
| 2 | top-middle-square | categorical | x, o, b |
| 3 | top-right-square | categorical | x, o, b |
| 4 | middle-left-square | categorical | x, o, b |
| 5 | middle-middle-square | categorical | x, o, b |
| 6 | middle-right-square | categorical | x, o, b |
| 7 | bottom-left-square | categorical | x, o, b |
| 8 | bottom-middle-square | categorical | x, o, b |
| 9 | bottom-right-square | categorical | x, o, b |
| 10 | class | categorical | positive, negative |

**Table 12.** Attribute description of the *Car* dataset.

| # | Attribute Name | Type | Possible Values |
|---|---|---|---|
| 1 | buying | categorical | vhigh, high, med, low |
| 2 | maint | categorical | vhigh, high, med, low |
| 3 | doors | categorical | 2, 3, 4, 5-more |
| 4 | persons | categorical | 2, 4, more |
| 5 | lug_boot | categorical | small, med, big |
| 6 | safety | categorical | low, med, high |
| 7 | class | categorical | unacc, acc, good, vgood |

**Table 13.** Attribute description of the *kr-vs-kp* dataset.

| # | Attribute Name | Type | Possible Values |
|---|---|---|---|
| 1 | bkblk | categorical | t , f |
| 2 | bknwy | categorical | t , f |
| 3 | bkon8 | categorical | t , f |
| 4 | bkona | categorical | t , f |
| 5 | bkspr | categorical | t , f |
| 6 | bkxbq | categorical | t , f |
| 7 | bkxcr | categorical | t , f |
| 8 | bkxwp | categorical | t , f |
| 9 | blxwp | categorical | t , f |
| 10 | bxqsq | categorical | t , f |
| 11 | cntxt | categorical | t , f |
| 12 | dsopp | categorical | t , f |
| 13 | dwipd | categorical | g , l |
| 14 | hdchk | categorical | t , f |
| 15 | katri | categorical | b , n , w |
| 16 | mulch | categorical | t , f |
| 17 | qxmsq | categorical | t , f |
| 18 | r2ar8 | categorical | t , f |
| 19 | reskd | categorical | t , f |
| 20 | reskr | categorical | t , f |
| 21 | rimmx | categorical | t , f |
| 22 | rkxwp | categorical | t , f |
| 23 | rxmsq | categorical | t , f |
| 24 | simpl | categorical | t , f |
| 25 | skach | categorical | t , f |
| 26 | skewr | categorical | t , f |
| 27 | skrxp | categorical | t , f |
| 28 | spcop | categorical | t , f |
| 29 | stlmt | categorical | t , f |
| 30 | thrsk | categorical | t , f |
| 31 | wkcti | categorical | t , f |
| 32 | wkna8 | categorical | t , f |
| 33 | wknck | categorical | t , f |
| 34 | wkovl | categorical | t , f |
| 35 | wkpos | categorical | t , f |
| 36 | wtoeg | categorical | n , t , f |
| 37 | class | categorical | won , nowin |

**Table 14.** Attribute description of the *Nursery* dataset.

| # | Attribute Name | Type | Possible Values |
|---|---|---|---|
| 1 | parents | categorical | usual, pretentious, great_pret |
| 2 | has_nurs | categorical | proper, less_proper, improper, critical, very_crit |
| 3 | form | categorical | complete, completed, incomplete, foster |
| 4 | children | categorical | 1, 2, 3, more |
| 5 | housing | categorical | convenient, less_conv, critical |
| 6 | finance | categorical | convenient, inconv |
| 7 | social | categorical | nonprob, slightly_prob, problematic |
| 8 | health | categorical | recommended, priority, not_recom |
| 9 | class | categorical | not_recom, recommend, very_recom, priority, spec_prior |

We have used the *ENORA* algorithm together with the $\mathcal{ACC_D}$ and $\mathcal{RMSE_D}$ objective functions in this case because these combinations have produced the best results for the *Breast Cancer* and *Monk's problem 2* datasets evaluated in 10-fold cross-validation (population size equal to 50, 20,000 generations and $M_{max} = 10 +$ number of classes). Table 15 shows the results of the best combination *ENORA-ACC* or *ENORA-RMSE* together with the results of the classical rule-based classifiers.

**Table 15.** Comparison of the performance of the learning models in 10-fold cross-validation mode—*Monk's Problem 2, Tic-Tac-Toe-Endgame, Car, kr-vs-kp* and *Nursery* datasets.

| Learning Model | Number of Rules | Percent Correct | TP Rate | FP Rate | Precision | Recall | F-Measure | MCC | ROC Area | PRC Area | RMSE |
|---|---|---|---|---|---|---|---|---|---|---|---|
| *Monk's problem 2* | | | | | | | | | | | |
| *ENORA-ACC* | 7 | 77.70 | 0.777 | 0.360 | 0.777 | 0.777 | 0.762 | 0.481 | 0.708 | 0.695 | 0.472 |
| *PART* | 47 | 79.53 | 0.795 | 0.253 | 0.795 | 0.795 | 0.795 | 0.544 | 0.884 | 0.893 | 0.380 |
| *JRip* | 1 | 62.90 | 0.629 | 0.646 | 0.526 | 0.629 | 0.535 | −0.034 | 0.478 | 0.537 | 0.482 |
| *OneR* | 1 | 65.72 | 0.657 | 0.657 | 0.432 | 0.657 | 0.521 | 0.000 | 0.500 | 0.549 | 0.586 |
| *ZeroR* | - | 65.72 | 0.657 | 0.657 | 0.432 | 0.657 | 0.521 | 0.000 | 0.491 | 0.545 | 0.457 |
| *Tic-Tac-Toe-Endgame* | | | | | | | | | | | |
| *ENORA-ACC/RMSE* | 2 | 98.33 | 0.983 | 0.031 | 0.984 | 0.983 | 0.983 | 0.963 | 0.976 | 0.973 | 0.129 |
| *PART* | 49 | 94.26 | 0.943 | 0.076 | 0.942 | 0.943 | 0.942 | 0.873 | 0.974 | 0.969 | 0.220 |
| *JRip* | 9 | 97.81 | 0.978 | 0.031 | 0.978 | 0.978 | 0.978 | 0.951 | 0.977 | 0.977 | 0.138 |
| *OneR* | 1 | 69.94 | 0.699 | 0.357 | 0.701 | 0.699 | 0.700 | 0.340 | 0.671 | 0.651 | 0.548 |
| *ZeroR* | - | 65.35 | 0.653 | 0.653 | 0.427 | 0.653 | 0.516 | 0.000 | 0.496 | 0.545 | 0.476 |
| *Car* | | | | | | | | | | | |
| *ENORA-RMSE* | 14 | 86.57 | 0.866 | 0.089 | 0.866 | 0.866 | 0.846 | 0.766 | 0.889 | 0.805 | 0.259 |
| *PART* | 68 | 95.78 | 0.958 | 0.016 | 0.959 | 0.958 | 0.958 | 0.929 | 0.990 | 0.979 | 0.1276 |
| *JRip* | 49 | 86.46 | 0.865 | 0.064 | 0.881 | 0.865 | 0.870 | 0.761 | 0.947 | 0.899 | 0.224 |
| *OneR* | 1 | 70.02 | 0.700 | 0.700 | 0.490 | 0.700 | 0.577 | 0.000 | 0.500 | 0.543 | 0.387 |
| *ZeroR* | - | 70.02 | 0.700 | 0.700 | 0.490 | 0.700 | 0.577 | 0.000 | 0.497 | 0.542 | 0.338 |
| *kr-vs-kp* | | | | | | | | | | | |
| *ENORA-RMSE* | 10 | 94.87 | 0.949 | 0.050 | 0.950 | 0.949 | 0.949 | 0.898 | 0.950 | 0.927 | 0.227 |
| *PART* | 23 | 99.06 | 0.991 | 0.010 | 0.991 | 0.991 | 0.991 | 0.981 | 0.997 | 0.996 | 0.088 |
| *JRip* | 16 | 99.19 | 0.992 | 0.008 | 0.992 | 0.992 | 0.992 | 0.984 | 0.995 | 0.993 | 0.088 |
| *OneR* | 1 | 66.46 | 0.665 | 0.350 | 0.675 | 0.665 | 0.655 | 0.334 | 0.657 | 0.607 | 0.579 |
| *ZeroR* | - | 52.22 | 0.522 | 0.522 | 0.273 | 0.522 | 0.358 | 0.000 | 0.499 | 0.500 | 0.500 |
| *Nursery* | | | | | | | | | | | |
| *ENORA-ACC* | 15 | 88.41 | 0.884 | 0.055 | 0.870 | 0.884 | 0.873 | 0.824 | 0.915 | 0.818 | 0.2153 |
| *PART* | 220 | 99.21 | 0.992 | 0.003 | 0.992 | 0.992 | 0.992 | 0.989 | 0.999 | 0.997 | 0.053 |
| *JRip* | 131 | 96.84 | 0.968 | 0.012 | 0.968 | 0.968 | 0.968 | 0.957 | 0.993 | 0.974 | 0.103 |
| *OneR* | 1 | 70.97 | 0.710 | 0.137 | 0.695 | 0.710 | 0.702 | 0.570 | 0.786 | 0.632 | 0.341 |
| *ZeroR* | - | 33.33 | 0.333 | 0.333 | 0.111 | 0.333 | 0.167 | 0.000 | 0.500 | 0.317 | 0.370 |

## 5. Analysis of Results and Discussion

The results of our tests allow for several considerations. The first interesting observation is that *NSGA-II* identifies fewer solutions than *ENORA* on the Pareto front, which implies less diversity and therefore a worse hypervolume ratio, as shown in Figures 3 and 4. This is not surprising: in several other occasions [19,34,60], it has been shown that *ENORA* maintains a higher diversity in the population than other well-known evolutionary algorithms, with generally positive influence on the final results. Comparing the results in full training mode against the results in cross-validation or in splitting mode makes it evident that our solution produces classification models that are more resilient to over-fitting. For example, the classifier learned by *PART* with *Monk's Problem 2* presents a 94.01% accuracy in full training mode that drops to 73.51% in splitting mode. A similar, although with a more contained drop in accuracy, is shown by the classifier learned with *Breast Cancer* dataset; at the same time, the classifier learned by *ENORA* driven by accuracy shows only a 5.57% drop in one case, and even an improvement in the other case (see Tables 5, 6, 9, and 10). This phenomenon is easily explained by looking at the number of rules: the more rules in a classifier, the higher the risk of over-fitting; *PART* produces very accurate classifiers, but at the price of adding many rules, which not only affects the interpretability of the model but also its resilience to over-fitting. Full training results seem to indicate that when the optimization model is driven by *RMSE* the classifiers are more accurate; nevertheless, they are also more prone to over-fitting, indicating that, on average, the optimization models driven by the accuracy are preferable.

From the statistical tests (whose results are shown in the Appendixes A and B) we conclude that among the six variants of the proposed optimization model there are no statistical significative differences, which suggests that the advantages of our method do not depend directly on a specific evolutionary algorithm or on the specific performance measure that is used to drive the evolutions. Significant statistical differences between our method and very simple classical methods such as *OneR* were expectable. Significant statistical differences between our method and a well-consolidated one such as *PART* have not been found, but the price to be paid for using *PART* in order to have similar results to ours is a very high number of rules (15 vs. 2 in one case and 47 vs. 7 in the other case).

We would like to highlight that both the *Breast Cancer* dataset and the *Monk's problem 2* dataset are difficult to approximate with interpretable classifiers and that none of the analyzed classifiers obtains high accuracy rates using the cross-validation technique. Even powerful black-box classifiers, such as *Random Forest* and *Logistic*, obtain success rates below 70% in 10-fold cross-validation for these datasets. However, *ENORA* obtains a better balance (trade-off) between precision and interpretability than the rest of the classifiers. For the rest of the analyzed datasets, the accuracy obtained using *ENORA* is substantially higher. For example, for the *Tic-Tac-Toe-Endgame* dataset, *ENORA* obtains a 98.3299% success percentage with only two rules in cross-validation, while *PART* obtains 94.2589% with 49 rules, and *JRip* obtains 97.8079% with nine rules. With respect to the results obtained in the datasets *Car*, *kr-vs-kp* and *Nursery*, we want to comment that better success percentage can be obtained if the maximum number of evaluations is increased. However, better success percentages imply a greater number of rules, which is to the detriment of the interpretability of the models.

## 6. Conclusions and Future Works

In this paper, we have proposed a novel technique for categorical classifier learning. Our proposal is based on defining the problem of learning a classifier as a multi-objective optimization problem, and solving it by suitably adapting an evolutionary algorithm to this task; our two objectives are minimizing the number of rules (for a better interpretability of the classifier) and maximizing a metric of performance. Depending on the particular metric that is chosen, (slightly) different optimization models arise. We have tested our proposal, in a first instance, on two different publicly available datasets, *Breast Cancer* (in which each instance represents a patient that has suffered from breast cancer and is described by nine attributes, and the class to be predicted represents the fact that the patient has suffered a recurring event) and *Monk's Problem 2* (which is an artificial, well-known dataset in which the class to be predicted represents a logical function), using two different evolutionary algorithms, namely *ENORA* and *NSGA-II*, and three different choices as a performance metric, i.e., accuracy, the area under the *ROC* curve, and the root mean square error. Additionally, we have shown the results of the evaluation in 10-fold cross-validation of the publicly available *Tic-Tac-Toe-Endgame*, *Car*, *kr-vs-kp* and *Nursery* datasets.

Our initial motivation was to design a classifier learning system that produces interpretable, yet accurate, classifiers: since interpretability is a direct function of the number of rules, we conclude that such an objective has been achieved. As an aside, observe that our approach allows the user to decide, beforehand, a maximum number of rules; this can also be done in *PART* and *JRip*, but only indirectly. Finally, the idea underlying our approach is that multiple classifiers are explored at the same time in the same execution, and this allows us to choose the best compromise between the performance and the interpretability of a classifier a posteriori.

As a future work, we envisage that our methodology can benefit from an *embedded* future selection mechanism. In fact, all attributes are (ideally) used in every rule of a classifier learned by our optimization model. By simply relaxing such a constraint, and by suitably re-defining the first objective in the optimization model (e.g., by minimizing the sum of the lengths of all rules, or similar measures), the resulting classifiers will naturally present rules that use more features as well as rules that use less (clearly, the implementation must be adapted to obtain an initial population in which the classifiers have rules of different lengths as well as mutation operators that allow a rule to grow or to shrink). Although this approach does not follow the classical definition of feature selection mechanisms (in which a subset of features is selected that reduces the dataset over which a classifier is learned), it is natural to imagine that it may produce even more accurate classifiers, and more interpretable at the same time.

*Entropy* **2018**, *20*, 684

Currently, we are implementing our own version of *multi-objective differential evolution* (*MODE*) for rule-based classification for inclusion in the Weka Open Source Software issued under the GNU General Public License. The implementation of other algorithms, such as *MOEA/D*, their adaptation in the Weka development platform and subsequent analysis and comparison are planned for future work.

**Author Contributions:** Conceptualization, F.J. and G.S. (Gracia Sánchez); Methodology, F.J. and G.S. (Guido Sciavicco); Software, G.S. (Gracia Sánchez) and C.M.; Validation, F.J., G.S. (Gracia Sánchez) and C.M.; Formal Analysis, F.J. and G.S. (Guido Sciavicco); Investigation, F.J. and G.S. (Gracia Sánchez); Resources, L.M.; Data Curation, L.M.; Writing—Original Draft Preparation, F.J., L.M. and G.S. (Guido Sciavicco); Writing—Review and Editing, F.J., L.M. and G.S. (Guido Sciavicco); Visualization, F.J.; Supervision, F.J.; Project Administration, F.J.; and Funding Acquisition, F.J., L.M., G.S. (Gracia Sánchez) and G.S. (Guido Sciavicco).

**Funding:** This research received no external funding.

**Acknowledgments:** This study was partially supported by computing facilities of Extremadura Research Centre for Advanced Technologies (CETA-CIEMAT), funded by the European Regional Development Fund (ERDF). CETA-CIEMAT belongs to CIEMAT and the Government of Spain.

**Conflicts of Interest:** The authors declare no conflict of interest.

## Abbreviations

The following abbreviations are used in this manuscript:

| | |
|---|---|
| *ML* | Machine learning |
| *ANN* | Artificial neural networks |
| *DLNN* | Deep learning neural networks |
| *CEO* | Chief executive officer |
| *SVM* | Support vector machines |
| *IBL* | Instance-based learning |
| *DT* | Decision trees |
| *RBC* | Rule-based classifiers |
| *ROC* | Receiver operating characteristic |
| *RMSE* | Root mean square error performance metric |
| *FL* | Fuzzy logic |
| *MOEA* | Multi-objective evolutionary algorithms |
| *NSGA-II* | Non-dominated sorting genetic algorithm, 2nd version |
| *ENORA* | Evolutionary non-dominated radial slots based algorithm |
| *PART* | Partial decision tree classifier |
| *JRip* | *RIPPER* classifier of *Weka* |
| *RIPPER* | Repeated incremental pruning to produce error reduction |
| *OneR* | One rule classifier |
| *ZeroR* | Zero rule classifier |
| *ENORA-ACC* | *ENORA* with objective function defined as accuracy |
| *ENORA-AUC* | *ENORA* with objective function defined as area under the *ROC* curve |
| *ENORA-RMSE* | *ENORA* with *RMSE* objective function |
| *NSGA-II-ACC* | *NSGA-II* with objective function defined as accuracy |
| *NSGA-II-AUC* | *NSGA-II* with objective function defined as area under the *ROC* curve |
| *NSGA-II-RMSE* | *NSGA-II* with *RMSE* objective function |
| *HVR* | Hypervolume ratio |
| *TP* | True positive |
| *FP* | False positive |
| *MCC* | Matthews correlation coefficient |
| *PRC* | Precision-recall curve |

## Appendix A. Statistical Tests for *Breast Cancer* Dataset

**Table A1.** Shapiro–Wilk normality test *p*-values for *percent correct* metric—*Breast Cancer* dataset.

| Algorithm | *p*-Value | Null Hypothesis |
|---|---|---|
| *ENORA-ACC* | 0.5316 | Not Rejected |
| *ENORA-AUC* | 0.3035 | Not Rejected |
| *ENORA-RMSE* | 0.7609 | Not Rejected |
| *NSGA-II-ACC* | 0.1734 | Not Rejected |
| *NSGA-II-AUC* | 0.3802 | Not Rejected |
| *NSGA-II-RMSE* | 0.6013 | Not Rejected |
| *PART* | 0.0711 | Not Rejected |
| *JRip* | 0.5477 | Not Rejected |
| *OneR* | 0.316 | Not Rejected |
| *ZeroR* | $\mathbf{3.818 \times 10^{-06}}$ | Rejected |

**Table A2.** Friedman *p*-value for *percent correct* metric—*Breast Cancer* dataset.

| | *p*-Value | Null Hypothesis |
|---|---|---|
| Friedman | $\mathbf{5.111 \times 10^{-04}}$ | Rejected |

**Table A3.** Nemenyi post-hoc procedure for *percent correct* metric—*Breast Cancer* dataset.

| | *ENORA-ACC* | *ENORA-AUC* | *ENORA-RMSE* | *NSGA-II-ACC* | *NSGA-II-AUC* | *NSGA-II-RMSE* | *PART* | *JRip* | *OneR* |
|---|---|---|---|---|---|---|---|---|---|
| *ENORA-AUC* | 0.2597 | - | - | - | - | - | - | - | - |
| *ENORA-RMSE* | 0.9627 | 0.9627 | - | - | - | - | - | - | - |
| *NSGA-II-ACC* | 0.9981 | 0.8047 | 1.0000 | - | - | - | - | - | - |
| *NSGA-II-AUC* | 0.2951 | 1.0000 | 0.9735 | 0.8386 | - | - | - | - | - |
| *NSGA-II-RMSE* | 1.0000 | 0.2169 | 0.9436 | 0.9960 | 0.2486 | - | - | - | - |
| *PART* | 0.1790 | 1.0000 | 0.9186 | 0.6997 | 1.0000 | 0.1461 | - | - | - |
| *JRip* | 0.9909 | 0.8956 | 1.0000 | 1.0000 | 0.9186 | 0.9840 | 0.8164 | - | - |
| *OneR* | **0.0004** | 0.6414 | **0.0451** | **0.0108** | 0.5961 | **0.0002** | 0.7546 | **0.0212** | - |
| *ZeroR* | 0.2377 | 1.0000 | 0.9538 | 0.7803 | 1.0000 | 0.1973 | 1.0000 | 0.8783 | 0.6709 |

**Table A4.** Summary of statistically significant differences for *percent correct* metric—*Breast Cancer* dataset.

| | *ENORA-ACC* | *ENORA-RMSE* | *NSGA-II-ACC* | *NSGA-II-RMSE* | *JRip* |
|---|---|---|---|---|---|
| *OneR* | *ENORA-ACC* | *ENORA-RMSE* | *NSGA-II-ACC* | *NSGA-II-RMSE* | *JRip* |

**Table A5.** Shapiro–Wilk normality test *p*-values for *area under the ROC curve* metric—*Breast Cancer* dataset.

| Algorithm | *p*-Value | Null Hypothesis |
|---|---|---|
| *ENORA-ACC* | 0.6807 | Not Rejected |
| *ENORA-AUC* | 0.3171 | Not Rejected |
| *ENORA-RMSE* | 0.6125 | Not Rejected |
| *NSGA-II-ACC* | 0.0871 | Not Rejected |
| *NSGA-II-AUC* | 0.5478 | Not Rejected |
| *NSGA-II-RMSE* | 0.6008 | Not Rejected |
| *PART* | 0.6066 | Not Rejected |
| *JRip* | 0.2978 | Not Rejected |
| *OneR* | 0.4531 | Not Rejected |
| *ZeroR* | **0.0000** | Rejected |

**Table A6.** Friedman *p*-value for *area under the ROC curve* metric—*Breast Cancer* dataset.

|  | *p*-Value | Null Hypothesis |
|---|---|---|
| Friedman | $8.232 \times 10^{-10}$ | Rejected |

**Table A7.** Nemenyi post-hoc procedure for *area under the ROC curve* metric—*Breast Cancer* dataset.

|  | ENORA-ACC | ENORA-AUC | ENORA-RMSE | NSGA-II-ACC | NSGA-II-AUC | NSGA-II-RMSE | PART | JRip | OneR |
|---|---|---|---|---|---|---|---|---|---|
| ENORA-AUC | 1.0000 | - | - | - | - | - | - | - | - |
| ENORA-RMSE | 0.9972 | 0.9990 | - | - | - | - | - | - | - |
| NSGA-II-ACC | 0.9999 | 1.0000 | 1.0000 | - | - | - | - | - | - |
| NSGA-II-AUC | 1.0000 | 1.0000 | 1.0000 | 1.0000 | - | - | - | - | - |
| NSGA-II-RMSE | 0.9990 | 0.9997 | 1.0000 | 1.0000 | 1.0000 | - | - | - | - |
| PART | 0.9999 | 1.0000 | 1.0000 | 1.0000 | 1.0000 | 1.0000 | - | - | - |
| JRip | 1.0000 | 1.0000 | 0.9992 | 1.0000 | 1.0000 | 0.9998 | 1.0000 | - | - |
| OneR | 0.0041 | 0.0062 | 0.0790 | 0.0323 | 0.0281 | 0.0582 | 0.0345 | 0.0067 | - |
| ZeroR | $3.8 \times 10^{-07}$ | $7.2 \times 10^{-07}$ | $4.6 \times 10^{-05}$ | $9.8 \times 10^{-06}$ | $7.8 \times 10^{-06}$ | $2.7 \times 10^{-05}$ | $1.1 \times 10^{-05}$ | $8.1 \times 10^{-07}$ | 0.6854 |

**Table A8.** Summary of statistically significant differences for *area under the ROC curve* metric—*Breast Cancer* dataset.

|  | ENORA-ACC | ENORA-AUC | ENORA-RMSE | NSGA-II-ACC | NSGA-II-AUC | NSGA-II-RMSE | PART | JRip |
|---|---|---|---|---|---|---|---|---|
| OneR | ENORA-ACC | ENORA-AUC | - | NSGA-II-ACC | NSGA-II-AUC | - | PART | JRip |
| ZeroR | ENORA-ACC | ENORA-AUC | ENORA-RMSE | NSGA-II-ACC | NSGA-II-AUC | NSGA-II-RMSE | PART | JRip |

**Table A9.** Shapiro–Wilk normality test *p*-values for *serialized model size* metric—*Breast Cancer* dataset.

| Algorithm | *p*-Value | Null Hypothesis |
|---|---|---|
| ENORA-ACC | $5.042 \times 10^{-05}$ | Rejected |
| ENORA-AUC | $2.997 \times 10^{-07}$ | Rejected |
| ENORA-RMSE | $4.762 \times 10^{-04}$ | Rejected |
| NSGA-II-ACC | $4.88 \times 10^{-06}$ | Rejected |
| NSGA-II-AUC | $2.339 \times 10^{-07}$ | Rejected |
| NSGA-II-RMSE | $2.708 \times 10^{-06}$ | Rejected |
| PART | 0.3585 | Not Rejected |
| JRip | $9.086 \times 10^{-03}$ | Rejected |
| OneR | $1.007 \times 10^{-07}$ | Rejected |
| ZeroR | 0.0000 | Rejected |

**Table A10.** Friedman *p*-value for *serialized model size* metric—*Breast Cancer* dataset.

|  | *p*-Value | Null Hypothesis |
|---|---|---|
| Friedman | $2.2 \times 10^{-16}$ | Rejected |

**Table A11.** Nemenyi post-hoc procedure for *serialized model size* metric—*Breast Cancer* dataset.

|  | ENORA-ACC | ENORA-AUC | ENORA-RMSE | NSGA-II-ACC | NSGA-II-AUC | NSGA-II-RMSE | PART | JRip | OneR |
|---|---|---|---|---|---|---|---|---|---|
| ENORA-AUC | 0.9998 | - | - | - | - | - | - | - | - |
| ENORA-RMSE | 0.0053 | 0.0004 | - | - | - | - | - | - | - |
| NSGA-II-ACC | 0.3871 | 0.0942 | 0.8872 | - | - | - | - | - | - |
| NSGA-II-AUC | 0.8872 | 0.4894 | 0.3871 | 0.9988 | - | - | - | - | - |
| NSGA-II-RMSE | $4.1 \times 10^{-05}$ | $1.3 \times 10^{-06}$ | 0.9860 | 0.2169 | 0.0244 | - | - | - | - |
| PART | $4.7 \times 10^{-09}$ | $5.6 \times 10^{-11}$ | 0.1973 | 0.0013 | $3.3 \times 10^{-05}$ | 0.8689 | - | - | - |
| JRip | 0.2712 | 0.6997 | $1.2 \times 10^{-08}$ | $7.0 \times 10^{-05}$ | 0.0025 | $6.3 \times 10^{-12}$ | $6.9 \times 10^{-14}$ | - | - |
| OneR | 0.0062 | 0.0546 | $1.5 \times 10^{-12}$ | $5.5 \times 10^{-08}$ | $5.5 \times 10^{-06}$ | $8.3 \times 10^{-14}$ | $8.3 \times 10^{-14}$ | 0.9584 | - |
| ZeroR | $1.9 \times 10^{-05}$ | 0.0004 | $7.3 \times 10^{-14}$ | $8.6 \times 10^{-12}$ | $2.3 \times 10^{-09}$ | $8.5 \times 10^{-14}$ | $<2 \times 10^{-16}$ | 0.2377 | 0.9584 |

**Table A12.** Summary of statistically significant differences for *serialized model size* metric—*Breast Cancer* dataset.

| | ENORA-ACC | ENORA-AUC | ENORA-RMSE | NSGA-II-ACC | NSGA-II-AUC | NSGA-II-RMSE | PART |
|---|---|---|---|---|---|---|---|
| ENORA-RMSE | ENORA-ACC | NSGA-II-AUC | - | - | - | - | - |
| NSGA-II-RMSE | ENORA-ACC | ENORA-AUC | - | - | NSGA-II-AUC | - | - |
| PART | ENORA-ACC | ENORA-AUC | - | NSGA-II-ACC | NSGA-II-AUC | - | - |
| JRip | - | - | JRip | JRip | JRip | JRip | JRip |
| OneR | OneR | - | OneR | OneR | OneR | OneR | OneR |
| ZeroR | ZeroR | ZeroR | ZeroR | ZeroR | ZeroR | ZeroR | ZeroR |

## Appendix B. Statistical Tests for *Monk's Problem* 2 Dataset

**Table A13.** Shapiro–Wilk normality test *p*-values for *percent correct* metric—*Monk's Problem 2* dataset.

| Algorithm | *p*-Value | Null Hypothesis |
|---|---|---|
| ENORA-ACC | 0.6543 | Not Rejected |
| ENORA-AUC | 0.6842 | Not Rejected |
| ENORA-RMSE | **0.0135** | Rejected |
| NSGA-II-ACC | 0.979 | Not Rejected |
| NSGA-II-AUC | 0.382 | Not Rejected |
| NSGA-II-RMSE | **0.0486** | Rejected |
| PART | 0.5671 | Not Rejected |
| JRip | **0.075** | Rejected |
| OneR | $\mathbf{4.672 \times 10^{-06}}$ | Rejected |
| ZeroR | $\mathbf{4.672 \times 10^{-06}}$ | Rejected |

**Table A14.** Friedman *p*-value for *percent correct* metric—*Monk's Problem 2* dataset.

| | *p*-Value | Null Hypothesis |
|---|---|---|
| Frideman | $\mathbf{1.292 \times 10^{-07}}$ | Rejected |

**Table A15.** Nemenyi post-hoc procedure for *percent correct* metric—*Monk's Problem 2* dataset.

| | ENORA-ACC | ENORA-AUC | ENORA-RMSE | NSGA-II-ACC | NSGA-II-AUC | NSGA-II-RMSE | PART | JRip | OneR |
|---|---|---|---|---|---|---|---|---|---|
| ENORA-AUC | 0.8363 | - | - | - | - | - | - | - | - |
| ENORA-RMSE | 1.0000 | 0.9471 | - | - | - | - | - | - | - |
| NSGA-II-ACC | 0.1907 | 0.9902 | 0.3481 | - | - | - | - | - | - |
| NSGA-II-AUC | **0.0126** | 0.6294 | **0.0342** | 0.9958 | - | - | - | - | - |
| NSGA-II-RMSE | **0.0126** | 0.6294 | **0.0342** | 0.9958 | 1.0000 | - | - | - | - |
| PART | 0.8714 | 1.0000 | 0.9631 | 0.9841 | 0.5769 | 0.5769 | - | - | - |
| JRip | $\mathbf{2.1 \times 10^{-06}}$ | **0.0048** | $\mathbf{1.0 \times 10^{-05}}$ | 0.1341 | 0.6806 | 0.6806 | **0.0036** | - | - |
| OneR | **0.0001** | 0.0743 | **0.0006** | 0.6032 | 0.9875 | 0.9875 | 0.0601 | 0.9984 | - |
| ZeroR | **0.0001** | 0.0743 | **0.0006** | 0.6032 | 0.9875 | 0.9875 | 0.0601 | 0.9984 | 1.0000 |

**Table A16.** Summary of statistically significant differences for *percent correct* metric—*Monk's Problem 2* dataset.

| | ENORA-ACC | ENORA-AUC | ENORA-RMSE | PART |
|---|---|---|---|---|
| NSGA-II-AUC | ENORA-ACC | - | ENORA-RMSE | - |
| NSGA-II-RMSE | ENORA-ACC | - | ENORA-RMSE | - |
| JRip | ENORA-ACC | ENORA-AUC | ENORA-RMSE | PART |
| OneR | ENORA-ACC | - | ENORA-RMSE | - |
| ZeroR | ENORA-ACC | - | ENORA-RMSE | - |

**Table A17.** Shapiro–Wilk normality test *p*-values for *area under the ROC curve* metric—*Monk's Problem 2* dataset.

| Algorithm | *p*-Value | Null Hypothesis |
|---|---|---|
| *ENORA-ACC* | 0.4318 | Not Rejected |
| *ENORA-AUC* | 0.7044 | Not Rejected |
| *ENORA-RMSE* | **0.0033** | Rejected |
| *NSGA-II-ACC* | 0.3082 | Not Rejected |
| *NSGA-II-AUC* | **0.0243** | Rejected |
| *NSGA-II-RMSE* | 0.7802 | Not Rejected |
| *PART* | 0.1641 | Not Rejected |
| *JRip* | 0.3581 | Not Rejected |
| *OneR* | **0.0000** | Rejected |
| *ZeroR* | **0.0000** | Rejected |

**Table A18.** Friedman *p*-value for *area under the ROC curve* metric—*Monk's Problem 2* dataset.

| | *p*-Value | Null Hypothesis |
|---|---|---|
| Frideman | $\mathbf{1.051 \times 10^{-08}}$ | Rejected |

**Table A19.** Nemenyi post-hoc procedure for *area under the ROC curve* metric—*Monk's Problem 2* dataset.

| | *ENORA-ACC* | *ENORA-AUC* | *ENORA-RMSE* | *NSGA-II-ACC* | *NSGA-II-AUC* | *NSGA-II-RMSE* | *PART* | *JRip* | *OneR* |
|---|---|---|---|---|---|---|---|---|---|
| *ENORA-AUC* | 0.8363 | - | - | - | - | - | - | - | - |
| *ENORA-RMSE* | 1.0000 | 0.7054 | - | - | - | - | - | - | - |
| *NSGA-II-ACC* | 0.8870 | 0.0539 | 0.9556 | - | - | - | - | - | - |
| *NSGA-II-AUC* | 1.0000 | 0.8544 | 1.0000 | 0.8713 | - | - | - | - | - |
| *NSGA-II-RMSE* | 0.5504 | **0.0084** | 0.7054 | 0.9999 | 0.5239 | - | - | - | - |
| *PART* | 0.7054 | 1.0000 | 0.5504 | **0.0269** | 0.7295 | **0.0036** | - | - | - |
| *JRip* | **0.0238** | $\mathbf{2.3 \times 10^{-05}}$ | **0.0482** | 0.6806 | **0.0211** | 0.9471 | $\mathbf{7.0 \times 10^{-06}}$ | - | - |
| *OneR* | **0.0084** | $\mathbf{4.7 \times 10^{-06}}$ | **0.0186** | 0.4715 | **0.0073** | 0.8363 | $\mathbf{1.4 \times 10^{-06}}$ | 1.0000 | - |
| *ZeroR* | **0.0084** | $\mathbf{4.7 \times 10^{-06}}$ | **0.0186** | 0.4715 | **0.0073** | 0.8363 | $\mathbf{1.4 \times 10^{-06}}$ | 1.0000 | 1.0000 |

**Table A20.** Summary of statistically significant differences for *area under the ROC curve* metric—*Monk's Problem 2* dataset.

| | *ENORA-ACC* | *ENORA-AUC* | *ENORA-RMSE* | *NSGA-II-ACC* | *NSGA-II-AUC* | *PART* | |
|---|---|---|---|---|---|---|---|
| *NSGA-II-RMSE* | - | *ENORA-AUC* | - | - | - | - | - |
| *PART* | - | - | - | *PART* | - | *PART* | - |
| *JRip* | *ENORA-ACC* | *ENORA-AUC* | *ENORA-RMSE* | - | *NSGA-II-AUC* | - | *PART* |
| *OneR* | *ENORA-ACC* | *ENORA-AUC* | *ENORA-RMSE* | - | *NSGA-II-AUC* | - | *PART* |
| *ZeroR* | *ENORA-ACC* | *ENORA-AUC* | *ENORA-RMSE* | - | *NSGA-II-AUC* | - | *PART* |

**Table A21.** Shapiro–Wilk normality test *p*-values for *serialized model size* metric—*Monk's Problem 2* dataset.

| Algorithm | *p*-Value | Null Hypothesis |
|---|---|---|
| *ENORA-ACC* | $\mathbf{4.08 \times 10^{-05}}$ | Rejected |
| *ENORA-AUC* | **0.0002** | Rejected |
| *ENORA-RMSE* | **0.0094** | Rejected |
| *NSGA-II-ACC* | **0.0192** | Rejected |
| *NSGA-II-AUC* | **0.0846** | Rejected |
| *NSGA-II-RMSE* | **0.0037** | Rejected |
| *PART* | 0.9721 | Not Rejected |
| *JRip* | **0.0068** | Rejected |
| *OneR* | **0.0000** | Rejected |
| *ZeroR* | **0.0000** | Rejected |

**Table A22.** Friedman *p*-value for *serialized model size* metric—*Monk's Problem 2* dataset.

|  | *p*-**Value** | **Null Hypothesis** |
|---|---|---|
| Frideman | $\mathbf{2.657 \times 10^{-13}}$ | Rejected |

**Table A23.** Nemenyi post-hoc procedure for *serialized model size* metric—*Monk's Problem 2* dataset.

|  | *ENORA-ACC* | *ENORA-AUC* | *ENORA-RMSE* | *NSGA-II-ACC* | *NSGA-II-AUC* | *NSGA-II-RMSE* | *PART* | *JRip* | *OneR* |
|---|---|---|---|---|---|---|---|---|---|
| *ENORA-AUC* | 1.0000 | - | - | - | - | - | - | - | - |
| *ENORA-RMSE* | 1.0000 | 1.0000 | - | - | - | - | - | - | - |
| *NSGA-II-ACC* | 1.0000 | 1.0000 | 1.0000 | - | - | - | - | - | - |
| *NSGA-II-AUC* | 0.9925 | 0.9696 | 0.9984 | 0.9841 | - | - | - | - | - |
| *NSGA-II-RMSE* | 0.8870 | 0.9556 | 0.7966 | 0.9267 | 0.2622 | - | - | - | - |
| *PART* | 0.2824 | 0.1752 | 0.3957 | 0.2246 | 0.9015 | 0.0027 | - | - | - |
| *JRip* | 0.1752 | 0.2824 | 0.1110 | 0.2246 | 0.0084 | 0.9752 | $1.0 \times 10^{-05}$ | - | - |
| *OneR* | 0.0211 | 0.0431 | 0.0110 | 0.0304 | 0.0004 | 0.6552 | $1.5 \times 10^{-07}$ | 0.9993 | - |
| *ZeroR* | 0.0012 | 0.0031 | 0.0006 | 0.0020 | $1.0 \times 10^{-05}$ | 0.1907 | $1.3 \times 10^{-09}$ | 0.9015 | 0.9993 |

**Table A24.** Summary of statistically significant differences for *serialized model size* metric—*Monk's Problem 2* dataset.

|  | *ENORA-ACC* | *ENORA-AUC* | *ENORA-RMSE* | *NSGA-II-ACC* | *NSGA-II-AUC* | *NSGA-II-RMSE* | *PART* |
|---|---|---|---|---|---|---|---|
| *PART* | - | - | - | - | - | NSGA-II-RMSE | - |
| *JRip* | - | - | - | - | *JRip* | - | *JRip* |
| *OneR* | *OneR* | *OneR* | *OneR* | *OneR* | *OneR* | - | *OneR* |
| *ZeroR* | *ZeroR* | *ZeroR* | *ZeroR* | *ZeroR* | *ZeroR* | - | *ZeroR* |

## Appendix C. Nomenclature

**Table A25.** Nomenclature table (Part I).

| Symbol | Definition |
|---|---|
| | *Equation (1): Multi-objective constrained optimization* |
| $x_k$ | k-th decision variable |
| $\mathbf{x}$ | Set of decision variables |
| $f_i(\mathbf{x})$ | i-th objective function |
| $g_j(\mathbf{x})$ | j-th constraint |
| $l > 0$ | Number of objectives |
| $m > 0$ | Number of constraints |
| $w > 0$ | Number of decision variables |
| $\mathcal{X}$ | Domain for each each decision variable $x_k$ |
| $\mathcal{X}^w$ | Domain for the set of decision variables |
| $\mathcal{F}$ | Set of all feasible solutions |
| $\mathcal{S}$ | Set of non-dominated solutions or Pareto optimal set |
| $\mathcal{D}(\mathbf{x}',\mathbf{x})$ | Pareto domination function |
| | *Equation (2): Rule-based classification for categorical data* |
| $\mathcal{D}$ | Dataset |
| $x_i$ | *ith* categorical input attribute in the dataset $\mathcal{D}$ |
| $\mathbf{x}$ | Categorical input attributes in the dataset $\mathcal{D}$ |
| $y$ | Categorical output attribute in the dataset $\mathcal{D}$ |
| $\{1,\ldots,v_i\}$ | Domain of i-th categorical input attribute in the dataset $\mathcal{D}$ |
| $\{1,\ldots,w\}$ | Domain of categorical output attribute in the dataset $\mathcal{D}$ |
| $p \geq 0$ | Number of categorical input attributes in the dataset $\mathcal{D}$ |
| $\Gamma$ | Rule-based classifier |
| $R_i^\Gamma$ | *ith* rule of classifier $\Gamma$ |
| $b_{ij}^\Gamma$ | Category for *jth* categorical input attribute and *ith* rule of classifier $\Gamma$ |

**Table A25.** *Cont.*

| Symbol | Definition |
| --- | --- |
| $c_i^\Gamma$ | Category for categorical output attribute and *ith* rule of classifier $\Gamma$ |
| $\varphi_i^\Gamma(\mathbf{x})$ | Compatibility degree of the *ith* rule of classifier $\Gamma$ for the example $\mathbf{x}$ |
| $\mu_{ij}^\Gamma(\mathbf{x})$ | Result of the *ith* rule of classifier $\Gamma$ and *jth* categorical input attribute $x_j$ |
| $\lambda_c^\Gamma(\mathbf{x})$ | Association degree of classifier $\Gamma$ for the example $\mathbf{x}$ with the class $c$ |
| $\eta_{ic}^\Gamma(\mathbf{x})$ | Result of of the *ith* rule of classifier $\Gamma$ for the example $\mathbf{x}$ with the class $c$ |
| $f_\Gamma(\mathbf{x})$ | Classification or output of the classifier $\Gamma$ for the example $\mathbf{x}$ |
| *Equation (3): Multi-objective constrained optimization problem for rule-based classification* | |
| $\mathcal{F}_\mathcal{D}(\Gamma)$ | Performance objective function of the classifier $\Gamma$ in the dataset $\mathcal{D}$ |
| $\mathcal{NR}(\Gamma)$ | Number of rules of the classifier $\Gamma$ |
| $M_{max}$ | Maximum number of rules allowed for classifiers |
| *Equations (4)–(6): Optimization models* | |
| $\mathcal{ACC}_\mathcal{D}(\Gamma)$ | *Acurracy*: proportion of correctly classified instances with the classifier $\Gamma$ in the dataset $\mathcal{D}$ |
| $K$ | Number of instances in the dataset $\mathcal{D}$ |
| $T_\mathcal{D}(\Gamma, i)$ | Result of the classification of the *ith* instance in the dataset $\mathcal{D}$ with the classifier $\Gamma$ |
| $\hat{c}_i^\Gamma$ | Predicted value of the *ith* instance in the dataset $\mathcal{D}$ with the classifier $\Gamma$ |
| $c_\mathcal{D}^i$ | Corresponding true value for the *ith* instance in the dataset $\mathcal{D}$. |
| $\mathcal{AUC}_\mathcal{D}(\Gamma)$ | Area under the *ROC* curve obtained with the classifier $\Gamma$ in the dataset $\mathcal{D}$. |
| $S_\mathcal{D}(\Gamma, t)$ | *Sensitivity*: proportion of positive instances classified as positive with the classifier $\Gamma$ in the dataset $\mathcal{D}$ |
| $1 - E_\mathcal{D}(\Gamma, t)$ | *Specificity*: proportion of negative instances classified as negative with the classifier $\Gamma$ in the dataset $\mathcal{D}$ |
| $t$ | Discrimination threshold |
| $\mathcal{RMSE}_\mathcal{D}(\Gamma)$ | Square root of the *mean square error* obtained with the classifier $\Gamma$ in the dataset $\mathcal{D}$ |

**Table A26.** Nomenclature table (Part II).

| *Equations (7) and (8): Hypervolume metric* | |
| --- | --- |
| $P$ | Population |
| $Q \subseteq P$ | Set of non-dominated individuals of $P$ |
| $v_i$ | Volume of the search space dominated by the individual $i$ |
| $HV(P)$ | Hypervolume: volume of the search space dominated by population $P$ |
| $H(P)$ | Volume of the search space non-dominated by population $P$ |
| $HVR(P)$ | Hypervolume ratio: ratio of $H(P)$ over the volume of the entire search space |
| $VS$ | Volume of the search space |
| $\mathcal{F}_\mathcal{D}^{lower}$ | Minimum value for objective $F_D$ |
| $\mathcal{F}_\mathcal{D}^{upper}$ | Maximum value for objective $F_D$ |
| $\mathcal{NR}^{lower}$ | Minimum value for objective $NR$ |
| $\mathcal{NR}^{upper}$ | Maximum value for objective $NR$ |

## References


1. Bishop, C.M. *Pattern Recognition and Machine Learning (Information Science and Statistics)*; Springer: Berlin/Heidelberg, Germany, 2006.
2. Russell, S.; Norvig, P. *Artificial Intelligence: A Modern Approach*, 3rd ed.; Prentice Hall Press: Upper Saddle River, NJ, USA, 2009.
3. Davalo, É. *Neural Networks*; MacMillan Computer Science; Macmillan Education: London, UK, 1991.
4. Goodfellow, I.; Bengio, Y.; Courville, A. *Deep Learning*; MIT Press: Cambridge, MA, USA, 2016.
5. Cortes, C.; Vapnik, V. Support-vector networks. *Mach. Learn.* **1995**, *20*, 273–297. [CrossRef]
6. Aha, D.W.; Kibler, D.; Albert, M.K. Instance-based learning algorithms. *Mach. Learn.* **1991**, *6*, 37–66. [CrossRef]
7. Gacto, M.; Alcalá, R.; Herrera, F. Interpretability of linguistic fuzzy rule-based systems: An overview of interpretability measures. *Inf. Sci.* **2011**, *181*, 4340–4360. [CrossRef]
8. Cano, A.; Zafra, A.; Ventura, S. An EP algorithm for learning highly interpretable classifiers. In Proceedings of the 11th International Conference on Intelligent Systems Design and Applications, Cordoba, Spain, 22–24 November 2011; pp. 325–330.

9.  Liu, H.; Gegov, A. Collaborative Decision Making by Ensemble Rule Based Classification Systems. In *Granular Computing and Decision-Making: Interactive and Iterative Approaches*; Pedrycz, W., Chen, S.-M., Eds.; Springer International Publishing: Cham, Switzerland, 2015; pp. 245–264.

10. Sulzmann, J.N.; Fürnkranz, J. *Rule Stacking: An Approach for Compressing an Ensemble of Rule Sets into a Single Classifier*; Elomaa, T., Hollmén, J., Mannila, H., Eds.; Discovery Science; Springer: Berlin/Heidelberg, Germany, 2011; pp. 323–334.

11. Jin, Y. Fuzzy Modeling of High-Dimensional Systems: Complexity Reduction and Interpretability Improvement. *IEEE Trans. Fuzzy Syst.* **2000**, *8*, 212–220. [CrossRef]

12. Breiman, L.; Friedman, J.H.; Olshen, R.A.; Stone, C.J. *Classification and Regression Trees*; Wadsworth and Brooks: Monterey, CA, USA, 1984.

13. Novák, V.; Perfilieva, I.; Mockor, J. *Mathematical Principles of Fuzzy Logic*; Springer Science + Business Media: Heidelberg, Germany, 2012.

14. Freund, Y.; Schapire, R.E. A Short Introduction to Boosting. In Proceedings of the Sixteenth International Joint Conference on Artificial Intelligence, Stockholm, Sweden, 31 July–6 August 1999; pp. 1401–1406.

15. Deb, K. *Multi-Objective Optimization Using Evolutionary Algorithms*; John Wiley and Sons: London, UK, 2001.

16. Coello, C.A.C.; van Veldhuizen, D.A.; Lamont, G.B. *Evolutionary Algorithms for Solving Multi-Objective Problems*; Kluwer Academic/Plenum Publishers: New York, NY, USA, 2002.

17. Jiménez, F.; Gómez-Skarmeta, A.; Sánchez, G.; Deb, K. An evolutionary algorithm for constrained multi-objective optimization. In Proceedings of the 2002 Congress on Evolutionary Computation, Honolulu, HI, USA, 12–17 May 2002; pp. 1133–1138.

18. Jiménez, F.; Sánchez, G.; Juárez, J.M. Multi-objective evolutionary algorithms for fuzzy classification in survival prediction. *Artif. Intell. Med.* **2014**, *60*, 197–219. [CrossRef] [PubMed]

19. Jiménez, F.; Marzano, E.; Sánchez, G.; Sciavicco, G.; Vitacolonna, N. Attribute selection via multi-objective evolutionary computation applied to multi-skill contact center data classification. In Proceedings of the 2015 IEEE Symposium Series on Computational Intelligence, Cape Town, South Africa, 7–10 December 2015; pp. 488–495.

20. Jiménez, F.; Jódar, R.; del Pilar Martín, M.; Sánchez, G.; Sciavicco, G. Unsupervised feature selection for interpretable classification in behavioral assessment of children. *Expert Syst.* **2017**, *34*, e12173. [CrossRef]

21. Rey, M.; Galende, M.; Fuente, M.; Sainz-Palmero, G. Multi-objective based Fuzzy Rule Based Systems (FRBSs) for trade-off improvement in accuracy and interpretability: A rule relevance point of view. *Knowl.-Based Syst.* **2017**, *127*, 67–84. doi:10.1016/j.knosys.2016.12.028. [CrossRef]

22. Ducange, P.; Lazzerini, B.; Marcelloni, F. Multi-objective genetic fuzzy classifiers for imbalanced and cost-sensitive datasets. *Soft Comput.* **2010**, *14*, 713–728. doi:10.1007/s00500-009-0460-y. [CrossRef]

23. Gorzalczany, M.B.; Rudzinski, F. A multi-objective genetic optimization for fast, fuzzy rule-based credit classification with balanced accuracy and interpretability. *Appl. Soft Comput.* **2016**, *40*, 206–220. doi:10.1016/j.asoc.2015.11.037. [CrossRef]

24. Ducange, P.; Mannara, G.; Marcelloni, F.; Pecori, R.; Vecchio, M. A novel approach for internet traffic classification based on multi-objective evolutionary fuzzy classifiers. In Proceedings of the 2017 IEEE International Conference on Fuzzy Systems (FUZZ-IEEE), Naples, Italy, 9–12 July 2017; pp. 1–6.

25. Antonelli, M.; Bernardo, D.; Hagras, H.; Marcelloni, F. Multiobjective Evolutionary Optimization of Type-2 Fuzzy Rule-Based Systems for Financial Data Classification. *IEEE Trans. Fuzzy Syst.* **2017**, *25*, 249–264. doi:10.1109/TFUZZ.2016.2578341. [CrossRef]

26. Carmona, C.J.; González, P.; Deljesus, M.J.; Herrera, F. NMEEF-SD: Non-dominated multiobjective evolutionary algorithm for extracting fuzzy rules in subgroup discovery. *IEEE Trans. Fuzzy Syst.* **2010**, *18*, 958–970. [CrossRef]

27. Hubertus, T.; Klaus, M.; Eberhard, T. *Optimization Theory*; Kluwer Academic Publishers: Dordrecht, The Netherlands, 2004.

28. Sinha, S. *Mathematical Programming: Theory and Methods*; Elsevier: Amsterdam, The Netherlands, 2006.

29. Collette, Y.; Siarry, P. *Multiobjective Optimization: Principles and Case Studies*; Springer-Verlag Berlin Heidelberg: New York, NY, USA, 2004.

30. Karloff, H. *Linear Programming*; Birkhauser Basel: Boston, MA, USA, 1991.

31. Maros, I.; Mitra, G. Simplex algorithms. In *Advances in Linear and Integer Programming*; Beasley, J.E., Ed.; Oxford University Press: Oxford, UK, 1996; pp. 1–46.

32. Bertsekas, D. *Nonlinear Programming*, 2nd ed.; Athena Scientific: Cambridge, MA, USA, 1999.

33. Jiménez, F.; Verdegay, J.L. Computational Intelligence in Theory and Practice. In *Advances in Soft Computing*; Reusch, B., Temme, K.-H., Eds.; Springer: Heidelberg, Germany, 2001; pp. 167–182.

34. Jiménez, F.; Sánchez, G.; García, J.; Sciavicco, G.; Miralles, L. Multi-objective evolutionary feature selection for online sales forecasting. *Neurocomputing* **2017**, *234*, 75–92. [CrossRef]

35. Deb, K.; Pratab, A.; Agarwal, S.; Meyarivan, T. A fast and elitist multiobjective genetic algorithm: NSGA-II. *IEEE Trans. Evol. Comput.* **2002**, *6*, 182–197. [CrossRef]

36. Bao, C.; Xu, L.; Goodman, E.D.; Cao, L. A novel non-dominated sorting algorithm for evolutionary multi-objective optimization. *J. Comput. Sci.* **2017**, *23*, 31–43. doi:10.1016/j.jocs.2017.09.015. [CrossRef]

37. Jiménez, F.; Sánchez, G.; Vasant, P. A Multi-objective Evolutionary Approach for Fuzzy Optimization in Production Planning. *J. Intell. Fuzzy Syst.* **2013**, *25*, 441–455.

38. Deb, K.; Jain, H. An Evolutionary Many-Objective Optimization Algorithm Using Reference-Point-Based Nondominated Sorting Approach, Part I: Solving Problems With Box Constraints. *IEEE Trans. Evol. Comput.* **2014**, *18*, 577–601. doi:10.1109/TEVC.2013.2281535. [CrossRef]

39. Frank, E.; Witten, I.H. *Generating Accurate Rule Sets without Global Optimization*; Department of Computer Science, University of Waikato: Waikato, New Zealand, 1998; pp. 144–151.

40. Witten, I.H.; Frank, E.; Hall, M.A. Introduction to Weka. In *Data Mining: Practical Machine Learning Tools and Techniques*, 3rd ed.; Witten, I.H., Frank, E., Hall, M.A., Eds.; The Morgan Kaufmann Series in Data Management Systems, Morgan Kaufmann: Boston, MA, USA, 2011; pp. 403–406.

41. Michalski, R.S. On the quasi-minimal solution of the general covering problem. In Proceedings of the V International Symposium on Information Processing (FCIP 69), Bled, Yugoslavia, 8–11 October 1969; pp. 125–128.

42. Quinlan, J.R. *C4. 5: Programs for Machine Learning*; Elsevier: Amsterdam, The Netherlands, 2014.

43. Rajput, A.; Aharwal, R.P.; Dubey, M.; Saxena, S.; Raghuvanshi, M. J48 and JRIP rules for e-governance data. *IJCSS* **2011**, *5*, 201.

44. Cohen, W.W. Fast effective rule induction. In Proceedings of the Twelfth International Conference on Machine Learning, Tahoe City, CA, USA, 9–12 July 1995; pp. 115–123.

45. Fürnkranz, J.; Widmer, G. Incremental reduced error pruning. In Proceedings of the Eleventh International Conference, New Brunswick, NJ, USA, 10–13 July 1994; pp. 70–77.

46. Holte, R.C. Very simple classification rules perform well on most commonly used datasets. *Mach. Learn.* **1993**, *11*, 63–90. [CrossRef]

47. Mukhopadhyay, A.; Maulik, U.; Bandyopadhyay, S.; Coello, C.A.C. A Survey of Multiobjective Evolutionary Algorithms for Data Mining: Part I. *IEEE Trans. Evol. Comput.* **2014**, *18*, 4–19. doi:10.1109/TEVC.2013.2290086. [CrossRef]

48. Mukhopadhyay, A.; Maulik, U.; Bandyopadhyay, S.; Coello, C.A.C. Survey of Multiobjective Evolutionary Algorithms for Data Mining: Part II. *IEEE Trans. Evol. Comput.* **2014**, *18*, 20–35. doi:10.1109/TEVC.2013.2290082. [CrossRef]

49. Ishibuchi, H.; Murata, T.; Turksen, I. Single-objective and two-objective genetic algorithms for selecting linguistic rules for pattern classification problems. *Fuzzy Sets Syst.* **1997**, *89*, 135–150. [CrossRef]

50. Srinivas, M.; Patnaik, L. Adaptive probabilities of crossover and mutation in genetic algorithms. *IEEE Trans. Syst. Man Cybern B Cybern.* **1994**, *24*, 656–667. [CrossRef]

51. Zwitter, M.; Soklic, M. Breast Cancer Data Set. Yugoslavia. Available online: http://archive.ics.uci.edu/ml/datasets/Breast+Cancer (accessed on 5 September 2018).

52. Thrun, S. MONK's Problem 2 Data Set. Available online: https://www.openml.org/d/334 (accessed on 5 September 2018).

53. Thrun, S.B.; Bala, J.; Bloedorn, E.; Bratko, I.; Cestnik, B.; Cheng, J.; Jong, K.D.; Dzeroski, S.; Fahlman, S.E.; Fisher, D.; et al. The MONK's Problems A Performance Comparison of Different LearningAlgorithms. Available online: http://digilib.gmu.edu/jspui/bitstream/handle/1920/1685/91-46.pdf?sequence=1 (accessed on 5 September 2018).

54. Metz, C.E. Basic principles of ROC analysis. *Semin. Nucl. Med.* **1978**, *8*, 283–298. [CrossRef]

55. Fawcett, T. An Introduction to ROC Analysis. *Pattern Recogn. Lett.* **2006**, *27*, 861–874. [CrossRef]

56. Hand, D.J. Measuring classifier performance: A coherent alternative to the area under the ROC curve. *Mach. Learn.* **2009**, *77*, 103–123. doi:10.1007/s10994-009-5119-5. [CrossRef]

57. Zitzler, E.; Deb, K.; Thiele, L. Comparison of multiobjective evolutionary algorithms: Empirical results. *Evol. Comput.* **2000**, *8*, 173–195. [CrossRef] [PubMed]

58. Zitzler, E.; Thiele, L.; Laumanns, M.; Fonseca, C.; Grunert da Fonseca, V. Performance Assessment of Multiobjective Optimizers: An Analysis and Review. *IEEE Trans. Evol. Comput.* **2002**, *7*, 117–132. [CrossRef]

59. Kohavi, R. A study of cross-validation and bootstrap for accuracy estimation and model selection. In Proceedings of the 14th International Joint Conference on Artificial Intelligence (II), Montreal, QC, Canada, 20–25 August 1995; pp. 1137–1143.

60. Jiménez, F.; Jodár, R.; Sánchez, G.; Martín, M.; Sciavicco, G. Multi-Objective Evolutionary Computation Based Feature Selection Applied to Behaviour Assessment of Children. In Proceedings of the 9th International Conference on Educational Data Mining EDM 2016, Raleigh, NC, USA, 29 June–2 July 2016; pp. 1888–1897.

MDPI
St. Alban-Anlage 66
4052 Basel
Switzerland
Tel. +41 61 683 77 34
Fax +41 61 302 89 18
www.mdpi.com

*Entropy* Editorial Office
E-mail: entropy@mdpi.com
www.mdpi.com/journal/entropy